HIGH-PRESSURE RHEOLOGY FOR QUANTITATIVE ELASTOHYDRODYNAMICS

TRIBOLOGY AND INTERFACE ENGINEERING SERIES

HIGH-PRESSURE RHEOLOGY
FOR QUANTITATIVE
ELASTOHYDRODYNAMICS

Aims & Scope

The Tribology Book Series is well established as a major and seminal archival source for definitive books on the subject of classical tribology. The scope of the Series has been widened to include other facets of the now-recognised and expanding topic of Interface Engineering.

The expanded content will now include:
• colloid and multiphase systems; • rheology; • colloids; • tribology and erosion; • processing systems; • machining; • interfaces and adhesion; as well as the classical tribology content which will continue to include • friction; contact damage; • lubrication; and • wear at all length scales.

TRIBOLOGY AND INTERFACE ENGINEERING SERIES, 54
EDITOR: B.J. BRISCOE

HIGH-PRESSURE RHEOLOGY FOR QUANTITATIVE ELASTOHYDRODYNAMICS

Scott Bair, Ph.D., P.E.

Principal Research Engineer
Director, Center for High-Pressure Rheology
George W. Woodruff School of Mechanical Engineering
Georgia Institute of Technology
Atlanta, GA
USA

With a section by Clare McCabe, Vanderbilt University

ELSEVIER

Amsterdam • Boston • Heidelberg • London • New York • Oxford
Paris • San Diego • San Francisco • Singapore • Sydney • Tokyo

Elsevier
The Boulevard, Langford Lane, Kidlington, Oxford OX5 1GB UK
Radarweg 29, PO Box 211, 1000 AE, Amsterdam, The Netherlands

First edition 2007

British Library Cataloguing in Publication Data
A catalogue record for this book is available from the British Library

Library of Congress Cataloging-in-Publication Data
A catalog record for this book is available from the Library of Congress

ISBN-13: 978-0-444-52243-6
ISBN-10: 0-444-52243-3
ISSN: 0167-8922 (series)

Transferred to digital print 2007

Printed and bound in Great Britain by CPI Antony Rowe, Eastbourne

CONTENTS

PREFACE

The reader will find this book to be very different from books on the topic of rheology or works dealing with elastohydrodynamics. It is not a rigorous treatment of either of these subjects and the reader may wish to consult the bibliography for a thorough development of the underlying topics. This book is instead intended to provide a much needed link between the two fields. The reader may also be surprised to find that the author's specialty is machine design and that he has been issued patents on machines unrelated to either rheology or tribology research such as a transmission for a self-propelled vacuum cleaner and instruments for cataract surgery. The utility of a machine design background should become apparent as the difficulty of performing rheological experiments under high pressure and with sufficient shear stress to yield nonlinear response in low-molecular-weight materials is appreciated. The author's laboratory at the Woodruff School of Mechanical Engineering, Georgia Institute of Technology, possesses many unique, custom-built instruments for the measurement of properties of liquids under pressure including viscosity, normal stress coefficients, volume, dielectric properties, refractive index, phase transformations and heat capacity.

Research into the high-pressure rheology of liquids has been motivated within the physics community by an interest in data that supply an additional state variable (pressure or density in addition to temperature) to evaluate theories of viscous flow and the dynamics of the glass transition. The tribology community has a need of a description of the temperature, pressure and shear dependence of the viscosity of the liquids used to generate lubricating films in the high-pressure contacts between nonconforming surfaces of many machine parts. With such similar interests, it is amazing that the descriptions of liquid behavior adopted by the two communities should differ by so much.

Computational elastohydrodynamics, a part of tribology, has existed happily enough for about fifty years without the use of accurate models for the rheology of the liquids used as lubricants. For low-molecular-weight liquids, such as low-viscosity mineral oils, it has been possible to calculate, with precision, the film thickness in a concentrated contact provided that the pressure and temperature are relatively low, even when the pressure variation of viscosity is not accurately modeled in detail. Other successes have been more qualitative in nature, using effective properties which come from the fitting of parameters used in calculations to experimental measurements of the contact behavior, friction or film thickness. This book is intended to provide a sufficiently accurate framework for the rheology of liquids at elevated pressure that it may be possible for computational elastohydrodynamics to discover the relationships between the behavior (film thickness, friction or other) of a lubricated concentrated contact and the (measurable) properties of the liquid lubricant. The rheological models will be mainly empirical, a phenomenological description being of more use to the tribologist than theory except where theory provides an improvement in the ability to extrapolate. The need for a book of this type should become apparent when considering that of the at least five monographs on the subject of elastohydrodynamic lubrication, not even one contains a complete description of the pressure and shear dependence of viscosity that is necessary for quantitative predictions of the tribological behavior of elastohydrodynamic contacts. With the current need for energy-efficient lubricants, the opportunity for improved understanding and prediction of the wear protection and friction reduction from lubricating films has never been more welcome. Rheologists may also find the description of pressure effects presented here useful for the polymer processing that sometimes requires elevated pressure.

"Quantitative prediction is the aim of every science" (Kagan, *Pure Appl. Geophys.*, Vol. 155, 1999, pp. 233–258).

NOMENCLATURE

a	the Yasuda parameter for the Carreau–Yasuda model
a_g	thermal expansivity of the glass, K^{-1}
a_V	thermal expansivity defined for volume linear with temperature, K^{-1}
a_ρ	thermal expansivity defined for density linear with temperature, K^{-1}
a_l	linear thermal expansivity of a solid, K^{-1}
a_t	horizontal shift factor
B	Doolittle parameter
B	Na/Wi^2
b	contact half-width, m
c	concentration by mass fraction, kg/kg
c_L	longitudinal sound velocity, m/s
c_T	transverse sound velocity, m/s
c_p	constant pressure specific heat, J/kg/K
c_V	constant volume specific heat, J/kg/K
C_{rb}	calibration factor for a rolling ball viscometer, Nm/kg
C_{fc}	calibration factor for a falling cylinder viscometer, Nm/kg
C_i	various constants, units given in the text
C_1	first WLF constant
C_2	second WLF constant, K
C_F	fragility parameter in the Johari and Whalley equation
C_E	parameter multiplying Σ in the traction sinh law
d	interaction parameter in a mixing rule
$\mathbf{D}$	rate of deformation tensor, s^{-1}
D	working diameter, m
D_F	fragility parameter in the VTF equation
D_i	inside diameter, m
D_o	outside diameter, m
e	ellipticity ratio
E	Young's modulus, Pa
E_a	thermal activation energy, J/kmol

E'	composite Young's modulus, Pa
F	vertical shift factor
Fo	heat transfer Fourier number
f	frequency, Hertz
g	acceleration of gravity = 9.807 m/s^2
G	shear modulus, liquid critical shear stress or material modulus associated with λ, Pa
G''	component of the dynamic shear modulus in phase with the shear strain, Pa
G_R	material modulus associated with λ at the reference state, T_R, $p = 0$, Pa
G_s	roller elastic shear modulus, Pa
G_∞	limiting high rate shear modulus of the liquid, Pa
h	film thickness, m
h_c	central film thickness, m
h_m	minimum film thickness, m
h_N	central film thickness from a Newtonian calculation, m
h_{NN}	central film thickness from a shear-thinning calculation, m
h_0	film thickness for $dp/dx = 0$, m
I	refractive index
$\mathbf{I}$	unit tensor
K	isothermal bulk modulus, Pa
K_0	isothermal bulk modulus at $p = 0$, Pa
K_0'	pressure rate of change of isothermal bulk modulus at $p = 0$
K_∞	K_0 at infinite temperature, Pa
$\dot{K}_0$	rate of change of K_0 with reciprocal temperature, Pa K
K_{00}	K_0 at zero absolute temperature, Pa
k	liquid thermal conductivity, W/m/K
k_B	Boltzmann constant = 1.380622×10^{-23} J/K
k_s	solid thermal conductivity, W/m/K
l	fall distance, m
L	length, m
M	molecular weight, kg/kmol
m	exponent in the shifting rule for G
m_s	the linear slope of a dry traction curve representing roller elasticity
m_T	temperature fragility index
m_p	pressure fragility index, Pa^{-1}
n	power–law exponent
Na	Nahme–Griffith number
N_1	first normal stress difference, Pa
N_2	second normal stress difference, Pa
p	pressure, Pa
$\bar{p}$	average contact pressure, Pa
p_e	maximum internal pressure for elastic behavior of a cylinder, Pa
p_i	minimum internal pressure for complete inelastic behavior of a cylinder, Pa
p_{inf}	pressure at $\partial\alpha/\partial p = 0$, the inflection pressure, Pa
p_m	mechanical pressure, Pa

p_p	viscosity pole pressure in the Roelands model, Pa
p_∞	idealized glass transition pressure, the pressure at which viscosity diverges, Pa
$p_{-\infty}$	the tension at which the associated viscosity term vanishes, Pa
p_{iv}	isoviscous pressure, Pa
p_{ai}	asymptotic isoviscous pressure ($=1/\alpha^*$), Pa
p_g	glass transition pressure, Pa
p_H	Hertz (maximum) pressure, Pa
P_+, P_-	parameters in the Bair and Kottke pressure–viscosity model, Pa
Q	flowrate, m^3/s, or flowrate per unit width, m^2/s
$\hat{Q}$	dimensionless flowrate per unit width $= Q\mu/h^2G$
q	exponent used in viscosity models
R_0	occupied volume fraction, V_∞/V, at the reference state, T_R, $p=0$
R	composite radius, m
R_g	universal gas constant = 8314.34 Pa m^3/kmol/K
r	dimensionless contact radius
s	integration variable
S	Roelands slope index
T	temperature, K
$\hat{T}$	dimensionless temperature
$\hat{T}_C$	dimensionless midplane temperature
T_g	glass transition temperature, K
T_R	reference temperature, K
T_0	the ambient pressure value of T_∞, K
T_{g0}	the ambient pressure value of T_g, K
T_0'	the ambient pressure value of dT_∞/dp, K/Pa
T_{g0}'	the ambient pressure value of dT_g/dp, K/Pa
T_0''	the ambient pressure value of d^2T_∞/dp^2, K/Pa^2
T_{g0}''	the ambient pressure value of d^2T_g/dp^2, K/Pa^2
T_∞	divergence temperature, K
t	time, s
t_0	time to reach test velocity after constant acceleration, s
$\bar{u}$	average surface velocity or rolling velocity, m/s
u_i	velocity of surface i, m/s
V	volume at T and p, m^3
V_g	volume at the glass transition, m^3
V_m	molecular volume, m^3
V_R	volume at reference state, T_R, $p=0$, m^3
V_0	volume at $p=0$, m^3
V_∞	occupied volume, m^3
$V_{\infty R}$	occupied volume at reference state, T_R, $p=0$, m^3
W	force normal to the surfaces at the contact, N for point contact, N/m for line contact
Wi	Weissenberg number
u	velocity in the x-direction (generally the rolling direction), m/s
v	velocity in the y-direction, m/s

w	velocity in the z-direction (generally the cross-film direction), m/s
x	coordinate in the direction of surface velocity, m
y	coordinate along the surface normal to the surface velocity, m
z	coordinate across the film, m
$\hat{z}$	dimensionless coordinate across the film, $= z/h$
Z	Roelands pressure–viscosity index
α	local pressure–viscosity coefficient, Pa^{-1}
α_0	initial pressure–viscosity coefficient, Pa^{-1}
α^*	reciprocal asymptotic isoviscous pressure coefficient ($=1/p_{ai}$), Pa^{-1}
α_B	secant pressure–viscosity coefficient, Pa^{-1}
α_{film}	general film-forming pressure–viscosity coefficient, Pa^{-1}
α_g	pressure–viscosity coefficient of the liquid at the glass transition, Pa^{-1}
α_n	parameter of the Irving and Barlow model, Pa^{-1}
α_p	parameter of the Irving and Barlow model, Pa^{-1}
α_s	limiting local pressure–viscosity coefficient of the Skinner equation, Pa^{-1}
$\alpha_{\eta,\tau}$	generalized pressure–viscosity coefficient for stress control, Pa^{-1}
$\alpha_{\eta,\dot{\gamma}}$	generalized pressure–viscosity coefficient for rate control, Pa^{-1}
β	temperature–viscosity coefficient, K^{-1}
β_g	temperature–viscosity coefficient of the liquid at the glass transition, K^{-1}
β_K	temperature coefficient of K_0, C^{-1}
Γ	inlet Weissenberg number, $\mu_0\bar{u}/(h_N G)$
$\dot{\gamma}$	shear rate, s^{-1}
$\hat{\dot{\gamma}}$	dimensionless shear rate $= \dot{\gamma}\mu/G$
$\dot{\gamma}*$	invariant equivalent shear rate, s^{-1}
$\dot{\gamma}_{app}$	apparent shear rate, s^{-1}
$\dot{\gamma}_{\max}$	maximum inlet zone shear rate, s^{-1}
$\dot{\gamma}_r$	apparent shear rate at the rim of a parallel disc rheometer, s^{-1}
γ	oscillatory shear strain
γ_A	oscillatory shear amplitude
ε	occupied volume thermal expansivity, K^{-1}
ς	$N_1/2p$
η	rate-dependent shear viscosity, Pa s
η'	dynamic shear viscosity, Pa s
η_e	rate-dependent elongational viscosity, Pa s
Θ	thrust, N
θ	inclination angle, deg
κ	h_N/h_{NN}
Λ	limiting stress–pressure coefficient
λ	characteristic time in shear-thinning models, s
λ_{EB}	Einstein–Debye rotational relaxation time of the molecule, s
λ_D	dielectric relaxation time, s
λ_R	characteristic time at the reference state, T_R, $p = 0$, s
μ	limiting low-shear viscosity and Newtonian viscosity, Pa s
μ_0	low-shear viscosity at $p = 0$, Pa s

μ_2	second Newtonian viscosity, Pa s
μ_∞	viscosity extrapolated to infinite temperature, Pa s
$\mu_{-\infty}$	viscosity extrapolated to infinite tension, Pa s
μ_a	limiting low-shear viscosity of an additive, Pa s
μ_b	limiting low-shear viscosity of a base oil or solvent in a polymer solution, Pa s
μ_i	viscosity at $t = 0$, Pa s
μ_g	viscosity at the glass transition, Pa s
μ_R	low-shear viscosity at reference state, T_R, $p = 0$, Pa s
μ_s	parameter in the Skinner equation, Pa s
μ_p	pole viscosity in the Roelands model, Pa s
ν	Poisson's ratio
ρ	mass density, kg/m^3
ρ_R	mass density, at reference state, T_R, $p = 0$, kg/m^3
ρ_s	average density of a falling body, kg/m^3
Σ	sliding/rolling velocity or slide-to-roll ratio
Σ_l	$\dot{\gamma} h/\bar{u}$, liquid shear component of slide-to-roll ratio
Σ_L	$(u_1 - u_2)/\bar{u}$, longitudinal slide-to-roll ratio
Σ_T	$(v_1 - v_2)/\bar{u}$, transverse slide-to-roll ratio
σ_{ij}	i,j component of the stress tensor, Pa
σ_θ	normal stress at inclination angle, θ, Pa
σ	stress tensor, Pa
T	torque, Nm
τ	shear stress, Pa
$\hat{\tau}$	dimensionless shear stress $= \tau/G$
τ'	oscillatory shear component of the shear stress in phase with shear rate, Pa
τ''	oscillatory shear component of the shear stress in phase with shear strain, Pa
$\bar{\tau}$	average contact shear stress, Pa
τ^*	invariant equivalent shear stress, Pa
τ_c	critical shear stress, Pa
τ_m	shear stress at the midplane of the film, Pa
$\tau_{\max}$	maximum inlet zone shear stress, Pa
τ_r	apparent shear stress at the rim of a parallel disc rheometer, Pa
τ_w	wall shear stress, Pa
τ_E	representative shear stress, the rate of change of the averaged shear stress with respect to the natural logarithm of the apparent shear rate, Pa
τ_θ	critical shear stress for slip which occurs at inclination angle, θ, Pa
υ	dilatational viscosity, Pa s
ϕ	internal friction coefficient of a liquid
Ψ_1	first normal stress coefficient, Pa s^2
ω	angular frequency, rad/s
Ω	rotation rate, rad/s

FIGURE CREDITS

Fig. 1.3 is adapted from Figure 11, page 102 of *Wear Control Handbook*, Peterson and Winer (Eds.), *Am. Soc. Mech. Eng*., 1980.

Fig. 2.2 is adapted from Figure 6a of *J. Non-Newt. Fluid Mech.*, Elsevier Science, Vol. 44, 1992, pp. 113–126.

Fig. 2.3 is adapted from Figure 13 of *J. Non-Newt. Fluid Mech.*, Elsevier Science, Vol. 6, 1980, pp. 303–324.

Figs. 4.5 and 4.6 are Figures 5 and 6 from *Lubricants and Lubrication*, Elsevier Science, 1995, pp. 169–187.

Figs. 4.7 and 4.9 are Figures 7 and 8 from *Tribol. Int.*, Elsevier Science, Vol. 34, No. 7, 2001, pp. 461–468.

Fig. 5.1 is adapted from Figure 9 of *Phil. Trans. Roy. Soc. London*, Series A, Vol. 255, No. 1056, 1963, pp. 281–312.

Fig. 5.2 is Figure 2 from *ASME J. Lubr. Techn.*, Vol. 90, Ser. F, No. 3, 1968, pp. 580–591.

Fig. 5.3 is Figure 1 from *Japanese J. Appl. Phys.*, Vol. 25, No. 7, Part 1, 1986, pp. 1091–1096.

Fig. 5.4 is Figure 1, page 2073, from *J. Appl. Phys.*, Vol. 71, No. 5, 1992, pp. 2071–2081.

Fig. 5.5 is Figure 2 from *Viscosity and Density of Over Forty Lubricating Fluids of Known Composition at Pressures to 150,000 psi and Temperatures to 425 F, ASME*, New York, 1953.

Fig. 5.6 is Figure 1, page 232, from *J. Phys. E.*, Vol. 4, No. 3, 1971, pp. 232–236.

Figs. 5.7 and 5.13 are Figures 1 and 4 from *STLE Tribol. Trans.*, Vol. 47, No. 3, 2004, pp. 356–360.

Fig. 5.9 is Figure 1, page 463, from *Tribol. Int.*, Elsevier Science, Vol. 34, No. 7, 2001, pp. 461–468.

Figs. 6.12, 6.13 and 6.14 are Figures 1, 4 and 5 from *ASME J. Tribol.*, Vol. 126, No. 4, 2004, pp. 697–702.

Fig. 7.1 is adapted from Figure 1 of *Tribological Research and Design for Engineering Systems*, Dowson, D. et al. (Eds.), Elsevier, 2002, pp. 693–699.

Fig. 7.6 is adapted from Figure 2 in *A. I. Ch. E. Journal*, Vol. 51, No. 8, 2005, pp. 2150–2170.

Fig. 7.7 is adapted from Figure 11 in *ASME J. Lubr. Techn.*, Vol. 90, Ser. F, No. 3, 1968, pp. 580–591.

Fig. 7.12 is Figure 3 from *Proc. Instn. Mech. Engrs.*, Vol. 216, Part J, 2002, pp. 139–149.

Fig. 7.14 is Figure 3 from *Rheologica Acta*, Vol. 35, 1996, pp. 13–23.

Fig. 7.15 is Figure 1 from *Tribol. Int.*, Elsevier Science, Vol. 31, No. 10, 1999, pp. 573–586.

Fig. 8.8 is Figure 6 from *ASME J. Tribol.*, Vol. 126, No. 4, 2004, pp. 697–702.

Fig. 8.9 is Figure 2 from *Tribol. Lett.*, Vol.18, No. 2, 2005, pp. 145–152.

Figs. 10.2 and 10.3 are Figures 4 and 5 from *Tribol. Int.*, Elsevier Science, Vol. 35, 2002, pp. 591–597.

Fig. 10.4 is Figure 3 from *Trans. ASME J. Tribol.*, Vol. 115, 1993, pp. 507–514.

Figs. 10.5 and 10.6 are adapted from Figures 1, 2, 3 and 4 in *Tribol. Int.*, Elsevier Science, Vol. 37, 2004, pp. 783–789.

Fig. 12.4 is Figures 4 and 5 from *Tribological Research and Design for Engineering Systems*, Dowson, D. et al. (Eds.), Elsevier, 2002, pp. 693–699.

Fig. 12.5 is Figure 2 from *Proc. I. Mech. E.*, Vol. 220, Part J, 2006, pp. 365–374.

Fig. 12.8 is adapted from Figure 6 in *STLE Tribol. Trans.*, Vol. 46, 2003, pp. 289–295.

Figs. 12.9 and 12.10 are Figures 3 and 2 from *Tribol. Lett.*, Vol. 13, No. 4, 2002, pp. 251–254.

Figs. 1.5 and 12.11 are Figures 1 and 2 from *STLE Tribol. Trans.*, Vol. 49, No. 2, 2006, pp. 218–224.

ACKNOWLEDGMENTS

The laboratory research that forms a large portion of this work would not have been possible without the long-term financial support of the author's work in the G.W. Woodruff School of Mechanical Engineering by the Office of Naval Research, Peter Schmidt, Program Officer and the Timken Company, Harvey Nixon, Manager Product R&D and many others.

The measurements of longitudinal sound velocity and some of the relative volume measurements were performed by Theodore Byer, graduate research assistant. Some of the C12 isomers for Fig. 5.11 were provided by Dr. Peter Gordon of ExxonMobil Corp.

The author is indebted to Dr. Simin Nasseri for her careful checking of Chapter 2 and many helpful suggestions.

CHAPTER 1

An Introduction to Elastohydrodynamic Lubrication

1.1 Lubrication

It would be difficult to imagine any type of machinery without lubrication. There have been numerous analyses showcasing the energy savings by the application of the best lubrication practices. However, if we had to live with the high friction coefficients and wear rates that go with bare metal contacts, heat engines, gear boxes and other common machines would be scarcely usable.

Elastohydrodynamics is a part of the broader field of lubrication. Lubricants are employed to generate a film of easily sheared material interposed between the surfaces of machine components to reduce friction and prevent wear. Early lubricants were often animal or vegetable fats and oils. Today, the majority of liquid lubricants are based on highly refined mineral oils and synthetic oils such as polyalphaolefins, polyolesters, polyglycols and others are becoming more important with time. The synthetics tend to have a reduced temperature dependence of viscosity as compared with mineral oils. Mineral oils (and sometimes synthetics as well) are often blended with high-molecular-weight polymer to reduce the temperature dependence. Greases are liquid lubricants that have been given a solid-like yield strength by the addition of a thickener which allows the grease to remain without flow in a desired location until needed.

Regimes of lubrication are identified by the mechanism of film formation. When a molecularly thin film formed from the oil itself or an additive is physically or chemically attached to the interacting surfaces, the regime is known as boundary lubrication. When the surfaces are separated by a film of pressurized liquid with sufficient thickness to completely prevent the contact between roughness features, full film lubrication is said to exist. If the liquid pressure is generated externally and applied through passages penetrating the surfaces into the space between the surfaces, the full film regime is called hydrostatic. If the generation of pressure is self-acting, that is, if the motion of the surfaces in combination with the viscous properties of the liquid and the shape of the conjunction acts to pressurize the film, the full film is termed hydrodynamic. Many machine components operate with the hydrodynamic mechanism active, but with an insufficiently thick film to prevent some interaction of the surface roughness. In this case the regime is said to be mixed, that is, mixed hydrodynamic and boundary lubrication.

For the liquid lubricants used for boundary lubrication, the base fluid is generally selected for its function as a solvent and carrier of the chemically active agent. Rheological properties such as viscosity are thought to be unimportant, although they often are. In contrast, the viscosity of a full film lubricant is the most important property.

Hydrodynamic lubrication is often classified by the shapes of the contacting surfaces. Journal bearings and planar slider bearings are examples of conformal conjunctions where the gap between the surfaces changes only slowly with distance along the surface in any direction. There does not seem to be a rigorous definition, but for conformal contacts the operating load is often insufficient to cause elastic deformations of the surfaces that result in a significant change in the film shape. On the other hand, nonconformal surfaces such as rollers, gear teeth and cams do not "fit" well against each other. Until a load is applied to form a contact patch, the gap varies rapidly with distance along the surface. When a load is applied the film shape can change appreciably. The conjunctions formed by nonconformal surfaces are known as *concentrated contacts* and an entire branch of hydrodynamic lubrication is devoted to the study of the lubrication of these contacts.

1.2 Concentrated Contact Lubrication

Contacts between nonconformal surfaces differ from those formed by conformal surfaces by the high contact pressure that can be generated. The elastic deformation of the surface may be greater than the thickness of any lubricant film formed and if the solid materials have a large modulus of elasticity, then the contact pressures may substantially change the viscosity of the liquid. The viscosity of organic liquids increases with pressure, roughly doubling for every 0.05 GPa increase in pressure. These liquids are *piezoviscous*. See Fig. 1.1 for a plot of the logarithm of viscosity versus pressure for a viscous organic liquid, clove oil, from data generated by Bridgman [1] nearly eighty years ago. For nonconformal steel surfaces, a common contact pressure may be 1 GPa so that incredibly large increases in viscosity are possible. For liquid metals, the pressure sensitivity is much less than for organic liquids (doubling for about a 3 GPa pressure) and for liquids that expand on freezing such as water and gallium, the response is for the viscosity to decrease with pressure for conditions approaching the melting temperature.

Concentrated contacts can be further classified by the geometry of the contact with zero load or perfectly rigid bodies. A cylinder on a plane or a pair of cylinders with parallel axes will contact along a line and these form what are known as *line contacts*. When a load is applied the contact patch is rectangular. A sphere on a sphere, a sphere on a plane or identical cylinders with perpendicular axes will contact at a point and these are known as *point contacts*. When a load is applied to this conjunction, the contact patch will be circular. Point contacts can also be formed by the conjunction of a sphere and a cylinder or more generally by the contact of crowned rollers where the radius about the axis of symmetry is different from the radius in the plane of the axis of symmetry

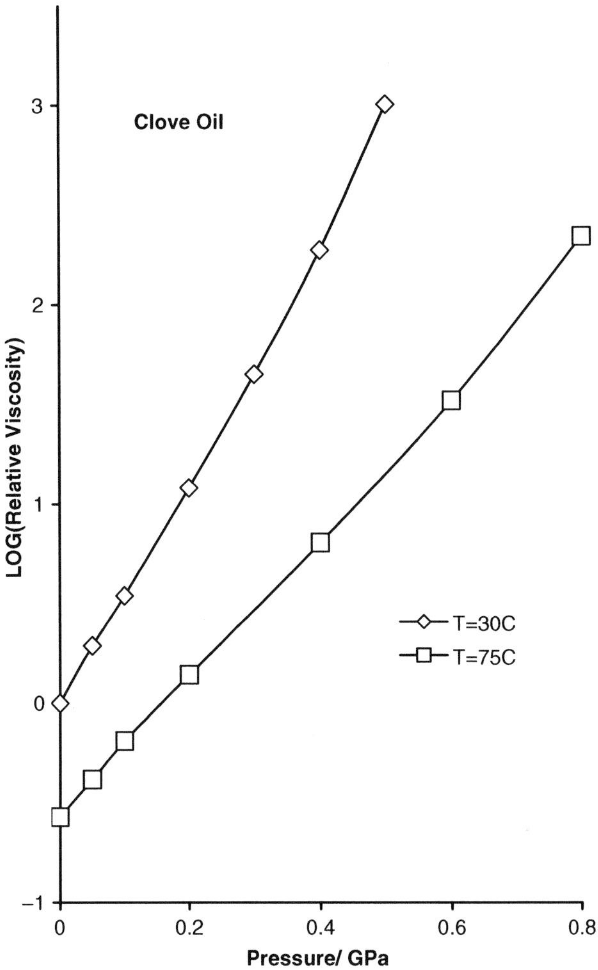

Fig. 1.1. The viscosity of clove oil (eugenol) relative to its viscosity at $p = 0$ and $T = 30°C$. Measurements from Bridgman, 1926 [1].

(the crown radius). These *elliptical point contacts*, when loaded, produce an elliptical contact patch.

The lubrication of nonconformal contacts can be further divided into regimes by the influence of the elastic deformation of the solid surfaces and the influence of pressure on the viscosity of the liquid. Johnson [2] identified four regimes of concentrated contact lubrication:

1. *Isoviscous rigid*: The pressure generated by the film is insufficient to substantially affect either the shape of the solid bodies relative to the thickness of the film or the

viscosity of the liquid. This regime may be encountered when ordinary lubricated machine parts are operating at high speed and very light load.

2. *Piezoviscous rigid*: The pressure generated by the film is sufficient to effectively increase the viscosity but not great enough to significantly alter the shape of the solid bodies. This regime requires a liquid with a viscosity that is very sensitive to pressure and surfaces moving at relatively high speed.
3. *Isoviscous elastic*: The pressure generated by the film alters the shape of the solid bodies but not the viscosity of the liquid. This regime, often referred to as *soft EHL*, is encountered when the machine components are made of low modulus material or when the viscosity of the liquid is relatively insensitive to pressure.
4. *Piezoviscous elastic*: The effect of the pressure within the film on both the shape of the bodies and the viscosity of the liquid cannot be neglected. This is the important *full elastohydrodynamic* regime that is commonly just known as elastohydrodynamics or EHL or EHD and is operating in rolling element bearings and gears lubricated by viscous organic liquids.

Johnson [2] identified the parameters that should characterize the particular regime of lubrication, the dimensionless viscosity and elasticity parameters, g_V and g_E, respectively. For line contact, these may be written as

$$g_V^2 = \frac{8\pi^3 \alpha^{*2} p_H^6 R}{E'^3 \bar{u} \mu_0} \tag{1.1}$$

$$g_E^2 = \frac{4\pi^2 p_H^4 R}{E'^3 \bar{u} \mu_0} \tag{1.2}$$

Here, μ_0 is the viscosity of the liquid at atmospheric pressure, $\mu(p = 0)$, and α^* is a pressure viscosity coefficient defined by

$$\frac{1}{\alpha^*} = p_{ai} = \int_0^\infty \frac{\mu(p = 0)}{\mu(p)} dp \tag{1.3}$$

Figure 1.2, showing two cylinders in contact, will be helpful in defining the other parameters. The composite radius is

$$R = \frac{1}{(1/R_1) + (1/R_2)} \tag{1.4}$$

and the reduced modulus of the solid bodies is

$$E' = \frac{2}{\left[(1 - v_1^2)/E_1\right] + \left[(1 - v_2^2)/E_2\right]} \tag{1.5}$$

The maximum contact pressure for a dry contact of width, $2b$, as shown in left portion of Fig. 1.2, is p_H and this pressure will not necessarily be the maximum pressure in

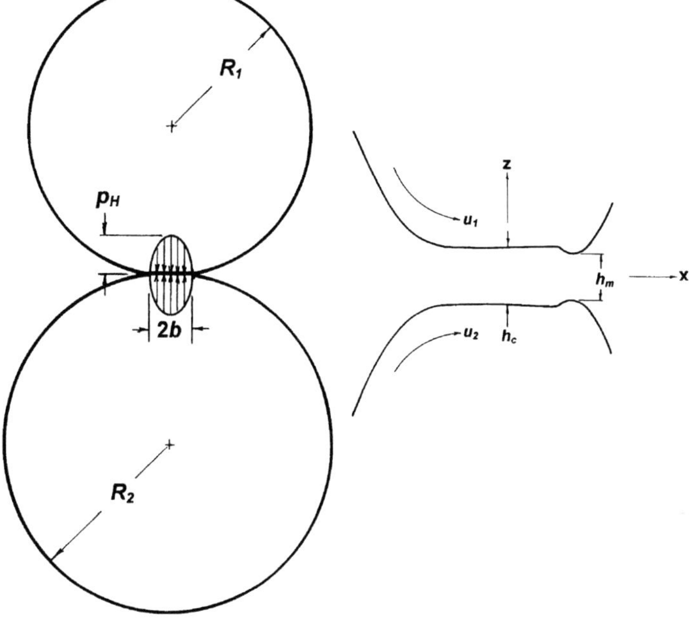

Fig. 1.2. A line contact formed by a pair of cylinders. The pressure distribution in dry contact on the left and the film shape for the piezoviscous-elastic lubrication regime on the right.

the liquid film shown in the right-hand portion of Fig. 1.2. The shape of the film will differ for the various regimes; the film shown in the right of Fig. 1.2 is the case of piezoviscous-elastic lubrication.

Johnson constructed a chart in g_V and g_E coordinates to delineate the four regimes of line contact lubrication. Hooke [3] adjusted the boundaries of Johnson's regime chart to fill in some missing portions. The boundaries in such charts are generally constructed by finding the relationships between g_V and g_E that make the theoretical solutions for film thickness equal for the two regimes at the boundary. Winer and Cheng [4] have shown the boundaries as "fuzzy" in Fig. 1.3. The regime chart in Fig. 1.3 indicates that the type of regime changes from isoviscous to piezoviscous as g_V increases, and changes from rigid to elastic as g_E increases. The arrows in the figure illustrate an example of a typical line contact. Here, $R = 0.02$ m, $E' = 211$ GPa, $\alpha^* = 15\ \text{GPa}^{-1}$ and $\mu_0 = 0.08\ \text{Pa} \cdot \text{s}$. The solid arrow in Fig. 1.3 represents a rolling velocity increasing from 0.1 to 10 m/s with $p_H = 1.0$ GPa and the theoretical film thickness grows from 0.081 to 2.04 μm as the velocity increases. The broken arrow indicates pressure increasing from 0.6 to 1.6 GPa, while fixing $\bar{u} = 1$ m/s and here the film diminishes from 0.46 to 0.36 μm for this pressure increase. These conditions are typical for steel lubricated by a mineral oil and illustrate the usual behavior for the piezoviscous-elastic regime or full elastohydrodynamic lubrication; the film thickness is much more sensitive to velocity (and viscosity and pressure–viscosity coefficient) than it is to pressure (or load or elastic modulus).

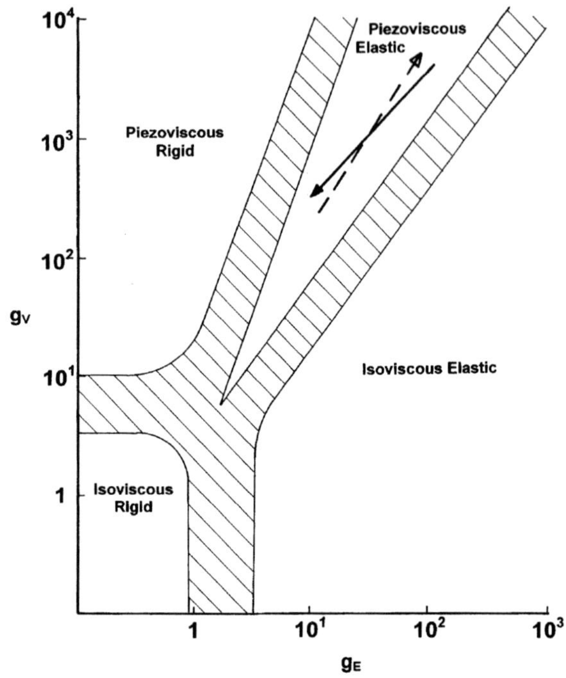

Fig. 1.3. The regimes of line contact lubrication, adapted from Winer and Cheng, "Film Thickness, Contact Stress and Surface Temperature", *Wear Control Handbook*, (Peterson and Winer, Eds.), American society of Mechanical Engineers, 1980, pp. 81–141, by permission of the American Society of Mechanical Engineers. The solid arrow represents an increase in $\bar{u}$ from 0.1 to 10 m/s at $p_H = 1.0$ GPa. The broken arrow is $\bar{u} = 1$ m/s and p_H increasing from 0.6 to 1.6 GPa.

Johnson [2] also contributed dimensionless viscosity and elasticity parameters for point contacts. These are written as

$$g_V = \frac{\left(\frac{2}{3}\pi^3\right)^3 \alpha^* p_H^9 R^2}{E'^6 \bar{u}^2 \mu_0^2} \tag{1.6}$$

$$g_E = \frac{\left(\frac{2}{3}\pi^3\right)^{\frac{8}{3}} p_H^8 R^2}{E'^6 \bar{u}^2 \mu_0^2} \tag{1.7}$$

Hamrock [5] has constructed regime charts for elliptical point contacts that function in the same way as Fig. 1.3 except that the dimensionless viscosity and elasticity parameters are defined by equations (1.6) and (1.7). These charts span ellipticity ratios from 0.5 to 6. Ellipticity ratio is the contact patch length in the direction normal to the motion divided by the length in the direction of motion. Large ellipticity ratio results in a thicker film by mitigating the flow of liquid out of the sides of the contact.

Now it must be pointed out that the formulations as they are given here and to follow, do not characterize the contact load by the distributed force, W, normal to the surfaces as is conventionally done, but by the maximum pressure, p_H, instead. Here the major concern is for the properties of the liquid in the conjunction and, obviously, the liquid properties depend upon the local pressure, not the force applied to the solid bodies.

1.3 Full Elastohydrodynamic Lubrication

The solution for the film shape and pressure in an elastohydrodynamic conjunction represents one of the more difficult analytical problems in engineering and the successful solution was one of the most significant accomplishments for the tribology field. Small changes in the film shape can produce large excursions in pressure, while the elastic deformations that result in the film shape can be orders of magnitude larger than the film thickness, $h(x)$. The problem requires the simultaneous solution of the film shape equation and the Reynolds equation. For a line contact, the elastic film shape equation is

$$h(x) = h_s + \frac{x^2}{2R} - \frac{2}{\pi E'} \int_{-b}^{b} p(s) \ln(x - s)^2 ds \tag{1.8}$$

The first term on the right, h_s, is a constant to be determined, the second is an approximation to the shape of the gap between the cylinders near the conjunction and the last term is the elastic deflection resulting from the pressure distribution, $p(x)$. The deformation at any position in the contact depends upon the pressure everywhere. For dry contact, the pressure distribution is the theoretical Hertz pressure distribution,

$$p(x) = p_H \left[1 - \left(\frac{x}{b} \right)^2 \right]^{\frac{1}{2}} \tag{1.9}$$

The Hertz contact half-width, b, shown in Fig. 1.2, is for dry contact

$$b = \frac{4 p_H R}{E'} \tag{1.10}$$

and substitution of the Hertz pressure distribution (1.9) into the film shape equation (1.8) and integrating yields a uniform thickness that can be made zero by the adjustment of h_s as expected.

The Newtonian assumption has been the basis for engineering film thickness formulas. The Reynolds equation that relates local pressure to the film shape, the Newtonian viscosity, μ, velocity, $\bar{u}$ and density, ρ, is for steady conditions,

$$\frac{d}{dx}\left(\frac{\rho h^3}{\mu} \frac{dp}{dx} \right) = 12\bar{u} \frac{d(\rho h)}{dx} \tag{1.11}$$

For the incompressible case, ρ is absent from the equation.

Equations (1.8) and (1.11) must be solved simultaneously for the pressure and film thickness. The limits of integration for equation (1.8) must be moved outside of the Hertz zone, the region of width $2b$, to capture the effect of the entire lubricant pressure distribution.

Since the pressure–viscosity effect in this regime is not trivial, expressions are required for the pressure dependence of viscosity and density. Referring to Fig. 1.1, the curves drawn through the data are of sigmoidal shape with an infection point but may be described, to a first approximation, as linear. Then, to a first approximation, the pressure variation of viscosity is exponential and can be described by the special case of an equation that apparently was introduced in 1938 by Skinner [6] motivated by a survey of available pressure–viscosity data for lubricating oils. Skinner's equation is

$$\mu = \mu_s + (\mu_0 - \mu_s)\exp(\alpha_s p) \tag{1.12}$$

With $\mu_s < 0$, equation (1.12) can be accurate from ambient to moderate pressures or, with $\mu_s = 0$, we have the very simple expression often used in elastohydrodynamics,

$$\mu = \mu_0 \exp(\alpha_s p) \tag{1.13}$$

Notice that for this simple relationship, $\alpha^* \equiv \alpha_s$, and for equation (1.12), $\alpha^* = \alpha_s(-\mu_s/\mu_0)\big/\ln(1-\mu_s/\mu_0)$.

Around 1960, Dowson and Higginson [7] solved the full elastohydrodynamic line contact problem using the simple Skinner equation (1.13) and an equation of state for $\rho(p)$ of a mineral oil at ambient temperature and relatively low pressure. Their results were fitted to an empirical expression for the minimum film thickness.

$$h_m = 1.26\alpha^{*0.6}(\mu_0\bar{u})^{0.7}E'^{0.16}R^{0.3}p_H^{-0.26} \tag{1.14}$$

Dowson and Toyoda [8] found an expression for the incompressible central film thickness.

$$h_c = 2.55\alpha^{*0.56}(\mu_0\bar{u})^{0.69}E'^{0.07}R^{0.31}p_H^{-0.20} \tag{1.15}$$

They found that given the same conditions, the minimum film thickness is about 80% of the central value. Much of the contact area can be characterized by a nearly uniform film thickness with the small variation resulting from compressibility. The minimum film thickness occurs at a restriction at the liquid's exit to the contact as in Fig. 1.2. The very large negative pressure gradient at the exit would generate a substantial flow out of the contact at this location. The appearance of the exit restriction avoids continuity of flow problems. A local spike in the pressure occurs just upstream of this restriction as a consequence of the pressure dependence of viscosity. Hamrock [5] found for the compressible central film thickness,

$$h_c = 2.154\alpha^{*0.47}(\mu_0\bar{u})^{0.692}E'^{0.110}R^{0.308}p_H^{-0.332} \tag{1.16}$$

The two-dimensional form of the steady Reynolds equation required for a point contact solution is

$$\frac{\partial}{\partial x}\left(\frac{\rho h^3}{\mu}\frac{\partial p}{\partial x}\right) + \frac{\partial}{\partial y}\left(\frac{\rho h^3}{\mu}\frac{\partial p}{\partial y}\right) = 12\bar{u}\frac{\partial(\rho h)}{\partial x} \tag{1.17}$$

The two-dimensional counterpart for the elastic film shape equation will not be given here. Hamrock and Dowson [9] have solved the film thickness for point contacts of variable ellipticity. For circular contacts and compressible liquids, the central film thickness is

$$h_c = 1.55\alpha^{*0.53}(\mu_0\bar{u})^{0.67}E'^{0.061}R^{0.33}p_H^{-0.201} \tag{1.18}$$

These solutions assume isothermal, fully flooded conditions and interestingly were obtained for very low (compared to practical machines) contact pressures, $p_H < 0.5$ GPa. Solutions were obtained later [10] for higher pressures but using some unrealistic assumptions for the pressure dependence of viscosity and density.

A feature of the elastohydrodynamic lubrication (EHL) problem is that the lubricated conjunction may be separated into zones that operate somewhat independently. The central film thickness is essentially determined by the fluid dynamics of the *inlet zone*, the region of the conjunction that the liquid encounters before entering the area that would be in contact if the liquid was not present. This area that would be in contact if the liquid was not present, the *Hertz zone*, is the region in which the traction force transferred across the film is essentially determined because the piezoviscous property and the high pressures in the Hertz zone combine to make the viscosity here orders of magnitude larger than that outside the zone. If viscous dissipation in the inlet zone raises the liquid temperature sufficiently to result in a reduction in viscosity, then the film will be thinner than predicted by the isothermal formulas above. A location along the inlet zone can be identified at which there exists a position across the gap where the liquid velocity is zero. Downstream of this *stagnation point*, the flow is always in the direction of the surface velocity and upstream there is a region of reverse flow. Unless the rollers are submerged in liquid there should be an air/oil meniscus in the inlet zone. When the meniscus is upstream of the stagnation point, the conjunction is *fully flooded* and the contact is being supplied with all the liquid needed to generate the theoretical film thickness. When the meniscus is downstream of the stagnation point, the contact is *starved* and the full theoretical film thickness will not be achieved. Correction formulas are available [4] for these thermal and starvation effects.

1.4 Experimental Elastohydrodynamics

The experimental investigation of elastohydrodynamic lubrication benefits from a geometry that, compared with say a journal bearing, is relatively easy to access so that the two important parameters of film thickness and traction may be measured in a bench-top setting. Two embodiments of what are known as EHL rigs are illustrated in Figs. 1.4 and 1.5. The most common today is the ball-on-disc machine shown in Fig. 1.4 where a steel ball is loaded against a disc which may be fabricated from steel or a transparent material such as glass or sapphire. When the disc is transparent, it is a simple matter to measure the film thickness of the liquid between the ball and disc by interferometry. The twin disc machine shown in Fig. 1.5 is better suited to traction measurement than film thickness; however this configuration has been used for film thickness measurement as well as by the electrical capacitance method.

The shear force transferred across the film is known as traction. Traction in the rolling direction, longitudinal traction, may be measured as the force shown as the bold arrow

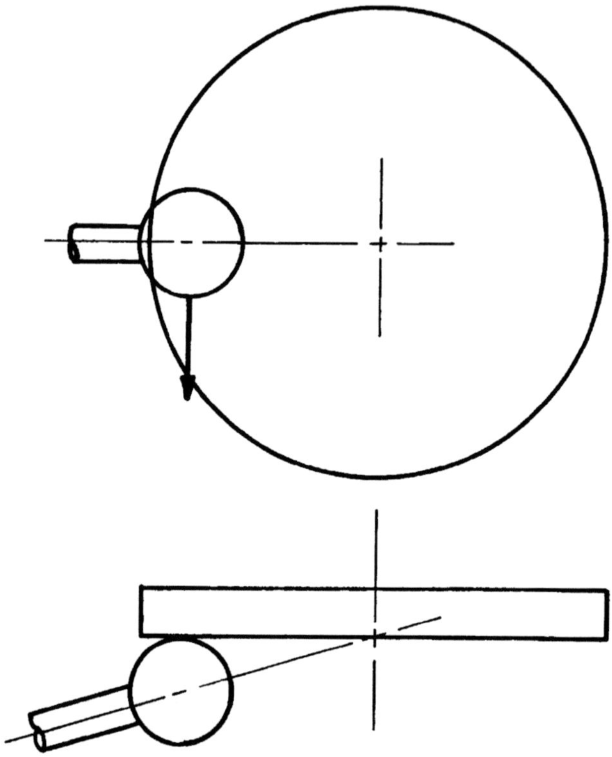

Fig. 1.4. The ball-on-disc machine.

in Fig. 1.4. The component of longitudinal traction due to shear of the liquid film results from a difference in velocities of the surfaces of the ball and disc in the rolling direction, say u_1 and u_2. The rolling velocity is $\bar{u} = (u_1 + u_2)/2$ and the longitudinal slide/roll ratio is $\Sigma_L = (u_1 - u_2)/\bar{u}$. The twin disc machine can also be used to measure longitudinal traction by varying the relative surface velocities of the discs. The twin disc machine is convenient for the measurement of the transverse traction that results from a transverse slide/roll ratio, $\Sigma_T = (v_1 - v_2)/\bar{u}$, where v is the component of surface velocity normal to the rolling direction. The transverse sliding condition can be imposed by skewing the rollers by an angle shown in the upper view of Fig. 1.4 and the associated transverse traction can be measured as the thrust indicated by the bold arrow in the lower view. Traction is generally reported as a traction coefficient, the ratio of traction force to normal force or load, W. Equivalently, the traction coefficient is the ratio of average shear stress to average pressure, $\bar{\tau}/\bar{p}$.

An additional kinematical condition is spin, the relative rotational velocity of the surfaces. Spin has been eliminated from the contact between the ball and disc in Fig. 1.4 by having the rotational axis of the ball intersect the axis of the disc at the contact plane. Other geometries may involve spin. Spin is generated by a nonzero tilt angle shown at the bottom of Fig. 1.5.

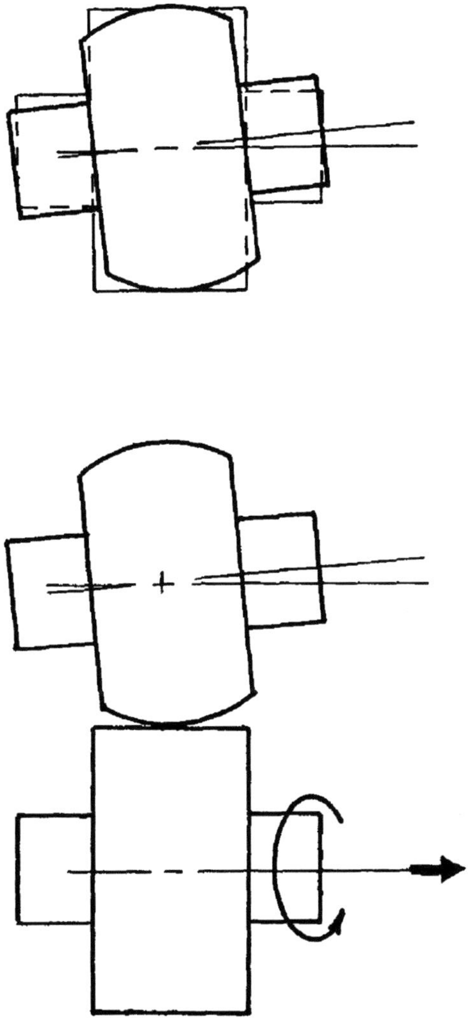

Fig. 1.5. The twin disc machine. Reproduced from Bair, S. and Kotzalas, M., "The Contribution of Roller Compliance to Elastohydrodynamic Traction", *STLE Tribol. Trans.*, Vol. 49, No. 2, 2006, pp. 218–224, with permission of the Society of Tribologists and Lubrication Engineers.

Some measurements of central film thickness are shown in Fig. 1.6. The data points have been reported by INSA, Lyon [11–13] using their colorimetric interferometry technique that has shown to be very repeatable. The contact pressure here was low, $p_H \approx 0.5\,\text{GPa}$, generated by the contact between a steel ball and glass disc. The film thickness of squalane, a low molecular weight (423 kg/kmol) branched alkane, shown as the squares, compares well to the theoretical Newtonian film thickness given by equation (1.18). The density and viscosity variation with pressure for squalane are similar to the assumptions used to arrive at this equation and squalane is Newtonian for the range of shear stress encountered in the inlet zone. The situation is different for the high

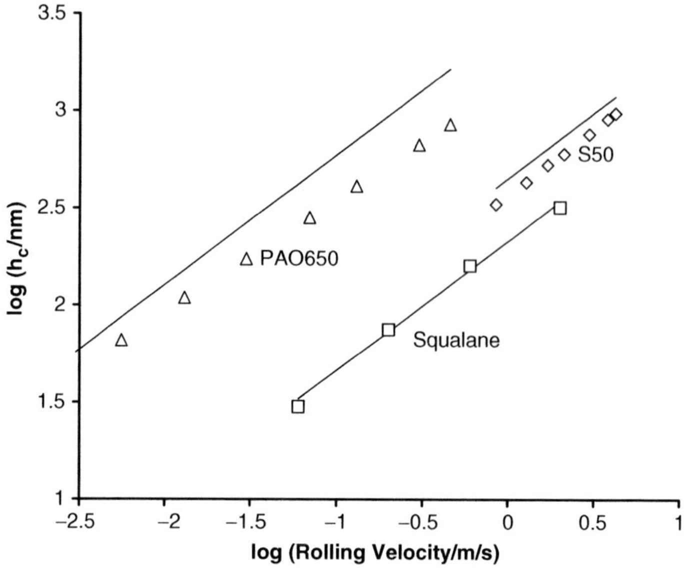

Fig. 1.6. The central film thickness for a high molecular weight polyalphaolefin, PAO650, triangles; a polymer-thickened traction fluid, S50, diamonds; and squalane, squares. The lines are the theoretical Newtonian film thickness given by equation (1.18).

molecular weight polyalphaolefin (PAO) and the polymer-thickened traction fluid (S50). For the S50, the film thickness is uniformly 20% less than theoretical and for the PAO, the film becomes progressively thinner than the Newtonian prediction as velocity increases declining to about one-half of the theoretical thickness at the highest speed. Actually this behavior can be explained by the nature of the shear dependence of viscosity. The S50 quickly shear-thins at low shear stress to a second Newtonian which governs the film thickness. The PAO continuously shear thins according to a power-law when the stress is greater than a critical value which is encountered in the inlet.

The traction behavior, $\bar{\tau}/\bar{p}$, of a lower molecular weight PAO, SHF 401, is shown in Fig. 1.7 for a Hertz pressure of 1.93 GPa or average contact pressure of $\bar{p} = 1.29$ GPa. The calculated central film thickness is 1.08 μm. The traction behavior of a dry contact is also shown where, of course, the film thickness is zero. As the contact pressure is increased, the initial slopes of the dry and lubricated traction curves will tend to converge. The data for both curves have been fitted to an empirical equation often used to describe traction at low slide to roll ratios,

$$\bar{\tau} = \tau_E \sinh^{-1}(C_E \Sigma) \tag{1.19}$$

Here, C_E is a constant for a given load, temperature and rolling velocity and τ_E is a representative stress often improperly associated with the Eyring theory of viscous flow. The value of $\tau_E = 11.6$ MPa for the PAO is approximately consistent with other published values [14]. A peculiar aspect of the field for thirty years has been the association of this

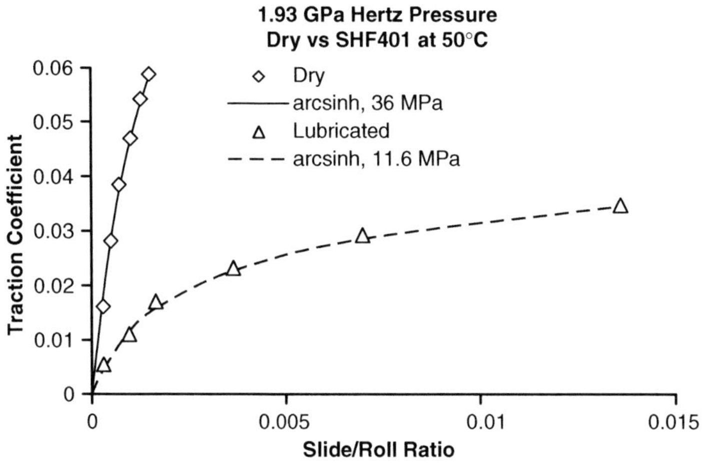

Fig. 1.7. Traction curves for unlubricated and lubricated contact. The lubricated central film thickness is 1.08 μm. Values of the representative stress, τ_E, are given in the legend.

particular form of the traction response at low sliding, that is, equation (1.19), with the constitutive behavior of the liquid film, the value of τ_E being associated with the limit of Newtonian response and the initial slope, $\tau_E C_E$, being used to calculate an average low shear viscosity. In fact, this has been the framework employed almost exclusively for the rheological characterization of the liquid in computational elastohydrodynamics. Now the quality of the fit of equation (1.19) to the dry traction data, with $\tau_E = 36$ MPa, where there is no liquid present, is at least as good as it is to the liquid film traction data. Then one may conclude that this technique of extracting rheological properties from traction curves has been capable of finding the rheological properties of nothing. (Pun intended.) We will see in Chapters 6 and 12 that the representative stress is instead related to the pressure fragility of the liquid.

At greater values of slide-to-roll ratio, a plateau is usually encountered at which the traction is essentially independent of sliding velocity. At still greater slide-to-roll ratio, the traction may decline with increasing sliding velocity.

1.5 Conclusion

The overview above is not intended to be an extensive review of the state of Newtonian elastohydrodynamics and the reader may benefit from the more detailed presentations by, for instance, Hamrock [5]. The following chapters will provide insight and practical information for the application of elastohydrodynamics to real liquids for which the piezoviscous response is more complicated and more severe than can be described by the simple case of the Skinner equation and to real liquids for which the Newtonian assumption seems to always fail in the Hertz zone and often fails in the inlet zone as well. This insight will be useful for understanding the film thickness and traction behavior in Figs. 1.6 and 1.7.

References

[1] Bridgman, P.W., "The Effect of Pressure on the Viscosity of Forty-Three Pure Liquids", *Proc. Amer. Acad.*, Vol. 61, 1926, pp. 57–99.

[2] Johnson, K.L., "Regimes of Elastohydrodynamic Lubrication", *J. Mech. Engr. Sci.*, Vol. 12, No. 1, 1970, pp. 9–16.

[3] Hooke, C.J., "The Elastohydrodynamic Lubrication of Heavily Loaded Contacts", *J. Mech. Engr. Sci.*, Vol. 19, No. 4, 1977, pp. 149–156.

[4] Winer, W.O. and Cheng, H.S., "Film Thickness, Contact Stress and Surface Temperature", *Wear Control Handbook*, Peterson and Winer (Eds.), American Society of Mechanical Engineers, New York, 1980, pp. 81–141.

[5] Hamrock, B.J., *Fluid Film Lubrication*, NASA Reference Publication 1255, 1991, pp. 461, 501–511.

[6] Skinner, S.M., "Film Lubrication of Finite Curved Surfaces", *J. Appl. Phys.*, Vol. 9, 1938, pp. 409–421.

[7] Dowson, D. and Higginson, G.R., *Elastohydrodynamic Lubrication*, Pergamon Press, 1966.

[8] Dowson, D. and Toyoda, S., "A Central Film Thickness Formula for Elastohydrodynamic Line Contacts", *Elastohydrodynamics and Related Topics*, Dowson, D., Taylor, C.M., Godet, M. and Berthe, D. (Eds.), 1978, pp. 60–67.

[9] Hamrock, B.J. and Dowson, D., *Ball Bearing Lubrication – The Elastohydrodynamics of Elliptical Contacts*, Wiley-Interscience, New York, 1981.

[10] Jalali-Vahid, D., Rahnejat, H., Gohar, R. and Jin, Z.M., Comparison between Experiments and Numerical Solutions for Isothermal Elastohydrodynamic Point Contacts", *J. Phys. D: Appl. Phys.*, Vol. 31, 1998, pp. 2725–2732.

[11] Hartl, M., Molimard, J., Krupka, I., Vergne, P., Querry, M., Poliscuk, R. and Liska, M., "Thin Film Lubrication Study by Colorimetric Interferometry", *Thinning Films and Tribological Interfaces*, Dowson et al. (Eds.), Elsevier Science, 2000, pp. 695–704.

[12] Bair, S., Vergne, P. and Querry, M., "A Unified Shear-Thinning Treatment of both Film Thickness and Traction in EHD", *Tribol. Lett.*, Vol. 18, No. 2, 2005, pp. 145–152.

[13] Villechaise, B., Dalmas, G., Makala, J. and Chaomleffel, J.P., "On the Traction Fluid Behavior in Concentrated Contacts", *Boundary and Mixed Lubrication: Science and Applications*, Dowson et al. (Eds.), Elsevier Science, 2002, pp. 401–411.

[14] LaFountain, A., Johnston, G.J. and Spikes, H.A., "Elastohydrodynamic Friction Behavior of Polyalphaolefin Blends", *Tribology for Energy Conservation*, Dowson et al. (Eds.), Elsevier Science, 1998, pp. 465–475.

CHAPTER 2

An Introduction to the Rheology of Polymeric Liquids

2.1 Background

The establishment, around 1930, of the field of rheology – the study of the deformation and flow of materials – was motivated by interest in the strange flow phenomena exhibited by polymeric liquids. A viscosity that depends upon the flow rate, liquids that "spring back" after a sudden deformation and liquids that climb from a free surface upward into a pipe are some examples of unusual flow behavior not associated with Newtonian liquids. Rheology generally excludes the study of low molecular weight liquids, mineral oil for example, as it has been accepted that the flow of these liquids can be successfully described in every case by the Newtonian model. Bird et al. [1, p. 1] placed a rough upper limit on Newtonian behavior at a molecular weight of 1000 kg/kmol. However, within elastohydrodynamics it has long been recognized that the Newtonian assumption must fail at the very high pressures and shear stress in the conjunction of the concentrated contacts described in the previous chapter. At typical pressures of 1–2 GPa, the viscosity may be 10^9 Pa·s and the shear rate within the film may be $10^3\ s^{-1}$. Thin elastohydrodynamic films of much less than one micron thickness are very effective at removing the heat of viscous dissipation. Then, for an isothermally sheared Newtonian liquid, the resulting shear stress of 10^3 GPa would easily exceed the strength of the strongest metals. Since sliding concentrated contacts operate successfully for years, with friction coefficients less than 0.1, lubricated by mineral oils of average molecular weight of about only 300 kg/kmol, some substantial departure from the Newtonian idealization must be occurring. Non-Newtonian flow in simple liquids, i.e. branched alkanes, has been observed with exciting new results from nonequilibrium molecular dynamics simulations [2]. These simulations and experiments in pressurized rheometers support the notion that the nonlinear aspects of the shear behavior of polymers, in the form of shear-thinning and normal stress differences, can be observed in simple liquids as well, provided that the shear stress is sufficiently large. Large shear stresses can be generated in thin compressed films of low molecular weight liquids.

Rheology can also been described as the search for the *constitutive equations* relating the stress to the deformation in a particular polymeric material. In this respect the field is incomplete; a constitutive equation can be written that is accurate over a limited range of kinematics only, and a part of the skill required in the application of rheology to practical problems is the recognition of which relation is appropriate for the present situation. Real flow problems are often so complex that the deformations cannot be known accurately without *a priori* knowledge of the constitutive relation. In order to test constitutive models and to provide rheological property data for their implementation, experiments are devised to produce flow kinematics that are simple and completely known. The apparatus for such experiments is a *rheometer*. If viscosity is the only property being measured, the apparatus is a *viscometer*.

The materials studied by rheologists include neat liquid polymers (polymer melts), polymers dissolved in low molecular weight liquids (polymer solutions), suspensions of a second immiscible liquid phase, suspensions of solid particles or gas and semi-solid materials displaying a yield stress. Here, because of the application to lubrication only the liquid polymer and polymer solution behavior is addressed.

2.2 The Newtonian Model

The constitutive equation for a compressible Newtonian liquid [3] with Newtonian shear viscosity, μ, is

$$\begin{bmatrix} \sigma_{xx} & \sigma_{xy} & \sigma_{xz} \\ \sigma_{yx} & \sigma_{yy} & \sigma_{yz} \\ \sigma_{zx} & \sigma_{zy} & \sigma_{zz} \end{bmatrix} = \tag{2.1}$$

$$\begin{bmatrix} \mu\left(\frac{\partial u}{\partial x}+\frac{\partial u}{\partial x}\right)+\upsilon\left(\frac{\partial u}{\partial x}+\frac{\partial v}{\partial y}+\frac{\partial w}{\partial z}\right)-p & \mu\left(\frac{\partial u}{\partial y}+\frac{\partial v}{\partial x}\right) & \mu\left(\frac{\partial u}{\partial z}+\frac{\partial w}{\partial x}\right) \\ \mu\left(\frac{\partial v}{\partial x}+\frac{\partial u}{\partial y}\right) & \mu\left(\frac{\partial v}{\partial y}+\frac{\partial v}{\partial y}\right)+\upsilon\left(\frac{\partial u}{\partial x}+\frac{\partial v}{\partial y}+\frac{\partial w}{\partial z}\right)-p & \mu\left(\frac{\partial v}{\partial z}+\frac{\partial w}{\partial y}\right) \\ \mu\left(\frac{\partial w}{\partial x}+\frac{\partial u}{\partial z}\right) & \mu\left(\frac{\partial w}{\partial y}+\frac{\partial v}{\partial z}\right) & \mu\left(\frac{\partial w}{\partial z}+\frac{\partial w}{\partial z}\right)+\upsilon\left(\frac{\partial u}{\partial x}+\frac{\partial v}{\partial y}+\frac{\partial w}{\partial z}\right)-p \end{bmatrix}$$

where σ_{xx} is the normal stress component in the x-direction and σ_{xy} is the (shear) stress component in the x,y-directions of the stress tensor, $\boldsymbol{\sigma}$ (see Fig. 2.1). For the normal stresses, tension is positive, leading to the minus sign on pressure. The dilation rate is $(\partial u/\partial x) + (\partial v/\partial y) + (\partial w/\partial z)$ and υ is a dilatational or bulk viscosity that together with μ results in a relaxation of the volume following an instantaneous change in the

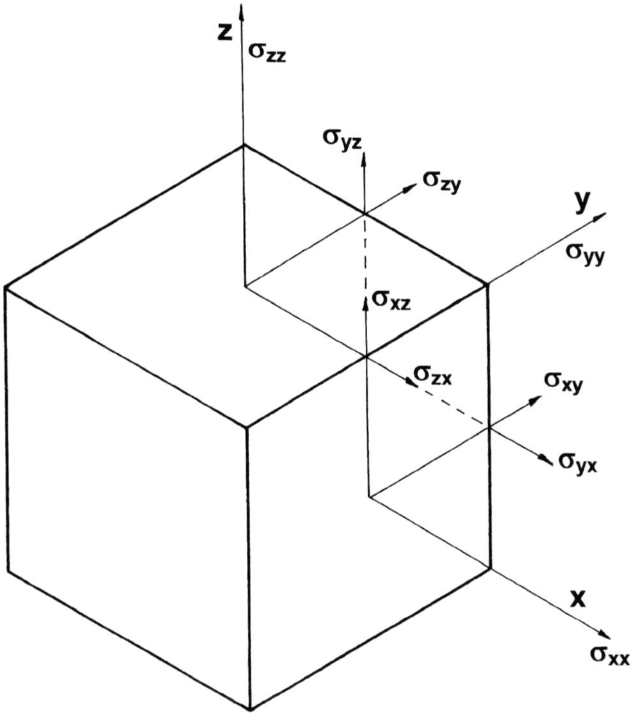

Fig. 2.1. Stresses defined.

mechanical pressure, defined as $p_m = -\left(\sigma_{xx} + \sigma_{yy} + \sigma_{zz}\right)/3$ [4]. Such volume relaxation can be experimentally observed at pressures near to the glass transition (Chapter 4). For compressible flow, a distinction must be made between p_m and the thermodynamic pressure, p, that is related to the density (or volume) and the temperature through an equation of state (Chapter 4). For compressible flow, p and p_m are related by

$$p_m = p - \left(\upsilon + \frac{2}{3}\mu\right)\left(\frac{\partial u}{\partial x} + \frac{\partial v}{\partial y} + \frac{\partial w}{\partial z}\right) \tag{2.2}$$

For an incompressible Newtonian liquid, $(\partial u/\partial x) + (\partial v/\partial y) + (\partial w/\partial z) = 0$, there is no distinction between p and p_m and the constitutive equation is then

$$\begin{bmatrix} \sigma_{xx} & \sigma_{xy} & \sigma_{xz} \\ \sigma_{yx} & \sigma_{yy} & \sigma_{yz} \\ \sigma_{zx} & \sigma_{zy} & \sigma_{zz} \end{bmatrix} = \begin{bmatrix} \mu\left(\frac{\partial u}{\partial x}+\frac{\partial u}{\partial x}\right)-p & \mu\left(\frac{\partial u}{\partial y}+\frac{\partial v}{\partial x}\right) & \mu\left(\frac{\partial u}{\partial z}+\frac{\partial w}{\partial x}\right) \\ \mu\left(\frac{\partial v}{\partial x}+\frac{\partial u}{\partial y}\right) & \mu\left(\frac{\partial v}{\partial y}+\frac{\partial v}{\partial y}\right)-p & \mu\left(\frac{\partial v}{\partial z}+\frac{\partial w}{\partial y}\right) \\ \mu\left(\frac{\partial w}{\partial x}+\frac{\partial u}{\partial z}\right) & \mu\left(\frac{\partial w}{\partial y}+\frac{\partial v}{\partial z}\right) & \mu\left(\frac{\partial w}{\partial z}+\frac{\partial w}{\partial z}\right)-p \end{bmatrix} \tag{2.3}$$

The quantities in parentheses on the right-hand side of equation (2.3) are the components of the rate of deformation tensor, **D**, multiplied by two. Equation (2.3) can be written in tensor "shorthand" as

$$\boldsymbol{\sigma} = 2\mu\mathbf{D} - \mathbf{I}p \tag{2.3a}$$

There are two simple ideal flows that are utilized for rheological experiments, simple uniaxial extension and simple shear. For simple uniaxial extension, the elongation strain rate is $\dot{\varepsilon} = \partial u/\partial x$, if the flow is incompressible, $\partial v/\partial y = \partial w/\partial z = -\dot{\varepsilon}/2$, and all other velocity gradients are zero. That is, for a cylindrical element of incompressible liquid being stretched at a particular elongation rate, the diameter will contract at half that rate. The off-diagonal (shear) stresses are zero and $\sigma_{yy} = \sigma_{zz}$. We shall ignore the body forces and acceleration and the only important stresses are the normal stress differences, $\sigma_{xx} - \sigma_{yy}$ or $\sigma_{xx} - \sigma_{zz}$. If we define the *elongational viscosity* as $\eta_e = \left(\sigma_{xx} - \sigma_{yy}\right)/\dot{\varepsilon}$, equation (2.3) gives for a Newtonian liquid,

$$\eta_e = 3\mu \tag{2.4}$$

For simple shear, $\partial u/\partial z = \dot{\gamma}$, the *shear rate*, and the other velocity gradients are zero. Equation (2.1) or (2.3) yields for a Newtonian liquid,

$$\sigma_{xz} = \sigma_{zx} = \tau = \mu\dot{\gamma} \tag{2.5}$$

as it should. The other shear stresses are zero and the first and second normal stress differences are

$$N_1 = \sigma_{xx} - \sigma_{zz} = 0 \quad \text{and} \quad N_2 = \sigma_{zz} - \sigma_{yy} = 0 \tag{2.6}$$

Simple shear has a special importance for rheological experiments related to lubrication.

Experimental measurements for polymeric liquids can yield very different results from the predictions of the Newtonian model. All of the data presented in this chapter have been obtained at ambient pressure.

2.3 Material Functions for Polymeric Liquids

2.3.1 Shear viscosity

For polymeric liquids the shear viscosity is a function of shear rate (or stress) and a generalized or non-Newtonian viscosity is defined in simple shear as $\eta(\dot{\gamma}) = \tau/\dot{\gamma}$ or $\eta(\tau) = \tau/\dot{\gamma}$. This is illustrated in Fig. 2.2 where the logarithm of the steady shear viscosity at 150°C of a molten low-density polyethylene is plotted against the logarithm of shear rate. This information is presented in the form of a graph called a *flow curve*. The steady elongational viscosity is also plotted against elongation rate. The shear viscosity is seen to monotonically decrease with shear rate in Fig. 2.2, a phenomenon known as *shear-thinning*. This effect is clearly most important for lubrication, as the property of viscosity provides the load-supporting pressure that is described by the Reynolds equation (1.17).

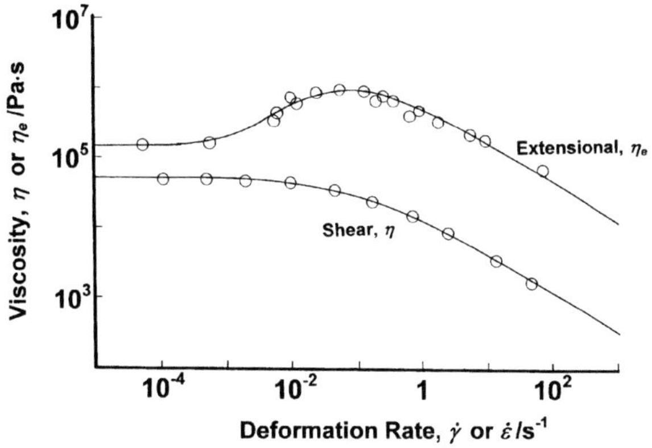

Fig. 2.2. The steady shear viscosity and steady elongational viscosity of a polyethylene melt from [5]. Reprinted with permission of Elsevier.

If the Newtonian viscosity, μ, is replaced by the shear-dependent viscosity, η, in the incompressible Newtonian constitutive equation (2.3) the result is the incompressible *generalized Newtonian fluid* model,

$$\boldsymbol{\sigma} = 2\eta\mathbf{D} - \mathbf{I}p \tag{2.7}$$

Of course for flows more complicated than simple shear, the argument of the viscosity function requires some consideration. The viscosity must depend upon a quantity that is independent of the way the coordinate system is specified. Then it should depend on some combination of the three scalar invariants of either the strain rate tensor, $\mathbf{D}$, or of the deviatoric stress tensor, $\boldsymbol{\sigma} + \mathbf{I}p$ [1]. The deviatoric stress is used in place of the total stress so that a hydrostatic pressure cannot affect flow other than through its effect on the low shear viscosity. If we define the magnitude of the strain rate tensor, $\dot{\gamma}^*$, so that it will be the argument for the viscosity function, one of the invariants is zero for shear flows and another is zero for incompressible flow [1, p. 170], leaving the second invariant as the choice for an argument of the function, $\eta(\dot{\gamma}^*)$, for general flows. There are a number of different ways to write the second invariant and for the purpose of replacing $\dot{\gamma} = \partial u/\partial z$ as the definition of $\dot{\gamma}^*$, the following relation is consistent with simple shear measurements where $\dot{\gamma}^* = \partial u/\partial z$.

$$\dot{\gamma}^* = \sqrt{\left(\frac{\partial u}{\partial y}+\frac{\partial v}{\partial x}\right)^2+\left(\frac{\partial u}{\partial z}+\frac{\partial w}{\partial x}\right)^2+\left(\frac{\partial w}{\partial y}+\frac{\partial v}{\partial z}\right)^2+2\left(\frac{\partial u}{\partial x}\right)^2+2\left(\frac{\partial v}{\partial y}\right)^2+2\left(\frac{\partial w}{\partial z}\right)^2} \tag{2.8}$$

Following a similar argument, if viscosity is to depend upon stress as $\eta(\tau^*)$, the appropriate invariant of the deviatoric stress tensor is [1]

$$\tau^* = \sqrt{\frac{(\sigma_{xx}+p)^2}{2}+\frac{(\sigma_{yy}+p)^2}{2}+\frac{(\sigma_{zz}+p)^2}{2}+(\sigma_{xy})^2+(\sigma_{xz})^2+(\sigma_{yz})^2} \tag{2.9}$$

where $\tau^* = \sigma_{xz} = \sigma_{zx}$ in simple shear. For a purely hydrostatic state of stress, $\sigma_{xx} = \sigma_{yy} = \sigma_{zz} = -p$ and $\tau^* = 0$ as expected. These relationships are useful for the incompressible generalized Newtonian fluid.

Examples of expressions for the viscosity function used in the generalized Newtonian fluid model are the Cross equation [6],

$$\eta(\dot{\gamma}) = \mu_2 + \frac{\mu - \mu_2}{1 + |\lambda\dot{\gamma}^*|^{1-n}} \tag{2.10}$$

and the Ellis equation [7],

$$\eta(\tau) = \mu_2 + \frac{\mu - \mu_2}{1 + \left|\dfrac{\tau^*}{\tau_c}\right|^{\left(\frac{1}{n}-1\right)}} \tag{2.11}$$

Here, n is the power-law exponent, λ is a characteristic time and τ_c is a characteristic stress. The dimensionless $\lambda\dot{\gamma}^*$ is the Weissenberg number for the flow. For the liquid in Fig. 2.2, at shear rate less than 10^{-3} s^{-1}, the viscosity is reasonably constant and approaches a limiting value we call the *limiting low shear viscosity*, μ, and the low shear regime for which $\eta \approx \mu$ is the *terminal regime*. In the rheology literature the low shear viscosity is typically given the term η_0, instead of μ, to differentiate the terminal behavior of a shear-thinning liquid from constant viscosity of a Newtonian liquid. In the elastohydrodynamic literature, the subscript, "0" is typically reserved for the value of a property at ambient pressure. Then, the same variable name, μ, is used here for both the Newtonian viscosity and the low shear viscosity to avoid confusion when considering the pressure variation of viscosity. The second Newtonian viscosity, μ_2, may appear in measurements as a limiting high shear viscosity, although no limiting high shear Newtonian behavior can be observed in Fig. 2.2.

The generalized Newtonian model is extremely useful in engineering calculations of flow but it has some interesting limitations to be seen in the following sections.

2.3.2 *Elongational viscosity*

The steady elongational viscosity, $\eta_e = (\sigma_{xx} - \sigma_{yy})/\dot{\varepsilon}$, is plotted in Fig. 2.2 for the polyethylene melt. In the terminal regime $\eta_e \approx 3\eta$, as the Newtonian model predicts but this relationship does not hold for higher deformation rates. The elongational viscosity first increases with elongation rate, and then decreases. The shear viscosity function when implemented in the generalized Newtonian model is not helpful for extensional viscosity.

There is another complication as well. If the direction of a simple shear is reversed, the same value of shear viscosity will be observed. The absolute value function appears in equations (2.10) and (2.11). Reversing the direction of elongation will not generally yield the same elongational viscosity.

For elastohydrodynamics, the elongation component of the flow should be insignificant compared with shear flow. Tanner [3] has investigated the effect of elongation on a simple squeeze film lubrication flow and found that the elongational contribution to the load was insignificant for the usual EHL thin films. We will not consider elongation further.

2.3.3 Normal stress differences

Shear-thinning is generally accompanied by the rapid development of differences in the normal stresses in simple shear. This nonintuitive effect is similar to the behavior of rubbery elastic solids that are subjected to a simple shear strain that is greater than the usual infinitesimal assumption. For rubber at large shear strain, the shear stress is linear with strain but the difference between the tension in the flow direction and the tension in the cross-flow direction increases with the square of the shear strain [8]. If unconstrained, the separation between the parallel shearing faces of a block of the solid will grow.

Normal stress differences for steady shear of a polymer solution of polyisobutylene in decalin, a low molecular weight cyclic hydrocarbon, are illustrated in the flow curves in Fig. 2.3. In this case the flow curves are in the form of the logarithm of stress plotted against the logarithm of shear rate. The shear stress, τ, can be seen to follow a power-law function of the shear rate,

$$\tau \propto \dot{\gamma}^n \tag{2.12}$$

usually known as the Ostwald–de Waele equation with $n = 0.53$. A weakness of this model is the lack of terminal behavior. The normal stress differences, $N_1 = \sigma_{xx} - \sigma_{zz}$ and $N_2 = \sigma_{zz} - \sigma_{yy}$, also follow a power-law but with much larger power-law exponents than the shear stress. Notice that $N_1 > 0$ and $N_2 < 0$ and $|N_2| \approx |N_1|/10$ for this liquid. The first normal stress difference in Fig. 2.3 is much larger than the shear stress and this behavior may be observed in polymer-thickened lubricants as well [9]. The generalized Newtonian fluid model (2.7) does not predict these stress differences.

The measurement of N_1 has become relatively routine, while N_2 requires difficult specialized laboratory technique. For Fig. 2.3, the shear stress and N_1 were obtained from the torsional flow between a rotating cone and a coaxial stationary circular disc. The torque to restrain the rotation of the stationary disc yields the shear stress and the normal force required to prevent the separation of the cone and plate yields N_1. A pressure profile is generated by N_1 that supports a normal load. Notice that the Newtonian based Reynolds equation (1.11) of Chapter 1 does not predict a pressure profile for this geometry. There is no change in film thickness in the flow direction, $dh/dx = 0$ in equation (1.11). Then the pressure gradient in the flow direction must be constant from (1.11) and must be zero for the pressure to be continuous around the gap.

The measurement of N_2 in Fig. 2.3 was provided by analysis of the shape of the free surface of liquid flowing downhill in an open channel. See Ref. [10] for a complete

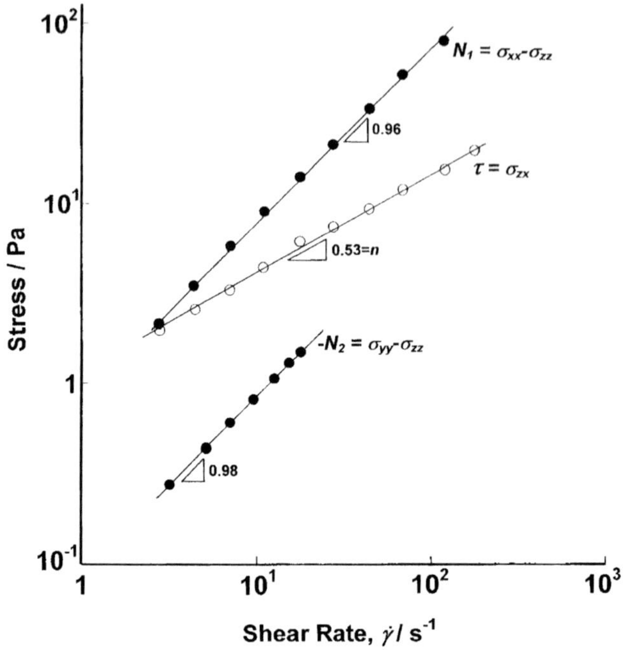

Fig. 2.3. The shear rate dependence of shear stress, τ, first normal stress difference, N_1, and second normal stress difference, N_2, for a 1.1% polyisobutylene in decalin solution from [10]. Reprinted with permission of Elsevier.

description of the technique. Until about 40 years ago it was supposed that N_2 was zero and this assumption is known as the Weissenberg hypothesis.

When the direction of shear, $\partial u/\partial z = \dot{\gamma}$, is reversed in a simple shear experiment, the direction of the shear stress, $\sigma_{zx} = \tau$, changes; however, symmetry requires that the normal stress differences $\sigma_{xx} - \sigma_{zz} = N_1$ and $\sigma_{zz} - \sigma_{yy} = N_2$ be independent of the direction of shear. That is, the normal stress differences must be even functions of $\dot{\gamma}$ while shear stress is an odd function. This requirement suggests that the definitions of the normal stress coefficients, ψ_1 and ψ_2, should be given by $N_1 = \psi_1\dot{\gamma}^2$ and $N_2 = \psi_2\dot{\gamma}^2$. It is usually found that in the terminal regime, N_1 varies with $\dot{\gamma}^2$ or τ^2, in other words, that like viscosity, ψ_1 is constant.

Some methods of calculating N_1 from the way that the viscosity varies with shear have been advanced. See [11] for examples. These methods should be limited to special cases because of the discovery of liquids which generate a large first normal stress difference while the viscosity appears to be independent of shear [12]. These *Boger liquids* are dilute solutions of high molecular weight polymers in a low molecular weight, but very viscous, solvent such as polybutene. Then the polymer contribution to the viscosity is small.

2.4 Rheological Models

2.4.1 Viscoelastic models

The experiments and models described to this point have considered steady-state deformation rates. If the shear rate in a polymeric liquid is imposed quickly enough, the stress may require some time to reach the steady value and may even overshoot the steady value if the strain rate is high. If the strain rate is applied periodically as a sinusoidally varying function of time, a component of stress will appear that is out of phase with the strain rate. These phenomena, like the normal stress differences, cannot be predicted by the generalized Newtonian model.

In classical elasticity, the stress depends on strain and not on strain rate. In classical fluid mechanics, the stress depends on strain rate and not on the strain. The behavior described above indicates that polymeric liquids can have the attributes of both viscosity and elasticity simultaneously. Models that combine an elastic modulus and a viscosity are time dependent. The first and perhaps the most simple viscoelastic model is the single mode Maxwell model that sums a strain rate due to elasticity, $\partial(\tau/G)/\partial t$, with a strain rate due to viscous flow, τ/μ, to obtain for constant shear modulus, G,

$$(\boldsymbol{\sigma} + \mathbf{I}p) + \frac{\mu}{G}\frac{\partial}{\partial t}(\boldsymbol{\sigma} + \mathbf{I}p) = 2\mu\mathbf{D} \tag{2.13}$$

The time derivative of the deviatoric stress in the equation above requires special care when the strains are not infinitesimal. There are three main objective time derivatives, the corotational (or Jaumann) derivative and the upper and lower convected derivatives, that may be substituted for $\partial/\partial t$ in equation (2.13). Joseph [13] has defined an interpolated invariant derivative that provides for a continuous mix of the three.

The Jaumann derivative has been employed extensively in lubrication, see Ref. [14] for example, and when used in the Maxwell model (2.13), yields some aspects of measurable steady shear response. The material derivative is

$$\frac{D}{Dt} = \frac{\partial}{\partial t} + u\frac{\partial}{\partial x} + v\frac{\partial}{\partial y} + w\frac{\partial}{\partial z} \tag{2.14}$$

The x,y element of the vorticity tensor is

$$\omega_{xy} = -\omega_{yx} = \frac{1}{2}\left(\frac{\partial v}{\partial x} - \frac{\partial u}{\partial y}\right) \tag{2.15}$$

and so on. Then the Jaumann derivative for the x,z component, σ_{xz}, of the stress tensor is, for an example,

$$\frac{\mathcal{D}\sigma_{xz}}{\mathcal{D}t} = \frac{D\sigma_{xz}}{Dt} + \{\omega_{xx}\sigma_{xz} + \omega_{xy}\sigma_{yz} + \omega_{xz}\sigma_{zz} - \sigma_{xx}\omega_{xz} - \sigma_{xy}\omega_{yz} - \sigma_{xz}\omega_{zz}\} \tag{2.16}$$

The derivatives of the other components follow the same rule.

For the steady simple shear experiment that has been employed as a previous example, $\partial u/\partial z = \dot{\gamma}$, and the other velocity gradients are zero. The stress gradients are zero and the stresses are steady. Then the material derivatives of the stress components with respect to time are zero. Solving then

$$(\boldsymbol{\sigma} + \mathbf{I}p) + \frac{\mu}{G}\frac{D}{Dt}(\boldsymbol{\sigma} + \mathbf{I}p) = 2\mu\mathbf{D} \tag{2.17}$$

for $\sigma_{xz} = \sigma_{zx} = \tau$ yields,

$$\sigma_{xz} + \frac{\mu}{G}\left\{\frac{\partial\sigma_{xz}}{\partial t} + u\frac{\partial\sigma_{xz}}{\partial x} + v\frac{\partial\sigma_{xz}}{\partial y} + w\frac{\partial\sigma_{xz}}{\partial z} + \frac{1}{2}\left[\sigma_{yz}\left(\frac{\partial v}{\partial x} - \frac{\partial u}{\partial y}\right) + \sigma_{xy}\left(\frac{\partial v}{\partial z} - \frac{\partial w}{\partial y}\right)\right.\right.$$
$$\left.\left. + (\sigma_{zz} - \sigma_{xx})\left(\frac{\partial w}{\partial x} - \frac{\partial u}{\partial z}\right)\right]\right\} = \mu\left(\frac{\partial u}{\partial z} + \frac{\partial w}{\partial x}\right) \tag{2.18}$$

The above simplifies to

$$\tau + \frac{\mu\dot{\gamma}}{2G}(\sigma_{xx} - \sigma_{zz}) = \mu\dot{\gamma} \tag{2.19}$$

Similarly solving (2.17) for σ_{xx}, σ_{zz} and σ_{yy},

$$\sigma_{xx} + p = \frac{\mu\dot{\gamma}}{G}\tau, \quad \sigma_{zz} + p = -\frac{\mu\dot{\gamma}}{G}\tau \quad \text{and} \quad \sigma_{yy} + p = 0 \tag{2.20}$$

Combining (2.19) and (2.20) gives,

$$\tau = \frac{\mu\dot{\gamma}}{1 + \left(\frac{\mu\dot{\gamma}}{G}\right)^2} \quad \text{and} \quad \eta = \frac{\mu}{1 + \left(\frac{\mu\dot{\gamma}}{G}\right)^2} \tag{2.21}$$

$$N_1 = \sigma_{xx} - \sigma_{zz} = \frac{2G\left(\frac{\mu\dot{\gamma}}{G}\right)^2}{1 + \left(\frac{\mu\dot{\gamma}}{G}\right)^2} \tag{2.22}$$

$$N_2 = \sigma_{zz} - \sigma_{yy} = -\frac{N_1}{2} \tag{2.23}$$

The elongational viscosity for the corotational Maxwell model is $\eta_e = 3\mu$.

These results may be compared to the data for a non-Newtonian standard reference material in Fig. 2.4. The National Institute for Science and Technology (NIST) has thoroughly characterized the steady shear and dynamic shear properties of two polymeric liquids. The shear stress and the first normal stress difference are plotted in Fig. 2.4 for the dimethyl silicone SRM 2491 in steady shear at two temperatures as shown in the figure. The data in Fig. 2.4 do not penetrate far into the nonlinear regime and using a value of $G = 9000$ Pa in equations (2.21) and (2.22) results in the curves plotted in the figure. First, the corotational Maxwell model represents the available shear stress

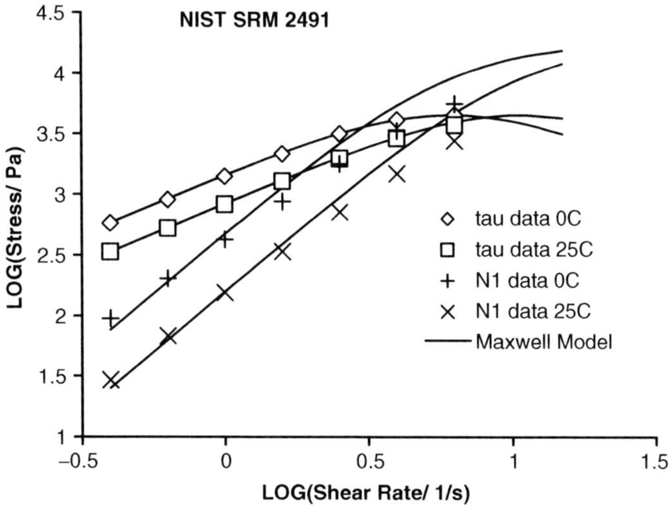

Fig. 2.4. The shear rate dependence of shear stress, τ, and first normal stress difference, N_1, at two temperatures, 0° and 25°C, for a non-Newtonian standard reference material, a dimethyl siloxane. The co-rotational Maxwell model with $G = 9$ kPa is shown as the curves. Data from [15].

data well, but at higher shear rates the shear stress displays a local maximum and then decreases with increasing rate as can be seen in Fig. 2.4. This response is not physically acceptable as nonmonotonic flow curves can result in a constitutive instability and banded shear flows where two shear rates can coexist in the same flow. Comparing, however, the N_1 data with the corotational Maxwell model shows good agreement in the terminal regime but not at higher shear rates. Then the corotational Maxwell model predicts shear-thinning that is overly severe but the prediction for the first normal stress difference is reasonable at low shear rates. The magnitude of the second normal stress difference, equation (2.23), is in general too large, but the sense is correct. The second normal stress difference is not shown in Fig. 2.4.

There is another interesting property of the single mode Maxwell model. For small strain oscillatory simple shear with $\gamma = \gamma_A \sin \omega t$ and $\dot{\gamma} = \gamma_A \omega \cos \omega t$, the time derivative in equation (2.13) is appropriate and we have,

$$\frac{d\tau}{dt} + \frac{G}{\mu}\tau = G\gamma_A \omega \cos \omega t \tag{2.24}$$

where γ_A is the shear strain amplitude and ω is the angular frequency. This linear first order equation has the solution

$$\tau = \frac{\dfrac{\mu^2 \omega \gamma_A}{G}}{1 + \left(\dfrac{\mu\omega}{G}\right)^2} \left[\frac{G}{\mu} \cos \omega t + \omega \sin \omega t\right] \tag{2.25}$$

where the shear stress in phase with the shear rate and the dynamic viscosity are,

$$\tau' = \frac{\mu\gamma_A\omega\cos\omega t}{1+\left(\frac{\mu\omega}{G}\right)^2} \quad \text{and} \quad \eta' = \frac{\tau'}{\dot{\gamma}} = \frac{\mu}{1+\left(\frac{\mu\omega}{G}\right)^2} \tag{2.26}$$

Compare this result for the dynamic viscosity, $\eta'(\omega)$, with the steady shear result for $\eta(\dot{\gamma})$ in equation (2.21). They are equivalent. This *analogy between continuous and oscillatory shear* has been used extensively in elastohydrodynamics although, experimentally, it sometimes fails to be accurate at large shear rates far beyond the terminal regime. Dyson [16] had some success in applying this analogy to the rheology of silicone oils and predicting the film thickness in line contact using viscosity measured at high frequency. There is experimental advantage to be gained through the use of small strains and high frequencies as a substitute for high shear rates. The viscous heating, which substantially limits the shear rates that can be attained isothermally, is rarely a problem in dynamic measurements. This technique was unfortunately abandoned by elastohydrodynamics about 30 years ago. An excellent exposition on the dynamic properties of liquids that includes high-pressure effects can be found in Harrison [17].

Other objective time derivatives can be utilized for the Maxwell model. The upper convected derivative results in the shear viscosity being constant and the elongational viscosity being unbounded at some strain rates. Also $N_1 = 2\mu^2\dot{\gamma}^2/G$ and $N_2 = 0$. There are other ways of combining elastic and viscous components in addition to the Maxwell approach.

It may be useful to substitute one of the shear-thinning viscosity functions, $\eta(\dot{\gamma}^*)$, into the Maxwell model (2.13). If a convected time derivative is employed, the White–Metzner model will be the result [1, p. 351]. Viscosity functions for this purpose can be found in the next section.

2.4.2 Generalized Newtonian fluid models

The generalized Newtonian model can be useful for engineering calculations of flow although we have seen that it lacks sufficient detail to describe many non-Newtonian effects. Since the Reynolds equation is essentially a flow conservation equation, the usefulness of the generalized Newtonian model for lubrication problems should be apparent. The two viscosity functions introduced in equations (2.10) and (2.11) require parameters, λ and τ_c, respectively, that characterize the limit of the terminal (constant viscosity) regime.

The relative viscosity and first normal stress difference of another NIST non-Newtonian standard reference liquid, SRM 2490, a polymer solution, is shown in Fig. 2.5 plotted against shear stress. The relative viscosity, η/μ, data for three temperatures are shown as superimposing to form a master curve at the top graph of Fig. 2.5. Plotting η/μ against shear rate does not superimpose the data in this way. The normal stress data for three temperatures are shown as superimposing to form a master curve at the bottom graph

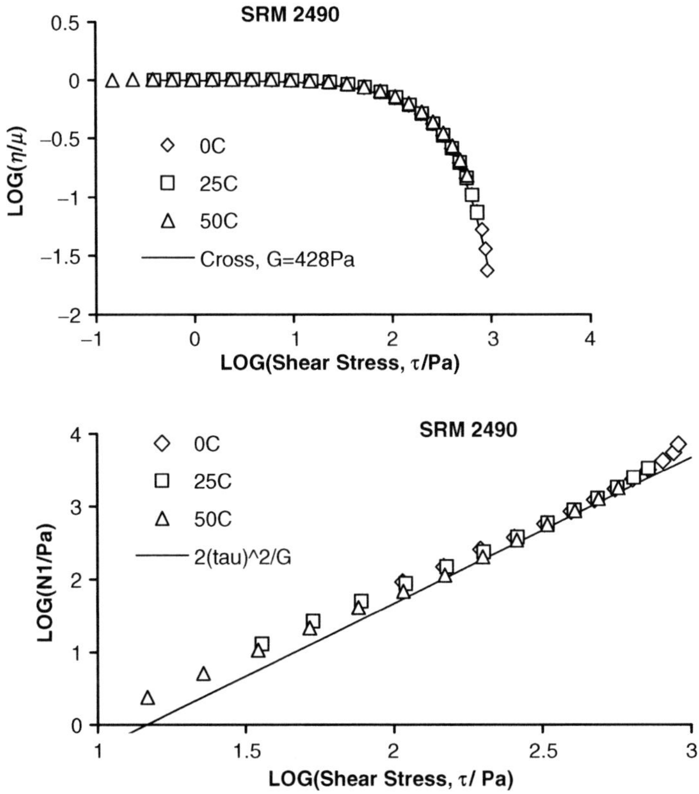

Fig. 2.5. The shear stress dependence of the relative viscosity, η/μ, top, and the first normal stress difference, N_1, bottom, for a non-Newtonian standard reference material, a solution of polyisobutylene in pristane, from [18].

of Fig. 2.5. This ability to normalize data by presenting properties as functions of stress (or as a function of $\mu\dot{\gamma}$) is a general feature of shear-thinning. The Cross equation (2.10) is plotted as the curve in the top of Fig. 2.5 using parameters supplied by NIST [18], $n = 0.18$ and $\mu_2 = 0$.

$$\eta\left(\dot{\gamma}\right) = \mu_2 + \frac{\mu - \mu_2}{1 + \left|\lambda\dot{\gamma}\right|^{1-n}} \tag{2.10}$$

Here, the characteristic time, λ, has been written as a Maxwell relaxation time,

$$\lambda = \frac{\mu}{G} \tag{2.27}$$

with $G = 428$ Pa. The approximation of the characteristic time with a viscoelastic relaxation time can be justified. Bird et al. [1, p. 234] state "... experimental data, many molecular theories, and many nonlinear viscoelastic constitutive equations do show a connection between the time constant determined from … the $\eta(\dot{\gamma})$ expression … and that determined from … linear viscoelasticity …". The Cross equation with $G = 428$ Pa is an excellent fit to the viscosity data in Fig. 2.5 for all three temperatures. The corotational Maxwell model gives for the terminal regime (see equation (2.22) for $\mu\dot{\gamma} \ll G$),

$$N_1 \approx \frac{2\tau^2}{G} \tag{2.28}$$

and the upper convected Maxwell model predicts this relationship for all shear stress. This relationship is plotted in the bottom graph of Fig. 2.5 and shows good agreement with the data. If we substitute $G = \tau_c$ in the $\eta(\tau)$ expressions then it will be shown in the following that we have a consistent way to write all of the ordinary viscosity functions.

The Cross equation (2.10) and the Ellis equation with G replacing τ_c,

$$\eta(\tau) = \mu_2 + \frac{\mu - \mu_2}{1 + \left|\frac{\tau}{G}\right|^{\left(\frac{1}{n}-1\right)}} \tag{2.29}$$

are plotted in Fig. 2.6 with $\mu = 10$, $\mu_2 = 10^{-3}$, $\lambda = 1$, $G = \mu/\lambda = 10$ and $n = 0.4$ using arbitrary units. The two viscosity functions are essentially the same in the terminal regime, $\tau = \mu\dot{\gamma}$, and power-law regime, $\tau = G(\lambda\dot{\gamma})^n$, where thin straight lines represent these relationships in Fig. 2.6. The lines intersect at $\lambda\dot{\gamma} = 1$ and $\tau/G = 1$. That is, $\lambda\dot{\gamma} = 1$

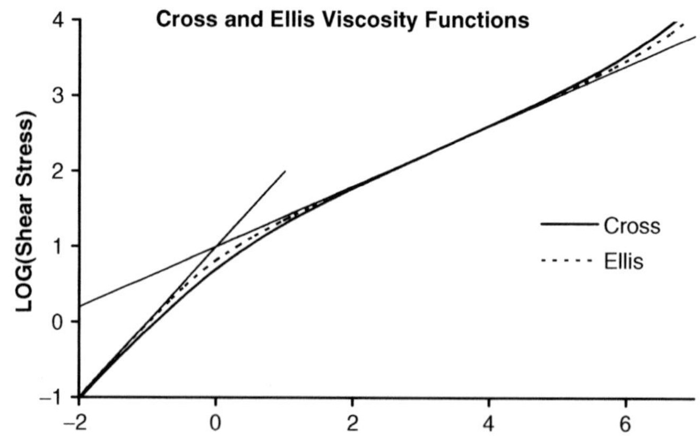

Fig. 2.6. The shear stress, τ, versus shear rate, $\dot{\gamma}$, for two viscosity functions using $\mu = 10$, $\mu_2 = 10^{-3}$, $\lambda = 1$, $G = \mu/\lambda = 10$ and $n = 0.4$ with arbitrary units. The fine lines represent the terminal regime, $\tau = \mu\dot{\gamma}$, and the power law regime, $\tau = G(\lambda\dot{\gamma})^n$, that are approximately tangent to the viscosity functions. The intersection occurs at $\tau = G$ and $\dot{\gamma} = 1/\lambda$.

and $\tau/G = 1$ establish approximately the limit of Newtonian response by specifying the rate, $\dot{\gamma}$, or stress, τ, respectively, at the intersection of lines approximately tangent to the terminal regime and to the power-law regime. All of the ordinary viscosity functions explicitly incorporating the power-law exponent, n, share this property provided that μ_2 is small compared with μ or is zero. When μ/μ_2 is not very much larger than one, then the power-law regime never fully develops.

Table 2.1 lists the viscosity functions already mentioned and, in addition, the Spriggs truncated power-law [19], Ferry [20], Rabinowitsch [21], Carreau [22], Carreau–Yasuda [3, p. 18], two models introduced by the author's laboratory [23,24] and the Ree–Eyring [25] model. The Ree–Eyring equation differs from the other ordinary viscosity functions in that it does not explicitly incorporate the power-law exponent, n. In the way that Eyring made use of this relation [25], it is clear that he intended it to be used to describe terminal, power-law, and possibly second Newtonian behavior.

Table 2.1. Viscosity functions for the generalized newtonian fluid model.

$\eta(\dot{\gamma})$		$\eta(\tau)$	
$\eta = \mu$	Newtonian	$\eta = \mu$	Newtonian
$\eta = \mu\,\lvert\lambda\dot{\gamma}\rvert^{n-1}$	Ostwald–deWaele	$\eta = \mu\left\lvert\frac{\tau}{G}\right\rvert^{1-\frac{1}{n}}$	Ostwald–deWaele
$\eta = \begin{cases}\mu, \lambda\dot{\gamma} \le 1 \\ \mu\,\lvert\lambda\dot{\gamma}\rvert^{n-1}, \lambda\dot{\gamma} \ge 1\end{cases}$	Spriggs [19]	$\eta = \begin{cases}\mu, {}^{\tau}/_{G} \le 1 \\ \mu\lvert{}^{\tau}/_{G}\rvert^{1-\frac{1}{n}}, {}^{\tau}/_{G} \ge 1\end{cases}$	Spriggs [19]
$\eta = \mu_2 + \dfrac{\mu - \mu_2}{1 + \lvert\lambda\dot{\gamma}\rvert^{1-n}}$	Cross [6]	$\eta = \mu_2 + \dfrac{\mu - \mu_2}{1 + \left\lvert\frac{\tau}{G}\right\rvert^{\left(\frac{1}{n}-1\right)}}$	Ellis [7]
		$\eta = \mu_2 + \dfrac{\mu - \mu_2}{1 + \left\lvert\frac{\tau}{G}\right\rvert}$	Ferry [20]
		$\eta = \mu_2 + \dfrac{\mu - \mu_2}{1 + \left(\frac{\tau}{G}\right)^2}$	Rabinowitsch [21]
$\eta = \mu_2 + \dfrac{\mu - \mu_2}{\left[1 + (\lambda\dot{\gamma})^2\right]^{\frac{1-n}{2}}}$	Carreau [22]	$\eta = \mu_2 + \dfrac{\mu - \mu_2}{\left[1 + \left(\frac{\tau}{G}\right)^2\right]^{\frac{\frac{1}{n}-1}{2}}}$	Bair, S. and Khonsari, M. [23]
$\eta = \mu_2 + \dfrac{\mu - \mu_2}{\left[1 + (\lambda\dot{\gamma})^a\right]^{\frac{1-n}{a}}}$	Carreau–Yasuda [3]	$\eta = \mu_2 + \dfrac{\mu - \mu_2}{\left[1 + \left(\frac{\tau}{G}\right)^a\right]^{\frac{\frac{1}{n}-1}{a}}}$	Bair, S. [24]
$\eta = \sum_{i=1}^{N} \dfrac{f_i\mu}{\lambda_i\dot{\gamma}} \sinh^{-1}(\lambda_i\dot{\gamma})$, $\sum_{i=1}^{N} f_i = 1, N > 1$	Ree–Eyring [25]		

2.5 Time–Temperature–Pressure Superposition

When flow curves obtained at different temperatures (or pressures) for one material are plotted as log η versus log $\dot{\gamma}$, the curves have the same shape. The relationships between the variables have the same functional form. The same holds true for log τ versus log $\dot{\gamma}$ or log η versus log τ. We have already seen in Fig. 2.5 that relative viscosity, η/μ, or the first normal stress difference measured for different temperatures can be reduced to essentially a single flow curve when the data are presented as a function of shear stress. This is a simple form of the tremendously useful principle known as *time–temperature–pressure superposition* or the *method of reduced variables*. More often "pressure" is omitted from the first name since most measurements do not include pressure as a variable. For the ordinary viscosity functions given in Table 2.1, and for the corotational Maxwell viscosity, equation (2.21), if n and G are held constant and only μ, and of course $\lambda = \mu/G$, vary with temperature and pressure, the simple shifting rule above will result. This concept allows the experimenter to generate rheological data under experimentally convenient conditions and obtain from that data accurate estimates of the response at conditions which may be too severe for an experimental measurement. The most obvious application is the determination of the viscosity function at high shear rates. The high shear measurement is generally a problem due to the viscous heating at high shear rate combined with the requirement that the stress be sufficiently large to penetrate into the shear-thinning regime. The alternative is to perform the measurement at low temperature where large shear stress can be obtained at low shear rate and with manageable heating, then shift the data to the desired temperature and shear rate.

Tanner [3, p. 458–464] provides a recipe for shifting of any viscosity function obtained at a reference state, T_R, p_R, that is written as $\tau = \dot{\gamma}\eta(\dot{\gamma}, \mu_R, \lambda_R, n_R)$. The shifting to another state is accomplished by replacing $\dot{\gamma}$ by $a_t\dot{\gamma}$, where a_t is the horizontal shift factor, $a_t = (\mu/\mu_R)((\rho_R T_R)/(\rho T))$, and τ is replaced by $F\tau$, where F is the vertical shift factor, $F = (\rho_R T_R)/(\rho T)$. The subscript "$R$" refers to the value of the property at the reference temperature and reference pressure. The power law exponent is not shifted, $n_R = n$. Then the shifted form is $F\tau = a_t\dot{\gamma}\eta(a_t\dot{\gamma}, \mu_R, \lambda_R, n)$. This is the standard form of time–temperature–pressure superposition, sometimes called the Ferry shifting rule [26]. The shifted Cross equation (2.10) becomes

$$\frac{\rho_R T_R}{\rho T}\tau = \frac{\mu}{\mu_R}\frac{\rho_R T_R}{\rho T}\dot{\gamma}\left\{\mu_{2_R} + \frac{(\mu_R - \mu_{2_R})}{1 + \left|\lambda_R \dfrac{\mu}{\mu_R}\dfrac{\rho_R T_R}{\rho T}\dot{\gamma}\right|^{1-n}}\right\} \tag{2.30}$$

Tanner [3, p. 459] points out that, compared with μ/μ_R, the quantity $(\rho_R T_R)/(\rho T)$ is close to unity for most changes of temperature (or pressure). Then if $\mu_2 = 0$, equation (2.30) can be approximated by

$$\eta = \frac{\tau}{\dot{\gamma}} = \frac{\mu}{1 + \left|\lambda_R \dfrac{\mu}{\mu_R}\dot{\gamma}\right|^{1-n}} \tag{2.31}$$

Since $\lambda_R = \mu_R/G_R$,

$$\frac{\eta}{\mu} = \frac{1}{1 + \left|\frac{\mu\dot{\gamma}}{G_R}\right|^{1-n}} \tag{2.32}$$

Now, a plot of η/μ versus $\mu\dot{\gamma}$ will yield the same curve regardless of temperature and pressure. This simple rule has been called the Vinogradov–Malkin rule [26] and is equivalent to correlating η/μ with shear stress.

The standard shifting rule can be obtained from theory [1, p. 139]. Molecularly based kinetic theory provides a theoretical foundation for the characteristic time, λ. The kinetic theory uses simple models of molecules to calculate the rheological behavior of polymers. The hydrodynamic force and the mass are concentrated at "beads" and the beads are connected by "springs" or rigid "rods". A thorough discussion of the subject can be found in Ref. [27]. For simple shearing of a solution of rigid dumbbells [27, p. 123],

$$\lambda = \frac{(\mu - \mu_s)M}{c\rho R_g T} \tag{2.33}$$

where μ_s is the viscosity of the solvent (zero for neat liquid polymer), c is the concentration (one for neat liquid polymer), M is the molecular weight of the polymer and R_g is the universal gas constant. Other models may yield a slightly different result, that is, with a leading coefficient on the right-hand side. For a neat liquid polymer, $c = 1$ and $\mu_s = 0$, and then $\lambda = \mu M/(\rho R_g T)$. Since $G = \mu/\lambda$,

$$G = \frac{\rho R_g T}{M} \tag{2.34}$$

a relationship that will be very useful later. Then the critical shear stress for the limit of Newtonian behavior varies inversely with molecular weight. Further, the ratio of characteristic times is

$$\frac{\lambda}{\lambda_R} = \frac{\mu \rho_R T_R}{\mu_R \rho T} = a_t \tag{2.35}$$

the horizontal shift factor used in temperature, pressure shifting.

Comparisons of equation (2.33) with tables of parameters for a viscosity function for a polymer solution, polystyrene [1, p. 174] for example, show that it is only roughly correct, with an error of as great as a factor of three as λ varies by two orders-of-magnitude. However, the molecular weight of a liquid polymer is not as straightforward as the molecular weight of a simple liquid such as, for example, squalane, hexamethyltetracosane, for which the precise formula molecular weight is known to be 422.8 kg/kmol. The molecular weight of a polymer is specified by an average value from a distribution of molecular weight. They are polydisperse rather than monodisperse like squalane. Different parts of the molecular weight distribution may contribute differently to the viscosity, shear-thinning and normal stress differences. The rheological techniques in the chapters that follow will provide an opportunity to evaluate expressions like (2.34) using well defined, monodisperse liquids.

There are situations in which time–temperature–pressure superposition may not superimpose flow curves from a rheological measurement. This may be the result of viscous heating so that the flow curve is not isothermal or the molecules may have been permanently degraded by the stress, or it may be the result of a failure of the liquid in somewhat the same manner that a solid can be said to fail.

2.6 Liquid Failure

Hutton [28] may have originated the term "liquid failure". He lists, in addition to molecular degradation, various instabilities such as edge fracture in torsional shear [29], fracture in hydrostatic tension [30], and fracture in elongational flow [29]. To these may be added, cavitation in shear flow [31], slip at the solid/liquid interface [29], slip within the liquid and banded shear flows where regions of different shear rates can coexist [32]. These effects are not included in any rheological theory and must be considered separately. For rheological measurements liquid failures are often an annoyance, but for elastohydrodynamics where the shear stress transferred across the film exhibits a plateau behavior just as banded flows do [32], the effect of liquid failure makes shear thinning at times appear to be a relatively insignificant part of the flow problem.

References

[1] Bird, R.B., Armstrong, R.C. and Hassager, O., *Dynamics of Polymeric Liquids Volume 1 Fluid Mechanics*, second edition, Wiley, New York, 1987, pp. 1, 170–171, 227–228, 351, 234, 174, 139.

[2] McCabe, C., Cui, S., Cummings, P.T., Gordon, P.A. and Saeger, R.B., "Examining the Rheology of 9-Octylheptadecane to Giga-Pascal Pressures", *J. Chem. Phys.*, Vol. 114, No. 4, 2001, pp. 1887–1891.

[3] Tanner, R.I., *Engineering Rheology*, second edition, Oxford Univ. Press, Oxford, 2000, pp. 55, 325, 18–19, 458–464.

[4] Currie, I.G., *Fundamental Mechanics of Fluids*, second edition, McGraw-Hill, New York, 1993, p. 27.

[5] Barnes, H.A. and Roberts, G.P., "A Simple Empirical Model Describing the Steady-State Shear and Extensional Viscosities of Polymer Melts", *J. Non-Newt. Fluid Mech.*, Vol. 44, 1992, pp. 113–126.

[6] Cross, M.M., "Rheology of Non-Newtonian Fluids: a New Flow Equation for Pseudoplastic Systems", *J. Colloid Sci.*, Vol. 13, 1965, pp. 417–437.

[7] Meter, D.M. and Bird, R.B., "Tube Flow of Non-Newtonian Polymer Solutions: Part 1. Laminar Flow and Rheological Models", *A.I.Ch.E. Journal*, Vol. 10, No. 6, 1964, pp. 878–881.

[8] Macosko, C.W., *Rheology Principles, Measurements, and Applications*, Wiley-VCH, New York, 1994, p. 7.

[9] Williamson, B.P., Walters, K., Bates, T.W., Coy, R.C. and Milton, A.L., The Viscoelastic Properties of Multigrade Oils and their Effect on Journal-Bearing Characteristics", *J. Non-Newt. Fluid Mech.*, Vol. 73, 1997, pp. 115–126.

[10] Keentok, M., Georgescu, A.G., Sherwood, A.A. and Tanner, R.I., The Measurement of the Second Normal Stress Difference for Some Polymer Solutions", *J. Non-Newt. Fluid Mech.*, Vol. 6, 1980, pp. 303–324.

[11] De Kee, D. and Stastna, J., "Primary Normal Stress Coefficient Predictions", *J. Rheology*, Vol. 30, No. 1, 1986, pp. 207–230.

[12] Binnington, R.J. and Boger, D.V., "Constant Viscosity Elastic Liquids", *J. Rheology*, Vol. 29, No. 6, 1985, pp. 887–904.

[13] Joseph, D.D., *Fluid Dynamics of Viscoelastic Liquids*, Springer-Verlag, New York, 1990, pp. 11–13.

[14] Johnson, K.L. and Tevaarwerk, J., "A Simple Non-Linear Constitutive Equation for Elastohydrodynamic Oil Films", *Wear*, Vol. 35, 1975, pp. 345–356.

[15] Schultheisz, C.R., Flynn, K.M. and Leigh, S.D., *Certification of the Rheological Behavior of SRM 2491, Polydimethylsiloxane*, NIST Special Publication 260–147, NIST, 2001.

[16] Dyson, A. and Wilson, A.R., "Film Thickness in Elastohydrodynamic Lubrication by Silicone Fluids", *Proc. Instn. Mech. Engrs.*, Vol. 180, Part 3K, 1965–66, pp. 97–105.

[17] Harrison, G., *The Dynamic Properties of Supercooled Liquids*, Academic Press, London, 1976.

[18] Schultheisz, C.R. and Leigh, S.D., *Certification of the Rheological Behavior of SRM 2490, Polyisobutylene Dissolved in 2,6,10,14-Tetramethylpentadecane*, NIST Special Publication 260–143, NIST, 2001.

[19] Florez, W.F., Power, H. and Janna, F.C., "Multi-Domain Dual Reciprocity for the Solution of Inelastic Non-Newtonian Flow Problems at Low Reynolds Number", *Comput. Mech.*, Vol. 27, 2001, pp. 396–411.

[20] Van Krevelen, D.W., *Properties of Polymers: Their Correlation with Chemical Structure: Their Numerical Estimation and Prediction*, Third Edition, Elsevier, Amsterdam, 1990, p. 477.

[21] Rotem, Z., "Non-Newtonian Flow in Annuli", *ASME J. Appl. Mech.*, Vol. 33, 1962, pp. 421–424.

[22] Carreau, P.J., "Rheological Equations from Molecular Network Theories", *Trans. Soc. Rheology*, Vol. 16, No. 1, 1972, pp. 99–127.

[23] Bair, S. and Khonsari, M. "An EHD Inlet Zone Analysis Incorporating the Second Newtonian", *ASME J. Tribol.*, Vol. 118, 1996, pp. 341–343.

[24] Bair, S., "A Rough Shear Thinning Correction for EHD Film Thickness", *STLE Tribol. Trans.*, Vol. 47, 2004, pp. 1–5.

[25] Ree, F.H., Ree, T. and Eyring, H., "Relaxation Theory of Transport Problems in Condensed Systems", *Ind. Eng. Chem.*, Vol. 50, 1958, pp. 1036–1038.

[26] Schurz, J., Vollrath-Rodiger, M. and Krassig, H., "Rheologische Untersuchungen an Polacrylnitril-Spinnlosungen: Erweiterung HV 6 fur hohere Temperaturen", *Rheologica Acta*, Vol. 20, 1981, pp. 569–578.

[27] Bird, R.B., Curtiss, C.F., Armstrong, R.C. and Hassager, O., *Dynamics of Polymeric Liquids Volume 2 Kinetic Theory*, Second Edition, Wiley, New York, 1987.

[28] Hutton, J.F., "Theory of Rheology", *Interdisciplinary Approach to Liquid Lubricant Technology*, Ku, P.M. (Ed.), NASA, Washington, 1973, pp. 187–261.

[29] Macosko, C.W., *Rheology Principles, Measurements, and Applications*, Wiley-VCH, New York, 1994, pp. 214–215, 286, 244–246, 194.

[30] Fisher, J.C., "The Fracture of Liquids", *J. Appl. Phys.* Vol. 19, 1948, pp. 1062–1067.

[31] Kottke, P.A., Bair, S.S. and Winer, W.O., "Cavitation in Creeping Shear Flows", *AIChE Journal*, Vol. 51, No. 8, 2005, pp. 2150–2170.

[32] Lee, J.Y., Fuller, G.G., Hudson, N.E. and Yuan, X.-F., "Investigation of Shear-Banding Structure in Worm-Like Micellar Solution by Point-Wise Flow Induced Birefringence Measurements", *J. Rheology*, Vol. 49, No. 2, 2005, pp. 537–550.

CHAPTER 3

General High-Pressure Experimental Technique

3.1 Background

There are several excellent resources for experimental techniques regarding the generation, containment and measurement of high pressures and methods of extracting information from a pressurized container. Bridgman [1], Comings [2], Eremets [3] and Sherman and Stadtmuller [4] are all very useful to anyone engaged in the design and construction of high-pressure dilatometers, piezometers, refractometers, viscometers and rheometers. All but Bridgman have a thorough treatment of the stresses and strains in thick-wall cylinders.

In high-pressure physics, the diamond-anvil cell has become the experimental mainstay. This device develops pressure in excess of 50 GPa by squeezing between two diamond gems a metal washer containing, in the central hole, a very small volume of sample. The gems are transparent to a wide spectrum of radiation making possible many spectroscopic techniques. The pressure within the cell is measured by placing a chip of synthetic ruby in the sample. The pressure-shift of one of the fluorescence peaks of ruby has been calibrated and from this shift in wavelength the pressure is determined. Shear viscosity measurements require that a known shear stress be applied to the sample and the resulting strain rate be measured, or the other way around. Gravity provides the only obvious means to drive a rheological experiment within a diamond anvil cell and the falling ball diamond anvil viscometer remains the only shear rheological measurement that is performed in this type of device, although this application is an important one.

The glass transition in liquid lubricants at normal operating temperature is generally reached by a pressure of about 2 GPa, so that if the aim is only to explore the viscous behavior of liquid lubricants, this pressure will be sufficient for all purposes. Single component or *monoblock* cylindrical vessels are acceptable for this pressure and thick-wall steel cylinders have been used to 5 GPa [4, p. 56], although a difficult and time consuming technique known as autofrettage is employed to achieve these pressures. Cylindrical vessels may have any number of access holes so that electrical feed-throughs, drive shafts, widows and connecting tubes may be employed. This is the type of apparatus employed by the author.

Some of the experimental techniques developed by the author that are specific to the study of high-pressure rheology along with some techniques that are more general in their applications but are not to be found in the above references are described in this chapter. The author has employed a philosophy of constructing the smallest possible apparatus to perform any measurement at hand. Besides the obvious safety benefits from having a small volume of compressed liquid for which the energy stored is small, the time required to recover from broken components is reduced. The failure of highly stressed components is not completely unexpected and should be considered in the design stage so that minimal disruption in the experimental schedule occurs. Smaller components can be fabricated more quickly to restore the operation of an instrument. Components that are part of the calibration of an apparatus should be protected so that they will not be consumed in a failure.

3.2 Pressure Containment

High-pressure rheological experiments generally utilize the same experimental techniques that are used in conventional ambient pressure experiments. Then the most obvious way to design a high-pressure rheological experiment is to simplify and miniaturize a conventional rheometer and enclose it within a pressure vessel. Vessels for this purpose are nearly always fabricated from high-strength steel. Some techniques employed by the author require that the vessel be nonmagnetic and some nonferrous alloys are useful for their low magnetic permeability.

For the thick-walled cylinders employed for high-pressure work, sometimes defined as pressures greater than 100 MPa, the usual hoop stress calculation is not helpful as the stress is nonuniformly distributed through the thickness of the wall with the highest tensile value being at the surface of the inner diameter, the bore. The ratio of outside diameter, D_o, to inside diameter, D_i, usually varies from about three to six. Assuming that the external pressure can be neglected, the maximum stress at the bore is tensile in the circumferential direction [2, p. 163].

$$\sigma = p\frac{(D_o/D_i)^2 + 1}{(D_o/D_i)^2 - 1} \tag{3.1}$$

In addition there is a compressive stress at the bore in the radial direction and equal to the internal pressure. If the closures at the ends of the vessel are supported by a structure different from the vessel, the stress in the axial direction is zero. However, if the closures are attached to the vessel, a small tensile stress is uniformly distributed through the wall in the axial direction [2, p. 163]. It is clear from equation (3.1) that for an elastic cylinder there is little to be gained in reduction of stress from diameter ratios greater than about five.

Sherman and Stadtmuller [4, p. 14–26] treat the elastic thick-walled cylinder and find that the greatest pressure that can be contained without exceeding the elastic limit of the material is

$$p_e = \tau_y \frac{(D_o/D_i)^2 - 1}{(D_o/D_i)^2} \tag{3.2}$$

where τ_y is the yield stress in shear. Then the greatest pressure that can be supported elastically approaches the magnitude of the yield shear stress as the diameter ratio becomes large. For, say, a cold-worked 300 series stainless steel commonly used for thick-wall, high-pressure tubing, the inelastic behavior begins at a pressure of only about 0.25 GPa. This tubing can be used for very many pressure cycles up to 1 GPa and for a limited number of cycles up to 1.4 GPa. Then, clearly, the elastic limit does not limit the pressure that may be contained by a steel cylinder.

When a cylinder is pressurized beyond its elastic limit, yielding of the material occurs first at the surface of the bore and then, progressively, extends outward as pressure increases. Upon release of internal pressure, an interesting event occurs. The material that was inelastically strained in tension now has an elastic compressive preload which will delay the pressure at which the elastic limit is reached in subsequent pressurizations. Sherman and Stadtmuller [4, p. 14–26] find that if there is no strain hardening, that is $\tau = \tau_y$ in the yielded material, the internal pressure required to overstrain the entire cylinder is

$$p_i = 2\tau_y \ln (D_o/D_i) \tag{3.3}$$

Thus for the tubing described above, if the diameter ratio is six, the entire wall would be overstrained at an internal pressure of about 0.9 GPa. Then even the value of p_i from equation (3.3) is not the maximum pressure that can be contained. Work hardening of the cylinder material raises the value of p_i in practice.

Also, the possibility exists that reverse yielding may occur at the bore surface on release of pressure from a highly overstrained condition. The author has applied a strain gage to the outside surface of a commercial high pressure (1.03 GPa rating) 304 stainless steel tube, $D_o/D_i = 4.7$, in an effort to use the gage as a pressure indicator. After an initial pressurization of 1.2 GPa the strain response was found to be linear with pressure for an interval of only 0.4 GPa. The pressure location of the linear interval was dependent upon the pressure history. On application of pressure from ambient the linear behavior extended from 0–0.4 GPa, and on a decrease in pressure the linear behavior extended from the maximum pressure to 0.4 GPa less than the maximum. Clearly, the material from which commercial tubing is made is unsatisfactory for a pressure indicator; although, with the proper choice of alloy the above arrangement can be useful for a pressure transducer, as will be shown later.

From the above it should be clear that for a pressure vessel to be safe and useful for pressures representative of elastohydrodynamics, the ductility or toughness is at least as important as strength. Ductility is measured by the elongation or reduction in area in a tensile test. For example, the author has constructed similar pressure vessels from O-1 and H-13 tool steels. When heat-treated to a hardness of Rc = 50 these alloys have nearly equal yield stress of 1.5 GPa. The H-13 alloy with a reduction in area property of 35% can survive multiple pressure cycles to 2.0 GPa, while O-1 with a reduction in area property of 5% failed (cleaved into two parts along the axis) at the first pressure of 1.2 GPa.

For high-pressure tubing, large strains at the bore of the tube do not affect the function of the component since the only place that the dimensions of the part are important to function is at the seal, which is generally provided by contact at an outside surface. For pressure-containment around a rheometer and for other vessels this is not generally the case. The closures at the ends of the vessel must seal on the inner surface and some remachining of the bore is necessary after each pressurization that substantially exceeds the present elastic limit. This process of pressurizing followed by removal of metal from the bore to restore a cylindrical internal shape is called *autofrettage*.

One cycle of the autofrettage process is shown in Fig. 3.1. The cylinder is fitted with a piston and a plug with small clearances and seals shown simplified in the figure. A metal slug (not shown) is usually placed inside the vessel to displace some of the liquid, thereby

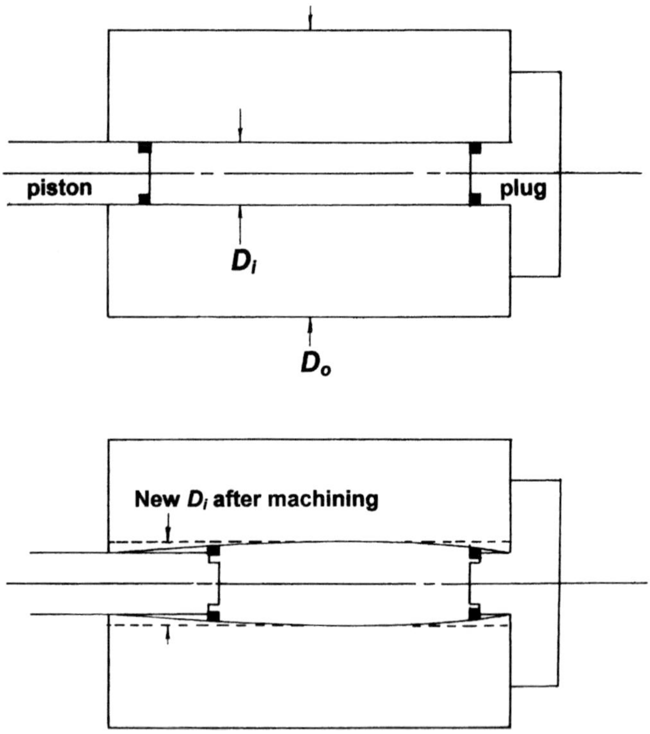

Fig. 3.1. A cylindrical pressure vessel with closures consisting of a piston on the left and a plug on the right. Seals are shown schematically at the ends of the closures as the blackened ring of square cross-section. Unpressurized, at top and under pressure, at bottom. Machining to the new bore diameter completes one autofrettage cycle.

Table 3.1. Materials for pressure containment assuming $D_o/D_i \geq 4$.

Alloy	Condition	Pressure at Bore Yield, p_e/GPa	Magnetic
316 Stainless Steel	Work Hardened	0.5–0.7	No
17-4 PH, AISI 630	480°C 1 h	0.95	Yes
H13 or 4340	Rc = 50	1.1	Yes
X-750	As Supplied	0.8	No
C 300	480°C 3 h	> 1.3	Yes
MP35N	500°C 4 h	> 1.3	No

reducing the overall compressibility of the contents and reducing the piston travel. The piston is driven into the cylinder to overstrain the material adjacent to the bore often until the seals fail by extruding into the large clearance shown in the bottom of Fig. 3.1. The bore is then remachined either to the largest existing diameter or to a sufficiently large diameter that the seals will now operate in a region of the bore of uniform diameter. A set of reamers of gradually increasing diameter is useful for this purpose. New closures and seals are fabricated to fit the new internal diameter and if necessary, the process is repeated until a pressure of about 10% greater than the maximum operating pressure is reached.

Some alloys useful for the construction of pressure vessels are listed in Table 3.1. The table also lists an approximate elastic limit pressure based upon experience, not analysis. Some of these alloys are nonmagnetic which makes available a technique for detecting the position of a cylinder of magnetic material in the high-pressure region using a linear variable different transformer, LVDT, positioned around outside of the vessel. Of the 300 series stainless steels, 304 becomes slightly magnetic with cold working while 316 does not and, also within this series, the mechanical properties vary significantly with degree of cold working making for large differences from lot to lot.

The 17-4 PH or AISI 630 alloy is useful as a general purpose material for high-pressure work. It machines easily and may be hardened by a convenient thermal aging. The X-750 is a nickel-based "super alloy" that may be precipitation hardened to a small extent. The MP35N is a relatively new and very expensive ternary alloy of nickel, cobalt, chromium and molybdenum, and this alloy responds to precipitation hardening.

3.3 Closures

Some method must be provided to contain the pressurized liquids at the two ends of the vessel and to close any access holes entering from the sides. These closures may be static or able to move axially to change the volume of the cylinder (a piston) or rotating to drive an experiment within the pressurized chamber (a drive shaft). Examples of closures that are useful in the construction of pressurized rheometers are shown in Fig. 3.2 arranged by the way that the closure is sealed.

The metal-to-metal sealed plug in Fig. 3.2(a) is constructed in a manner identical to commercial high-pressure tube fittings. Conical surfaces of different vertex angles initially

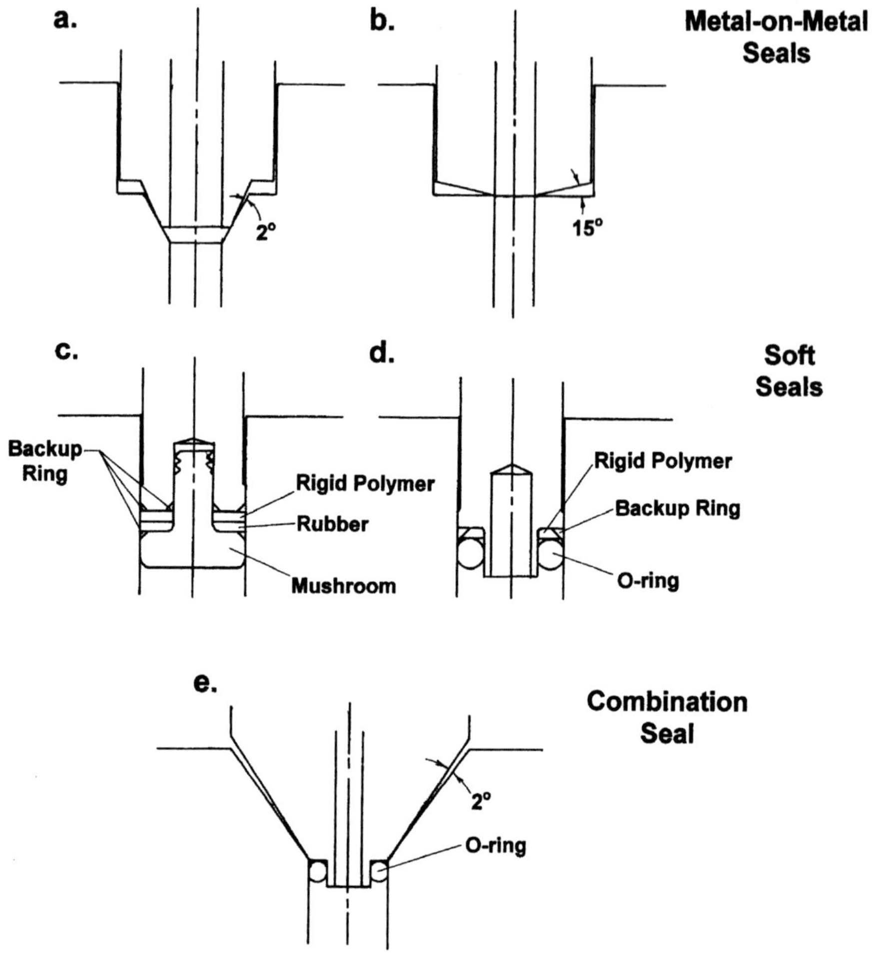

Fig. 3.2. Vessel closures arranged according to the seal material. Types (a), (b) and (e) are always static and types (c) and (d) may be static or translating as a piston.

contact along a circle at the small end of the plug. It will be shown that this closure can be of use for passing electrical conductors into the vessel. A similar arrangement that is useful for the connection of pressure transducers is shown in Fig. 3.2(b). The closure is driven toward the cylinder with sufficient force to plastically deform the material at the interface. The seal in Fig. 3.2(a) is somewhat self-acting since the pressure acting on the inside surface of the hole in the closure tightens the seal. All of the closures when used as a static plug may be retained by either a threaded nut or a bolted flange. The metal-to-metal seals and the combination seal, Fig. 3.2(e) require a greater retaining force than the soft seals in order to maintain a contact stress that substantially exceeds the sealed pressure. For the metal-on-metal sealed closures, the total cross-section area of the bolts or the nut threads should be about 30 times the area of the aperture for 0.7 GPa and twice that area ratio for 1.4 GPa. In the case of the combination seal these area ratios may be reduced by half. For the soft seals it is only necessary to have sufficient threaded cross section area to restrain the closure against pressure with a factor-of-safety. The concept of a self-acting seal should be utilized whenever possible.

The closure shown in Fig. 3.2(c) employs the well-known Bridgman [1] unsupported area seal. This is the most reliable and most complicated seal. The "mushroom" shaped steel piston has a stem that fits into a hole in the closure. The soft seal packing, the rubber washer and rigid polymer washer indicated in 3.2(c), resides between the head of the mushroom and the closure. The stem extends through a hole in the packing to provide the unsupported area. Pressure acting on the head of the mushroom applies a force to the packing. The packing is supported by the area of the closure minus the area of the hole. Then the hydrostatic stress in the packing must exceed the pressure being sealed. This seal cannot leak as long as some packing remains in place. Steel back-up rings are provided to prevent extrusion of the rigid polymer, often Teflon, toward the atmosphere. The rigid polymer prevents extrusion of the rubber toward the backup ring. Because the pressure in the rubber exceeds the liquid pressure, a backup ring is useful between the rubber washer and the liquid, particularly when the seal is used on a moving piston. The purpose of the rubber washer is to provide sealing at low pressures. This part may be eliminated, retaining only the rigid polymer as a packing; however, the seal will often leak as the pressure is initially applied, necessitating very fast compression to begin pressurization.

For the highest pressures, care must be taken in the selection of steels for the closure, the mushroom and the backup rings. The closure or piston is least critical with a high hardness being most important. For the closure or piston, precipitation-hardened C350 maraging steel or A2 tool steel at Rc = 56 are acceptable. For the mushroom, this laboratory has found A6 tool steel at Rc = 50 (480°C temper) to be successful. The radius at the junction of the stem and the head should be generous as failures of the mushroom always occur here. A thread on the end of the stem facilitates removal from the vessel. The backup rings can be a problem when the seal must translate. Too low a hardness will result in extrusion of the steel from the outer ring into a thin tube encasing the closure shank. Too high a hardness will result in cracking of the ring and loss of the packing by extrusion through the crack. The best compromise appears to be A6 tool steel tempered after air quenching at 565°C. Now it must be admitted that the selection of A6 may not be entirely optimum; however, A6 is unique in being an air quenching steel that does not require an inert atmosphere for thermal soaking. This makes possible rapid fabrication of replacement rings with a thermal treatment that can be accomplished in an hour.

The other variation of the soft seal shown in Fig. 3.2(d) utilizes an ordinary o-ring of the type used in household plumbing. In fact, the author has experimented with more exotic material formulations than the standard buna-n or nitrile rubber, but unless the sealed fluid is aggressively attacking the rubber, these ordinary compounds are superior in sealing. As above, the hardened steel backup ring prevents extrusion of the o-ring. The rigid polymer ring fills the volume inside the backup ring and urges the backup ring against the bore. The central hole pressurizes the inside of the closure closing the gap through which the backup ring may extrude. The o-ring must be provided with a large crush or squeeze, about 20%, since the volume will necessarily be reduced by pressure and large o-ring cross-sections perform better than small. This is a very reliable seal for pressures up to the glass transition pressure of the rubber o-ring which can be about 0.5–0.6 GPa at room temperature for 70 durometer (a hardness measure) buna-n. This glassy state pressure limit can be raised by heating the region around the seal. The glass transition temperature of buna-n increases roughly by about 5 MPa/°C. Another useful way to raise the glass transition pressure is to apply a plasticizing agent to the o-ring. This must be done in a clever way as the resulting swelling of the o-ring may prevent the closure from entering the vessel bore. A drop of dibutyl phthalate applied to the surface of the polymer ring will contact the o-ring in the process of installation. This type of closure has been used successfully to 1.0 GPa as a static plug and to 0.85 GPa as a pressurizing piston.

With all of the difficulty associated with the closure of Fig. 3.2(d) one may wonder why it may ever be used in preference to the other (Bridgman) soft sealed closure, Fig. 3.2(c). It does have a large central hole, however, that may be used for access for the feed-throughs to be discussed in the next section. In addition, when used as a translating piston the friction force amounts to about 2% of the force to move the piston compared to 4.5% for the Bridgman seal. Then the closure seal shown in Fig. 3.2(d) should result in a more repeatable estimation of the pressure from the piston force when that particular method of pressure measurement is chosen since the variability of the friction will then be a smaller proportion of the total force.

The combination seal in Fig. 3.2(e) is useful to close vessels that have been strained inelastically without resorting to the autofrettage process. Extrusion of the o-ring is prevented by the metal-to-metal contact. For some instruments, such as the cartridge type of falling body viscometer, the inside diameter of the vessel is not utilized to accurately establish a critical dimension and inelastic expansion of the bore is inconsequential to its function. For these instruments, the combination seal can be used to advantage for pressures of at least 1.4 GPa. The o-ring seal here is subject to the same problems and limitations discussed above. Again, buna-n seams to be the best compound. Experience has shown that 70 durometer hardness will operate to 1.4 GPa at 140°C but will extrude at only 0.7 GPa at 200°C, while 90 durometer hardness will seal 1.3 GPa at 165°C. A 70 durometer o-ring softened by dibutyl phthalate plasticizer has extruded at 1.3 GPa for 24°C and at 0.8 GPa for 100°C.

3.4 Feed-Throughs

Some means must be provided to communicate with the pressurized environment within the vessel to perform a rheological experiment. One way to do this is simply to place a transformer around a nonmagnetic vessel and inductively detect the position of a magnetic core. To measure forces within the vessel it is useful to detect the strain in an elastic component and this can be done electrically with strain gages or optically by laser reflection from a mirror applied to the element. These techniques require an electrical or optical feed-through, respectively. To impose a velocity in an experimental measurement, it is desirable to provide a mechanical shaft extending into the high-pressure chamber requiring a mechanical feed-through.

3.4.1 Rotating mechanical feed-throughs

This component is apparently unique to the author's laboratory. There are available, commercial mechanical feed-throughs that may be operated continuously at high speed and pressure to about 0.4 GPa. These permanent magnet drives have magnets arranged around and bonded to a shaft within a relatively thin nonmagnetic pressure vessel. Then magnets attached to a rotating armature outside of the vessel apply a driving torque to the internal shaft. These devices suffer from low-torque capability, low-pressure capability and very large space requirements making them unsuitable for high-pressure rheological measurements related to elastohydrodynamics.

The evolution of the rotating feed-through used in the author's laboratory is shown in Fig. 3.3. Both examples employ the soft seal of Fig. 3.2(d) to prevent leakage between the closure and the vessel, although the combination seal of Fig. 3.2(e) has been used [5] at the highest pressure, 1.0 GPa, for which this feed-through has been utilized. The essential difference between the two embodiments is that, at the top of Fig. 3.3 the thrust bearing is within the high pressure liquid while at the bottom it has been moved outside the vessel. Then for the first case, the bearing is lubricated by the high-pressure liquid and rotation of the bearing must result in churning of the high-pressure liquid.

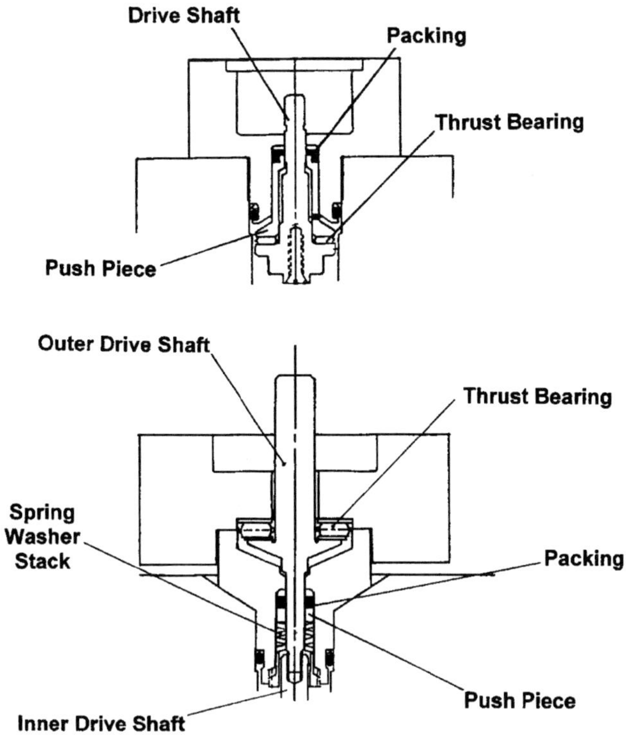

Fig. 3.3. Two types of rotating mechanical feed-throughs. Unsupported area (Bridgman) seal at top and spring energized seal at bottom.

This necessitates the introduction of another low-viscosity liquid in which to immerse the bearing as the viscosity of the liquids that are ordinarily the subject of investigation is so large as to require a torque in excess of the strength of the drive shaft. This is not so much of an inconvenience as might be expected since most high-pressure viscometers employ a pressurizing medium that is different from the liquid under investigation and this medium, to be discussed in a later section, is required to have a low viscosity under pressure. For the first application (not shown here) of the internal bearing version [6], the bearing was fitted with an integral shield and was packed with a silicone grease. This was not entirely satisfactory as the grease eventually migrated past the shield and contaminated the sample. The thrust bearing must often be used at a load close to the static load limit. Both of the bearings shown in Fig. 3.3 are of the caged needle type. Ball thrust bearings will have lower frictional torque.

There is, of course, a second seal in each embodiment in Fig. 3.3 in addition to the closure seal. The rotating shaft is sealed at the exit to the closure by the packing that is labeled in the figure. The shaft at the top of the figure uses an unsupported area seal for which the packing is energized in the same manner as the seal in Fig. 3.2(c). The packing pressure is automatically made larger than the liquid pressure. This feature is not available to the type shown at the bottom of Fig. 3.3, since the force to expel the shaft is reacted by the external bearing. Here, a stack of Belleville spring washers is used instead to provide a slightly greater pressure in the packing. The springs should be sized to provide about 10–60 MPa increase in pressure and a threaded screw, shown surrounding the internal drive shaft in the lower drawing, compresses the springs. The packing for both types of seal may be Teflon or glass-filled Teflon and the backup ring shown above the packing may be bearing bronze or for pressure up to 350 MPa it can be a rigid polymer such as Vespel. The shaft should be made of a high-strength material but must also be resistant to galling when sliding against the bronze backup ring. Tool steels of the O-1 and A-6 type are satisfactory with hardness of about Rc = 56.

The frictional torque at ambient pressure for a 3.5 mm diameter shaft is about 1 N·m and for a 6 mm shaft is about 2 N · m. This friction will increase with pressure, once nearly doubling at about 0.8 GPa, but in one case the friction showed a maximum whereupon it decreased with pressure. The drive motor can be sized for just the frictional torque as the torque requirement of the experiment is of a much lower magnitude. The purpose of these mechanical components is not to provide for continuous rotation. The frictional heating would probably be excessive and in any event for the present use, rotations of more than one or two revolutions are unnecessary.

The evolution to the external bearing in the bottom of Fig. 3.3 carried some additional complication. The drive shaft must terminate inside the vessel with a diameter that can pass through the seal. This requirement leaves little opportunity for a rotating coupling to the experimental setup. The solution has been to provide a hexagonal cross-section shape to the end of the outer shaft that connects with a mating hexagon on an internal shaft as shown at the bottom of the figure.

The development of these mechanical rotary feed-throughs made possible the operation of rotational viscometers and rheometers, often the measurement instruments of choice, within a high-pressure environment.

3.4.2 Electrical feed-throughs

For the measurement of many parameters and effects such as pressure, temperature, force and electrical properties a means to pass an electrical signal from the high-pressure region through the vessel to a recording device outside is useful if not essential. Generally, more than one electrical connection is required. For example, a full strain-gage bridge for strain measurement requires four connections and a manganin resister for pressure measurement requires at least two. Sherman and Stadtmuller [4, p. 266–280] have an excellent section dealing with electrical leads into a pressure vessel. Here, a number of electrical feed-throughs that have been useful to high-pressure rheology will be described.

The configuration shown in Fig. 3.4(a) is of the unsupported area (Bridgman) type. A stainless steel, 1.6 mm diameter, wire is passed through a hole in a short,

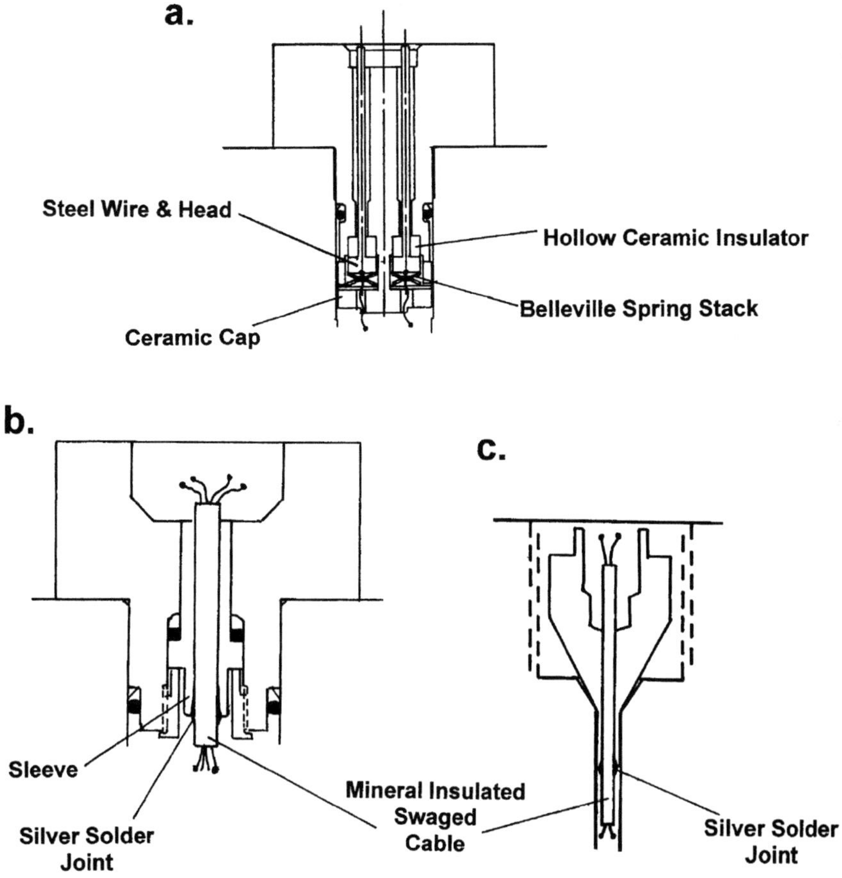

Fig. 3.4. Electrical feed-throughs employing: (a) rigid rods supported by alumina washers; (b) and (c) thermocouple type cable consisting of stainless steel tube swaged around wires embedded in magnesia insulation.

5–6 mm diameter, steel cylinder (the head) and is silver brazed in place. The head is supported by a ceramic hollow cylinder (the washer) between the head and the flat surface of the closure. The unsupported area is that of the hole in the closure so that the contact pressure between the washer and head and between the washer and closure exceeds the liquid pressure. These components are lapped together with a rotating motion and thread-locking compound is used as a sealant. These components are held in place by a stack of Belleville spring washers retained by a ceramic cap. As many as six assemblies may be arranged around a closure as shown in Fig. 3.4(a) and a plug using commercial pin receptacles can be constructed to fit the external ends of the rods. Internal lead wires may be soldered to a copper disc inserted into the spring stack. This feed-through is very rugged and reliable. On occasion, however, they show a peculiar trait. The ceramic insulating washer that is made of alumina may crack. Pressure tends to hold the crack closed so that no leak has been detected when this occurs but the crack becomes conductive making a closed circuit to the closure body.

The other types of feed-through shown in Figs. 3.4(b) and (c), utilize a commercial electrical cable that is generally used for thermocouple wire leads. It consists of one to six bare metal conducting wires packed into a stainless steel tube with magnesium oxide powder as an insulator. The wires may be thermocouple alloys or copper or nickel. This cable is provided with the tube having been tightly swaged around the powder and wires. If the tube is sufficiently long, perhaps ten times the diameter, the magnesia powder will resist pressures to at least 1.3 GPa, the highest pressure for which these feed-throughs have been used in the author's laboratory. Handling of the cable will cause powder to escape from the ends and this can be prevented by allowing the powder to absorb a mixture of acetone and two-part epoxy glue, of course, after the cable has been brazed into whatever component receives it as the temperature of the brazing process would decompose the glue.

In Fig. 3.4(c), the cable is simply placed into a through hole in a steel closure of the shape of a commercial high-pressure plug. These two are then brazed together using high-temperature silver solder. The silver solder joint should always be placed at the internal end so that pressure will tend to tighten the joint. The nut that is used with a commercial plug can be used to retain the feed-through.

In Fig. 3.4(b), the swaged cable is brazed in the same manner as above except that in this case it is received into a sleeve that functions as the mushroom in a Bridgman closure similar to Fig. 3.2(c). A threaded hollow screw preloads the sleeve to initially energize the seal. The ends of the wires emanating from the external end of the swaged cable are connected to a commercial electrical plug for strain relief and convenience.

3.4.3 Optical feed-throughs

There are often requirements for optical access to the internal environment of the pressure chamber. For flow visualization, flow birefringence measurements, refractive index measurements, for optical strain measurement and for simply verifying the internal operation of a rotational rheometer a window is useful. Actually, windows are relatively simple and

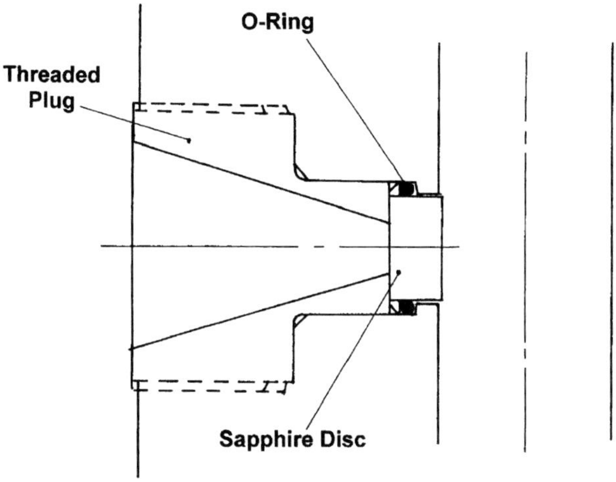

Fig. 3.5. Optical feed-through employing a sapphire window.

easy to implement. Sherman and Stadtmuller [4, p. 254–266] have an excellent section dealing with windows.

An example of a window used by the author for general purposes is shown in Fig. 3.5. The transparent material is a sapphire disc with a thickness of approximately half of the diameter and polished on the faces. These discs can be inexpensive if the design process begins by inquiring from commercial suppliers as to the dimensions of left over or odd lots that meet the thickness requirement. Then the instrument may be designed around the available widow discs. The disc is supported by a hardened steel plug with an aperture diameter about half of the disc diameter. The plug surface contacting the window must be flat and lapped smooth. The o-ring seal in Fig. 3.5 is similar to the seal in Fig. 3.2(d) except that there is no central hole to self-energize the seal.

3.5 Pressure Generation and Measurement

There are commercially available hand-pumps for pressures to about 0.4 GPa and pressure intensifiers for 1.4 GPa. The intensifier is a very simple device that uses a large swept volume from a low pressure source to generate a small swept volume at high-pressure to be supplied to an instrument. The commercial intensifiers are generally cyclic so that after a volume is delivered to the instrument, the intensifier and instrument are isolated, the intensifier is recharged and the cycle is repeated. This cyclic operation requires at least two high-pressure valves and a charging pump. For a small pressurized rheometer, the simplest alternative is a single-cycle intensifier sized to sweep a volume of one half to one times the internal volume of the pressurized rheometer volume. Such a pressure intensifier is shown in Fig. 3.6. It may be directly attached to the rheometer or connected by a short length of thick-wall tubing. A pressure transducer may be fitted to the high-pressure cylinder shown in Fig. 3.6 or in a standard high-pressure tee fitting between the intensifier and the rheometer.

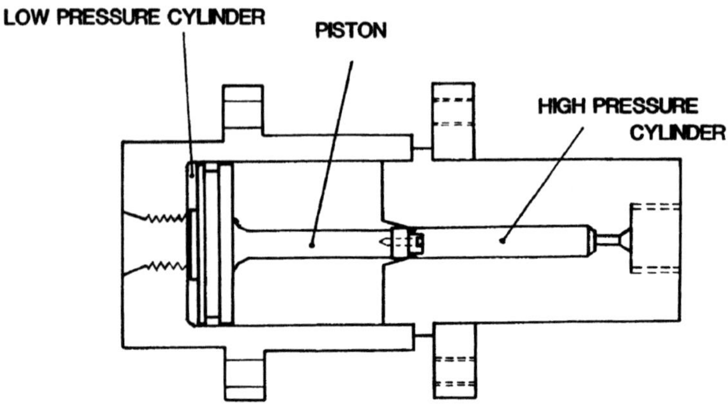

Fig. 3.6. Simplest form of a pressure intensifier.

The pressure intensifier in Fig. 3.6 consists primarily of three components; a small bore high-pressure cylinder, a large bore low-pressure cylinder and a double ended piston that translates axially between the two cylinders. The square of the ratio of the large bore diameter to the small bore diameter, the area ratio, is also the idealized pressure ratio. This laboratory has fabricated intensifiers with pressure ratios from 3 to 40. In practice the operating pressure ratio is about 1% to 10% less than the idealized ratio because of the friction of the seals. The piston seal at the low pressure end may be simply an o-ring for pressures to about 30 MPa. The harder 90 durometer compound resists extrusion and is best for this application. For the low pressure seal, pressures to about 70 MPa may reliably sealed if a split backup ring of rigid polymer is placed in the o-ring groove with the o-ring. For the high-pressure seal, the reinforced o-ring seal of Fig. 3.2(d) may be used with very low friction to about 0.85 GPa if the precautions outlined in Section 3.3 are followed. The Bridgman unsupported area seal of Fig. 3.2(c) can be used to 2.0 GPa. A useful rule is to size the swept volume of the high-pressure side of the intensifier to be roughly equal to the volume of the instrument to be pressurized.

Any high-pressure experiment requires a measurement of pressure. This can be satisfactorily accomplished by calibration of the relationship between the supplied low pressure, p_L, and the generated high pressure, p, of a pressure intensifier if great care is taken. First, it must be recognized that there will be a substantial difference in the pressure relationship for increasing pressure as opposed to decreasing pressure. For convenience, only the increasing pressure behavior is used for calibration and measurement. For this approach to be accurate, the same high-pressure medium must be used consistently. The calibration should be made only after several cycles of the intensifier to complete running-in of the sliding components. The o-ring seal of Fig. 3.2(d) is preferred to the Bridgman seal of Fig. 3.2(c) since the friction is lower and friction is the major source of error. Even so, intensifiers using Bridgman seals can be used for pressure measurement. As an example, a 1.5 GPa viscometer in the author's laboratory utilizes for pressure generation and measurement, an intensifier with a Bridgman seal. It was originally calibrated against a commercial 0.7 GPa Bourdon tube pressure gage (stated accuracy ±1 MPa) and a calibration function was arrived at that reads $p = 37.14p_L - 13$ MPa. The theoretical pressure ratio was $p/p_L = 38.26$. Two years later the calibration was checked against a commercial 1.4 GPa manganin gage (stated accuracy ±1 MPa). The largest difference between calculated pressure and pressure measured by the manganin cell occurred at 0.75 GPa where the difference was 0.02 GPa. The difference at 1.25 GPa was 0.007 GPa. Obviously, if possible, a separate pressure transducer should be provided but the intensifier may be used as a pressure measuring device as well and establishing the relationship between the input and output pressures provides a check on the health of the pressure transducer.

The stated accuracy of commercial pressure indicators needs some discussion. The two pressure measurement components employed above, the Bourdon tube gage and the manganin cell each have an advertised accuracy of ±1 MPa. They were connected together and to a pressure intensifier. Over a range of pressure from 0 to 0.7 GPa, a maximum difference in indicated pressure of 5 MPa was observed. Then it is unclear as to what can be understood from the ±1 MPa specification except that it is only a very rough estimate. The strategy employed in the author's laboratory to arrive at a pressure standard for calibration of a new instrument is to utilize at least two reliable pressure indicators and average the result.

The Bourdon tube dial type of gage is becoming rare in high-pressure work. At one time these were available for pressure to 1 GPa. Strain gage based transducers that indicate pressure by measuring the elastic strain of a high-strength pressure resisting component have become more reliable than they were 30 years ago, and today are more reliable than the Bourdon tube. Strain gage transducers can be purchased for pressures up to 1.5 GPa. Another type of pressure transducer is the manganin cell. The electrical resistance of the manganin alloy has been shown to be linear with pressure to well above 2 GPa and the measurement of resistance is straightforward. The disadvantage of using a commercial version of either type is that many of these have very large internal volumes, up to 3 ml, and when the internal volume of the transducer is large compared to the instrument, the requirement of compressing the liquid within the transducer determines the required size of the pressure intensifier.

Two pressure transducers that possess small internal or dead volume are shown in Fig. 3.7. The strain gage type at top consists simply of a cylinder of the highest yield strength alloy, C350 maraging steel in this case, with a blind internal hole. The two active strain gages are bonded with epoxy glue in the circumferential direction at a location where the diameter has been reduced to provide a diameter ratio of about $D_o/D_i = 4.4$. Inactive or dummy gages are bonded close by but outside of the strained region. A housing protects the delicate wiring and serves as a mounting for a commercial multi-pin electrical connecter. This transducer can resolve pressure changes of 1 MPa. Drift is the greatest problem and can be as much as 5 MPa per day. Commercial strain gage signal conditioners with digit readout are useful with this transducer.

The transducer shown at the bottom of Fig. 3.7 utilizes the pressure dependence of electrical resistance to indicate pressure. For most pure metals the resistance decreases with pressure and for most alloys the opposite is true. Manganin is an alloy that finds

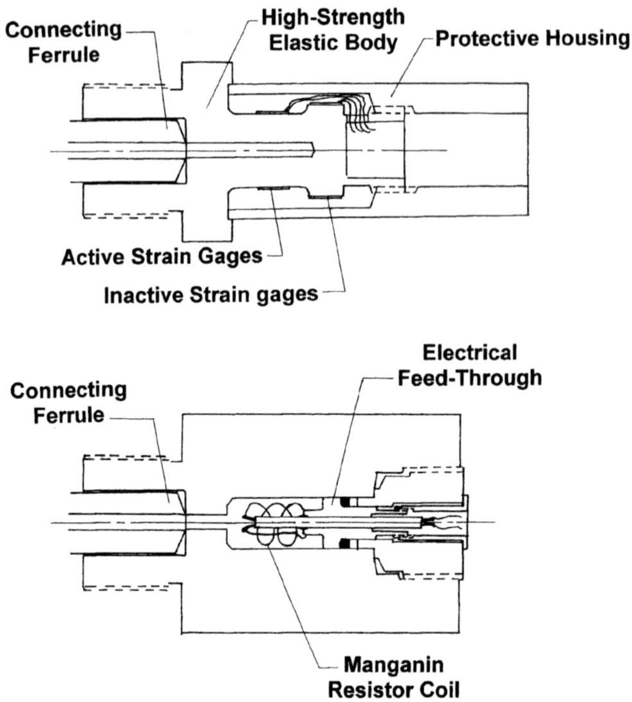

Fig. 3.7. Pressure transducers for small dead volume. Strain gage based, top and manganin resistance based, bottom.

use in electrical resistors because of its high resistivity and low temperature dependence of resistivity near room temperature. Manganin has been used as a pressure indicator for some time [7]. The resistance increases by about 2.5% per gigapascal and the exact value varies significantly from lot to lot. In Fig. 3.7, the manganin coil encircles the end of a mineral-insulated, swaged cable of the types shown in Figs. 3.4(b) and (c). The high sensitivity of the resistance to pressure results in an unexpected problem. The manganin resistor may be incorporated as one arm of a Wheatstone bridge and interfaced with a commercial strain gage signal conditioner with digit readout. However, the quarter bridge configuration is linear only for small signal levels. When the pressure exceeds about 0.8 GPa, the relatively large resistance change will lead to large errors in the indicated pressure. Then for higher pressures the transducer output must be read with something like an ohmmeter.

Two of the greatest problems encountered in constructing a reliable manganin cell are the insulation of the resistance wire and support of the wire. The usual polymer insulating coating will introduce a significant hysteresis, about 2%. Woven fiberglass insulation or simply wound thread will allow the liquid to communicate the pressure directly to the wire. To reduce the volume of wire and thereby minimize the internal volume of the transducer, while maintaining a useful resistance of at least 10 ohms, a very small diameter wire is required. Fifty micron diameter is used in this laboratory. This wire is extremely delicate and will be moved about by the flow due to compression of the liquid inside of the transducer unless restrained by a grooved spool or by bonding a fiberglass sleeve into the shape of a helix not shown in the figure. The lowest possible viscosity pressurizing medium must be used.

The two transducers of Fig. 3.7 may be integrated into the pressure generating intensifier unless the intensifier is heated to raise the glass transition pressure of the seal which will cause thermal drift of the transducer output. If the feed-through of the type of Fig. 3.4(c) is used to support the manganin resistance coil, the transducer may be installed directly into any port made for a standard high-pressure fitting including ports in the rheological instrument. Otherwise, the transducer must be installed in a connecting tube.

3.6 Hydrostatic Media and Volume Compensation

In almost all high-pressure viscometers and rheometers, there will be at least one liquid in addition to the liquid under study. There are several reasons for this and some have already been discussed. The manganin cell described above will not function reliably with a high-viscosity medium. If the pressure is to be indicated by the input pressure to the pressure intensifier then the same medium must be used consistently in the high-pressure output side of the intensifier regardless of the liquid being studied. If a thrust bearing of any type is operating inside of the instrument, it must be immersed in a low-viscosity liquid.

In addition, if a force measurement is to be made by sensing of the elastic strain of a component, then a displacement occurs in association with that strain. This displacement will be damped by the viscosity of the liquid surrounding the force sensing component. The damped response time may be quite long if the viscosity is that of the liquid being studied.

The time for the effect of a pressure change to be communicated along a length of high-pressure tubing may be very long for a viscous medium. For example, a 35 cm long tube of 1.5 mm inside diameter was used to connect an intensifier to a commercial manganin cell. The pressure medium was a diester, di(2-ethylhexyl)sebacate, a standard for high-pressure work [4]. For a pressure increase from 1.1 to 1.2 GPa, the gage required several minutes to fully respond to the increase. The addition of 10% of a light hydrocarbon, commercially known as Coleman Fuel, to the diester reduced the response time to a few seconds. In the author's laboratory, di(2-ethylhexyl)sebacate, di(2-ethylhexyl)adipate, and these diesters mixed with the light hydrocarbon are used as pressurizing media. In some special cases perfluorinated hydrocarbons find special use, to be discussed later.

The low-viscosity media must be separated from the liquid under study in a way that allows for the volume change of the liquid under study. This is accomplished in one of three ways:

1. A volume make-up cylinder can be attached to the pressure vessel of the instrument. This component houses a metal bellows or an isolating piston as shown

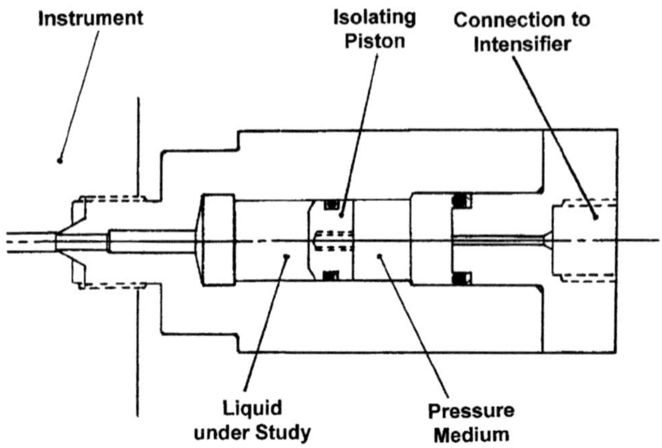

Fig. 3.8. A volume make-up cylinder attached directly to the instrument and using an isolating piston.

in Fig. 3.8 to separate and equalize the pressure of the medium and the liquid under study.

2. The rheological experiment can be confined to a cartridge that fits within the pressure vessel. One end of the cartridge is fitted with an isolating piston or metal bellows.
3. The pressure medium can be a material that is immiscible with the liquid under study and be of greater (or lesser) density than the liquid under study. Then the two liquids will naturally divide into two regions one above the other separated by a meniscus. The meniscus then functions as an isolating piston. Perfluorinated hydrocarbons are useful here.

The metal bellows is the most reliable method to assure that the liquids do not mix and that the difference in pressure is negligible. The bellows, however, adds a considerable dead volume since it cannot be fully collapsed and it is easily damaged when, for whatever reason, the volume displacement exceeds its design. The isolating piston is very simple to implement and this is now used exclusively for volume make-up devices in the author's laboratory. It does, however, offer some problems. The seal is generally an o-ring which suffers from glass transition under pressure that was mentioned previously. Then at some pressure, it will lose contact and no longer completely seal the two liquids. This is not a catastrophic failure since the seal is passive–the only pressure difference is due to seal friction which vanishes as contact is lost or due to the weight of the piston if the axis is vertical. The mixing that does occur is generally not a problem as long as the number of pressure cycles above about 0.5–0.6 GPa at room temperature is kept to a minimum.

The isolating piston shown in Fig. 3.8 incorporates a threaded hole for extraction. The bore of the cylinder can be seen to be enlarged at either end. The purpose of these enlargements is to break the seal if the piston should travel to either extreme of the cylinder. If the seal were maintained during a failure in some other part of the system the resulting pressure difference can damage other components as well.

References

[1] Bridgman, P.W., *The Physics of High Pressure*, Dover, New York, 1970, first published 1931.

[2] Comings, E.W., *High Pressure Technology*, McGraw-Hill, New York, 1956.

[3] Eremets, M.I., *High Pressure Experimental Methods*, Oxford Univ. Press, Oxford, 1996.

[4] Sherman, W.F. and Stadtmuller, A.A., *Experimental Techniques in High-Pressure Research*, Wiley, Chichester, 1987.

[5] Bair, S., "The High-Pressure Rheology of Some Simple Model Hydrocarbons", *Proc. Instn. Mech. Engrs.: J. Eng. Tribol.*, Vol. 216, Part J, 2002, pp. 139–149.

[6] Bair, S. and Winer, W.O., "The High Shear Stress Rheology of Liquid Lubricants at Pressures from 2 to 200 MPa", *ASME J. Tribol.*, Vol. 112, No. 2, 1990, pp. 246–253.

[7] Boren, M.D., Babb, S.E. and Scott, G.J., "Fixed Point Calibrations of Pressure Gauges", *Rev. Sci. Instruments.* Vol. 36, No. 10, 1965, pp. 1456–1459.

CHAPTER 4

Compressibility and the Equation of State

4.1 Background

For most engineering applications, liquids are treated as incompressible. In acoustics and low-pressure hydraulics, where the compressibility is certainly important, the compression, $V_0 - V$, may be regarded as linear with pressure and very small relative to the ambient pressure volume, V_0. Elastohydrodynamic lubrication, EHL, is quite different from these examples in the magnitude of pressure and the relative compression involved. A liquid lubricant pressurized to 1 GPa at 0°C may experience a relative compression, $(V_0 - V)/V_0$, of 20% and at 200°C of 30%. The compression is nonlinear for EHL pressures. The first half of the compression at 1 GPa comes at a pressure of 1/3 GPa at 0°C and at only 1/4 GPa at 200°C. Here and throughout this book, the distinction between $p = 0$ and $p = 1$ atmosphere is neglected owing to the relatively high pressures.

Now the compressibility of the lubricant is clearly a second order effect for the generation of a film thickness. Variations in velocity and viscosity can effect orders-of-magnitude changes in the film thickness while the effect of compressibility is to approximately reduce the central film by the amount of the relative compression [1],

$$\frac{h_c^{incom} - h_c^{compr}}{h_c^{incom}} \approx \frac{(V_0 - V)}{V_0} = 1 - \frac{V}{V_0} \tag{4.1}$$

An understanding of the compressibility of a liquid lubricant is required for the estimation of the refractive index, I, of the lubricant at the film pressure so that interferometry may be used to experimentally measure the thickness of the EHL film. The refractive index is related to the density or volume by the Lorenz–Lorentz equation [2]

$$\frac{(I^2 - 1)}{(I^2 + 2)} \frac{(I_R^2 + 2)}{(I_R^2 - 1)} = \frac{\rho}{\rho_R} = \frac{V_R}{V} \tag{4.2}$$

where subscript R refers to a reference state. The greatest use to be made of the pressure–volume–temperature, PVT, results obtained in this chapter will be for the free-volume correlations of viscosity in Chapter 5.

While the PVT measurement at first glance seems to be the most straightforward high-pressure experiment, it is in fact one of the most difficult to perform in a meaningful way. For example, a 30% random error in a viscosity measurement at very high-pressure will not result in a significant change in the interpretation of the strength of the piezoviscous effect as the viscosity is increasing by an order-of-magnitude every 25 to 250 MPa. In contrast, a 1% error in the measurement of volume under pressure will generally render the measurement useless for the correlation of viscosity; something around 0.1% being required. The variation in compressions among different organic liquids may only amount to less than 10% at 400 MPa so that, to resolve the differences among different chemical structures the resolution must be about 1%. The difficulty of precise measurement can be illustrated by noting that it was established experimentally by 1762 that water is compressible. However, it was not until one hundred years later that a precise value for the compressibility at even ambient pressure was found [3].

Compression measurements are the simplest means to arrive at the glass transition pressure and for this purpose accuracy is not an issue, except that of the pressure determination. The glass transition topic will arise often in the study of high-pressure rheology. The rapid increase in viscosity with pressure will eventually slow the structural changes that accompany a liquid-like compressibility. Eventually, there will be insufficient time for a liquid-like compression to occur and for increasing pressures the compression will follow a solid-like compressibility. For a given material the glass transition appears to occur at an isoviscous state although the glass transition viscosity, μ_g, seems to be material dependent.

4.2 PVT Measurement Techniques and Results

4.2.1 Dilatometers

Bridgman [3] provides an excellent review of experimental technique and an entertaining review of the history of the development of PVT measurements. The simplest device is the dilatometer, a piston-cylinder arrangement where the cylinder provides the pressure containment. Here a piston is driven into a cylinder containing a known initial volume of sample. See Fig. 4.1 for two examples. The displacement of the piston is measured, to find the compression, by means of the extensions shown in the figure. Displacement transducers (not shown) of the potentiometer or LVDT type are connected to the extensions. The piston force is used to calculate the pressure in the following manner: piston operation in reversed directions allows an estimate of the seal friction. Assuming that friction is symmetric with respect to direction, the reported pressure is the average of the theoretical pressures obtained in the two directions. Near to the glass transition, there is a large hysteresis in the material response so that extrapolation of the seal friction is required. Errors in both the volume compression and the pressure arise from the deformations of the cylinder, the piston and the seal and these corrections are both difficult to calculate and significant when compared with the quantities being measured.

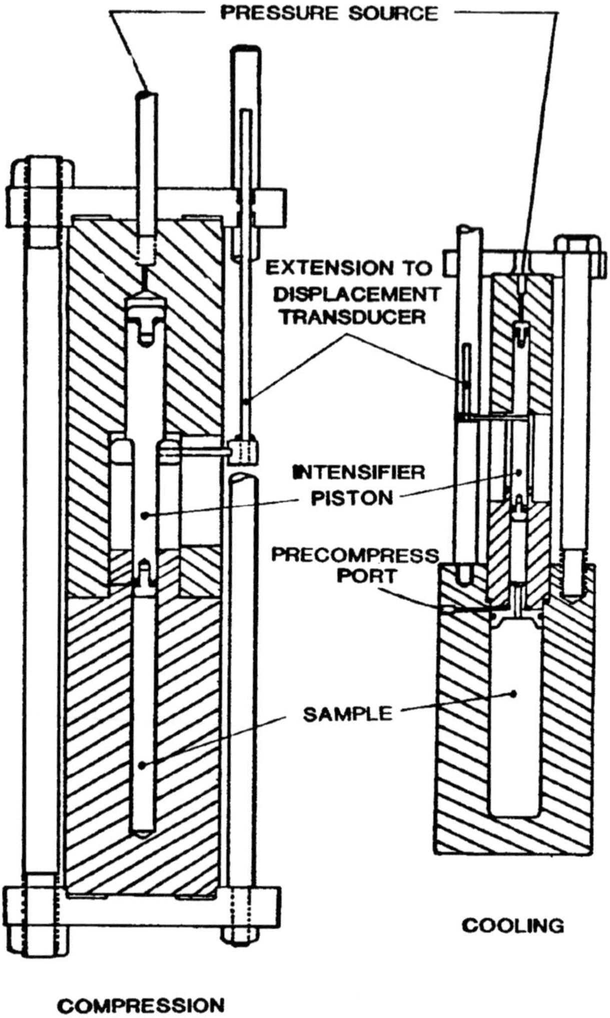

Fig. 4.1. Dilatometers for glass transition detection using isothermal compression and isobaric cooling histories.

Because of these shortcomings, the dilatometers shown in Fig. 4.1 only find use for measurement of the glass transition and solidification state variables.

Two types of dilatometer are illustrated in Fig. 4.1; the one on the left for isothermal compression to 2 GPa and on the right for isobaric cooling at up to 0.5 GPa. The much smaller volume change encountered in a cooling experiment requires a smaller diameter piston to be able to resolve the change in expansivity at the glass transition. This relatively small piston would be unable to effect a sufficient compression to reach the test pressure. To solve this problem, the dilatometer on the right has a separate isolating piston above the sample and a precompress port. Then pressurizing medium from the pressurizing source may be injected through the port into the space between the isolating piston and the intensifier piston to arrive at the test pressure. Once at the test pressure, the port is sealed by a valve and the temperature is reduced while the intensifier piston is advanced to maintain a constant pressure. Then a portion of the measured compression results from the pressure medium trapped between the intensifier piston and the isolator.

The dilatometer is immersed in a thermostat for temperature control. The displacement transducer and the pressure transducer outputs are sent to an analog plotter to record the compression curve. Some typical results are shown in Figs. 4.2 and 4.3 for the compression and cooling histories, respectively. The glass transition is obtained from the intersection of linear extrapolations of the liquid and glass portions of the curves. Clearly, some judgment is required for the extrapolation of the nonlinear liquid compression part and this judgment is aided by having available compression curves for increasing temperatures. In general, the cooling and compression histories yield consistent values, although the glass transition temperature (or pressure) is not a precise value, being somewhat different depending on the specific technique [4]. Not all liquids are glass-forming. Liquids for which the viscosity is low at the freezing temperature (pressure) do not easily supercool, for example, the normal alkanes like n-hexadecane and the liquid metals. See Fig. 4.4 where the compression curves for molten cadmium are shown. Cadmium has a melting temperature at ambient pressure of 321°C. The freezing pressure, in contrast to the glass transition pressure is indicated by a change in volume at the

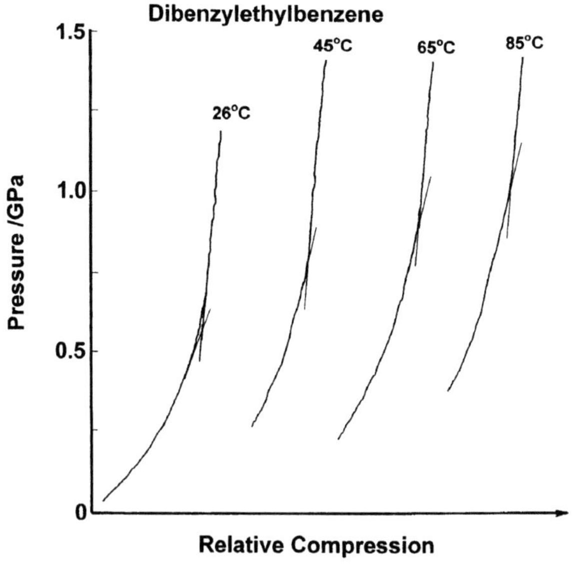

Fig. 4.2. Compression curves for a pure glass-forming hydrocarbon by the isothermal compression history at indicated temperatures. There is no volume change at the liquid-to-glass transition. The glass transition pressure is obtained from the intersection of the extrapolated compressions. Curves for different pressures have been shifted horizontally to avoid confusion.

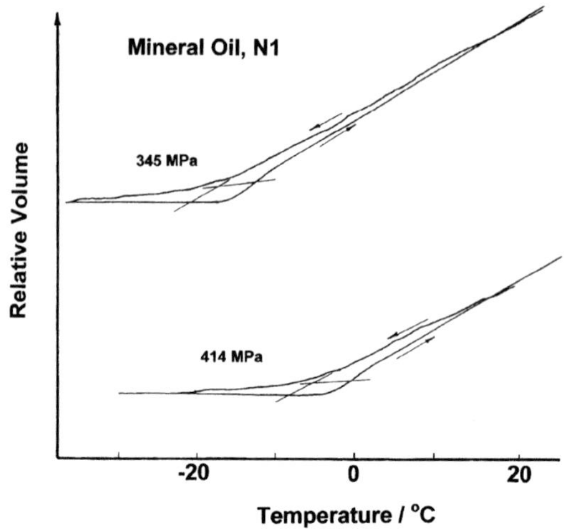

Fig. 4.3. Expansion and volume contraction curves for a naphthenic mineral oil by the isobaric cooling history at indicated pressures. There is no volume change at the liquid to glass transition. The glass transition temperature is obtained from the intersection of the extrapolated expansions. Curves for different pressures maintain continuity of relative volume.

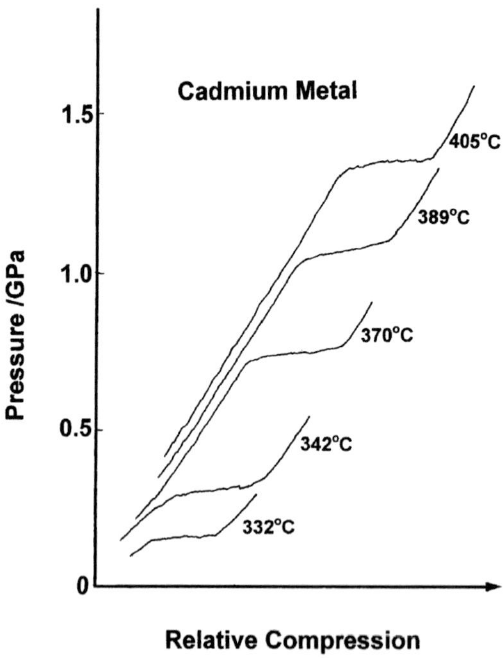

Fig. 4.4. Compression curves for a non-glass-forming liquid metal where there is a volume change at the liquid–solid transition. Compare with Fig. 4.2 where the volume is continuous at the liquid–glass transition.

freezing temperature. The freezing pressure for cadmium at each temperature may be defined as the average pressure along each of the plateaus in Fig. 4.4.

The dilatometer can be made more accurate for the measurement of volume compression by performing a direct measurement of the difference in compression of the sample and a material of known and low compressibility. A differential dilatometer for up to 0.7 GPa is shown in cross-section in Fig. 4.5. The temperature controlled block has two chambers containing a thin polymer capsule in each. The capsule in one contains a known volume of sample liquid and the capsule in the other contains the same volume of tungsten, a metal with a very high bulk modulus ($K_0 = 313$ GPa) and a low (cubic) thermal expansivity ($a_v = 1.38 \times 10^{-5}\text{C}^{-1}$). Each chamber communicates by high-pressure tubing with a separate high-pressure cylinder of a series intensifier. The intensifier vessel in Fig. 4.5 has two unconnected bores, each fitted with a piston, at opposite ends. The pistons are arranged in series so that the force on each is identical and therefore the pressure in the two bores should be essentially the same. The piston displacements are measured by potentiometers and their signals are electrically subtracted to give only the difference in piston displacement.

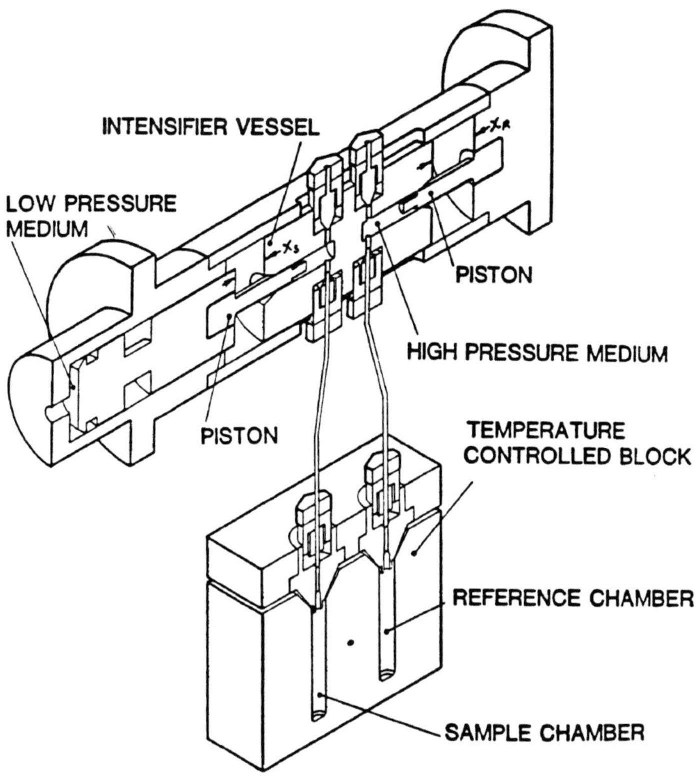

Fig. 4.5. A differential dilatometer for accurate measurements of compression to 700 MPa from [6]. Reprinted with permission of Elsevier.

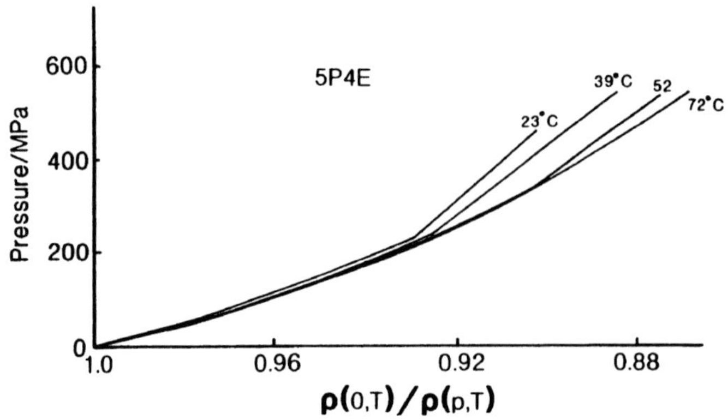

Fig. 4.6. The differential dilatometer result for pressure versus relative volume, $V(p, T)/V(0, T) = \rho(0, T)/\rho(p, T)$ for a synthetic lubricant from [6]. Reprinted with permission of Elsevier.

Water has become a standard reference material [5] for compression measurements and the differential dilatometer shown in Fig. 4.5 can repeat the accepted values for relative density within about 0.2% [6]. The relative volume, $V(p, T)/V(0, T) = \rho(0, T)/\rho(p, T)$ of a synthetic turbine lubricant, bis(phenoxyphenoxy)benzene called 5P4E, is shown in Fig. 4.6. The glass transition is clearly present in these compressions where the isothermal bulk modulus, $K(T, p) = \rho\,|\partial p/\partial \rho|_{T=const} = -V\,|\partial p/\partial V|_{T=const}$, which is related to the slope in Fig. 4.6, undergoes a transition from a low liquid-like value to a high solid-like value as in Fig. 4.2.

The compression of a liquid during a PVT measurement would result in a large temperature increase, perhaps 30°C for 0.7 GPa, if the heat of compression were not removed by conduction to the environment. In fact, following a rapid pressure increase, the volume will slowly relax to an equilibrium value as the temperature of the sample comes to the temperature of the thermostat. This means that the pressurization rate must be limited so that the compression may be assumed to be isothermal. Assuming that the container is made of a material with substantially greater thermal conductivity than the liquid, the pressurization rate for a 6 mm diameter liquid sample must be less than about 100 MPa/min to avoid serious error due to the temperature rise. A larger sample diameter requires a slower rate of pressurization.

4.2.2 Bellows piezometers

For accurate measurements of relative volume (relative density) to pressures important to EHL, the metal bellows piezometer that was originally devised by Bridgman [3] has become the standard instrument [7]. The bellows piezometer that is used in the author's laboratory is shown in Fig. 4.7. A commercial electro-deposited nickel bellows has been brazed to a stainless steel bulb that is fitted with a filling plug. The sample liquid is contained in the bellows and the bulb provides additional volume to improve

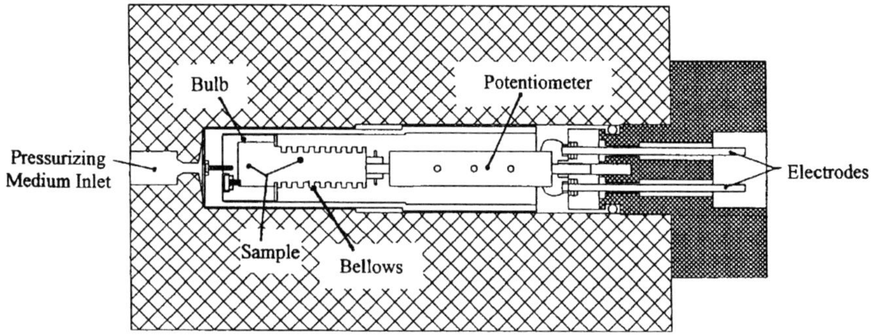

Fig. 4.7. A bellows piezometer for accurate measurements of compression to 450 MPa from [8]. Reprinted with permission of Elsevier.

the sensitivity. The change in length of the bellows is detected by a potentiometer that is immersed in the pressurizing medium along with the bellows. An LVDT may be substituted for the potentiometer [7]. The relationship between the length change and volume change was determined by the manufacturer and is independent of length. The difference between the internal and external pressures due to the elasticity of the bellows is less than 0.2 MPa. It is important that the sample be loaded under vacuum to be certain that bubbles are not present. Loading of liquid sample into any instrument under vacuum is good experimental practice and a glass syringe with a fitting that engages the same port as the fill plug is useful for this. Heating of the instrument during filling under vacuum will aid in removal of dissolved air that may present a problem if it leaves solution.

The relative volumes, V/V_R, of some liquids have been measured with the metal bellows piezometer of Fig. 4.7 and results are listed in Tables 4.1–4.9. The reference state for V_R is ambient pressure and the temperature for $V/V_R = 1$ at ambient pressure.

Table 4.1. The relative volume, V/V_R, of squalane, 2,6,10,15,19,23-hexamethyl tetracosane.

p/MPa	40°C	100°C
0.1	1.0000	1.0493
34.5	0.9758	1.0140
53.8	0.9651	1.0002
69	0.9572	0.9901
105	0.9412	0.9707
146	0.9277	0.9540
210	0.9095	0.9327
224	0.9052	0.9279
301	0.8874	0.9067
316	0.8831	0.9021
378	0.8707	0.8884
424	0.8625	0.8791

Table 4.2. The relative volume, V/V_R, of 1,1-(1,1,3-trimethyl-1,3-propanediyl) bis-cyclohexane, BCH.

p/MPa	40°C	100°C
0.1	1.0000	1.0482
34.5	0.9801	1.0162
69	0.9651	0.9969
146	0.9392	0.9642
224	0.9196	0.9396
301	0.9022	0.9192
378	0.8880	0.9024

Table 4.3. The relative volume, V/V_R, of dibenzylethylbenzene, DBEB.

p/MPa	40°C	100°C
0.1	1.0000	1.0393
34.5	0.9847	1.0159
69	0.9719	1.0002
146	0.9484	0.9720
224	0.9292	0.9491
301	0.9132	0.9303
378	0.8997	0.9150

Table 4.4. The relative volume, V/V_R, of trioctylmethane.

p/MPa	40°C	100°C
0.1	1.0000	1.0535
34.5	0.9761	1.0157
53.8	0.9654	1.0027
69	0.9571	0.9930
105	0.9418	0.9709
138	0.9304	0.9571
172	0.9207	0.9440
210	0.9093	0.9310
316	0.8829	0.9011
345	0.8768	0.8940
424	0.8614	0.8768

Some chemically well-defined liquids, squalane, 1,1-(1,1,3-trimethyl-1,3-propanediyl) bis-cyclohexane, dibenzylethylbenzene, polydecene trimer and trioctylmethane, are reported in Tables 4.1–4.5. Some lubricants are reported in Tables 4.6–4.9. These are the much-investigated mineral oils, HVI650 and LVI260, and the much-investigated perfluorinated polyalkylethers, the branched 143AD and the linear Z25. Additional data obtained from a similar instrument may be found in Ref. [9] for both chemically pure liquids and lubricants.

All of the instruments described above provide differential measurements of the changes in volume (density) with temperature and pressure. The values directly obtained

Table 4.5. The relative volume, V/V_R, of polydecene trimer.

p/MPa	40°C	100°C
0.1	1.0000	1.0512
34.5	0.9768	1.0113
69	0.9586	0.9893
146	0.9297	0.9524
224	0.9069	0.9252
301	0.8873	0.9031
378	0.8716	0.8841

Table 4.6. The relative volume, V/V_R, of HVI650 mineral oil.

p/MPa	20°C	70°C	120°C
0.1	1.000	1.033	1.071
34.5	0.980	1.007	1.036
69	0.966	0.991	1.016
103	0.954	0.976	1.000
138	0.943	0.965	0.987
207	0.925	0.944	0.964
276	0.909	0.927	0.944
345	0.896	0.913	0.929

Table 4.7. The relative volume, V/V_R, of LVI260 mineral oil.

p/MPa	20°C	70°C	120°C
0.1	1.000	1.035	1.070
34.5	0.982	1.010	1.039
69	0.969	0.994	1.020
103	0.957	0.980	1.003
138	0.948	0.969	0.990
207	0.931	0.949	0.968
276	0.916	0.932	0.949
345	0.902	0.918	0.934

Table 4.8. The relative volume, V/V_R, of 143AD branched perfluorinated oil.

p/MPa	0°C	25°C	50°C
0.1	0.976	1.000	1.0226
34.5	0.949	0.967	0.982
69	0.931	0.946	0.959
103	0.916	0.929	0.941
138	0.904	0.916	0.927
207	0.883	0.894	0.903
276	0.865	0.877	0.884
345	0.857	0.860	0.870

Table 4.9. The relative volume, V/V_R, of Z25 linear perfluorinated oil.

p/MPa	25°C	60°C	100°C
0.1	1.000	1.034	1.084
34.5	0.966	0.994	1.026
69	0.946	0.969	0.997
146	0.914	0.932	0.952
301	0.868	0.882	0.896
456	0.836	0.848	0.859

(compressibility or expansivity) are the difference in the response of the liquid sample and the response of the container material. Some correction must be applied although it is generally quite small. It should be possible, however, to measure other properties that depend directly upon the density or its variation with pressure and temperature and calculate the PVT response from this measurement. The measurement of the refractive index has a use for its own sake for EHL film thickness measurements but can be employed to generate PVT data.

4.2.3 *Refractive index measurements*

The ordinary approach to construct a refractometer is to place two nonparallel windows in a pressure vessel so that light entering through one window and exiting through the other window will pass through a wedge of liquid [2]. Then the light will be refracted through an angle by the liquid prism and the refractive index may be obtained using Snell's law [2] and a measurement of the angular displacement of the light exiting the cell. A slight variation on this approach is shown in Fig. 4.8. A mirror is retained in a fixture behind a single sapphire window. The fixture is held tightly against the window by a spring and the mirror makes a nonzero angle (determined by the fixture) with the surface of the window. Now the mirror can move with the window to accommodate the distortion of the cell body with pressure. The window will still deform with pressure so that care should be taken that the light passes close to the axis of the window where the spherical distortion is small. Refractometers operating at ambient pressure can be accurate to five decimal places. The problems of cell distortion with increasing pressure, limit the accuracy of the instrument shown in Fig. 4.8 to about three decimal places. The refractive index for a chemically well-defined liquid, bis(2-ethylhexyl)sebacate is listed in Table 4.10. Extracting a density change from a refractive index change using the Lorenz–Lorentz equation (4.2) tends to increase the error. Then the use of the refractometer to generate PVT data constitutes a poorly conditioned measurement.

4.2.4 *Longitudinal sound velocity*

The longitudinal sound velocity,

$$c_L = \sqrt{\frac{c_p}{c_V}\frac{K}{\rho}} \tag{4.3}$$

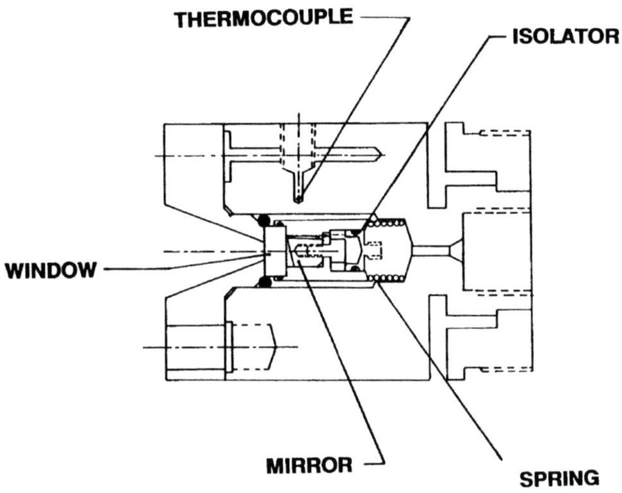

Fig. 4.8. A refractometer for measurement of the pressure variation of the refractive index to 700 MPa.

Table 4.10. The refractive index, *I*, of bis(2-ethylhexyl)sebacate at room temperature.

p/MPa	*I*
0.1	1.451
34.5	1.460
69	1.468
103	1.482
207	1.493
276	1.505
345	1.516
414	1.521
483	1.530
552	1.534
621	1.540

where c_p/c_V is the ratio of specific heats and $Kc_p/c_V = K_s$, the adiabatic bulk modulus, is a sensitive measure of K_s provided that ρ is known. The sound velocity may be measured under pressure by determining the time for an acoustic wave to travel from a source transducer to a receiving transducer. An instrument for this purpose is shown in Fig. 4.9. Two parallel quartz discs that are piezoelectrically active along the thickness are separated by a distance shown as the acoustic path in the figure. An isolating piston separates the sample from the pressure medium. The sound velocity is determined by launching a longitudinal wave, at the natural frequency of the disc, from one transducer and detecting the wave at the opposite transducer after a time, t. The sound velocity

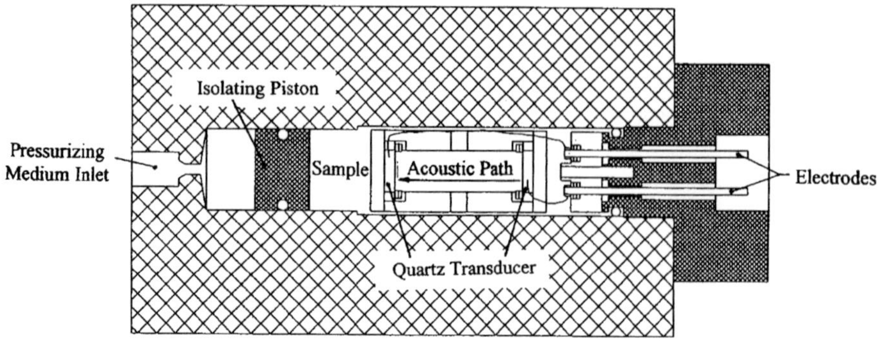

Fig. 4.9. An acoustic interferometer for the measurement of longitudinal sound velocity from [8]. Reprinted with permission of Elsevier.

is calculated from the length of the acoustic path divided by the measured t that has been corrected for electronic delay and the phase shift from diffraction. The operating frequency is about 1.5 MHz.

Some values of longitudinal sound velocity obtained in the instrument shown in Fig. 4.9 for a chemically well-defined liquid, bis(2-ethylhexyl)sebacate, and a mineral oil, HVI650, are listed in Tables 4.11 and 4.12, respectively. If the density at ambient pressure is known and the pressure dependence of volume and sound velocity is also known, it is possible to obtain the ratio of specific heats, c_p/c_V, as a function of pressure from equation (4.4). This has been done for the bis(2-ethylhexyl)sebacate and values are listed in Table 4.13 for c_p/c_V, K_s and K. The isothermal bulk modulus, K, was obtained by fitting the relative volume obtained with the bellows piezometer to an equation of state, the Tait equation, and analytically differentiating the equation. Equations of state are extremely useful for elastohydrodynamics and these are investigated in the following section.

Table 4.11. The longitudinal sound velocity, c_L, of bis(2-ethylhexyl)sebacate in m/s.

p/MPa	0°C	38°C	99°C
0.1	1495.3	1350.6	1147.5
34.5	1640.0	1521.1	1337.9
69	1744.9	1636.4	1482.7
103	1842.4	1743.9	1587.9
138	1930.1	1826.7	1686.1
172	2001.0	1911.7	1768.6
207	2070.0	1987.7	1856.1
241	2137.8	2055.8	1932.6
276	2223.4	2136.3	2001.0
310	2284.3	2200.6	2078.6
345	2326.9	2268.8	2142.4

Table 4.12. The longitudinal sound velocity, c_L, of the mineral oil, HVI650, in m/s.

p/MPa	70°C	120°C
0.1	1387.4	1189.9
34.5	1520.3	1403.6
69	1683.2	1535.9
103	1773.8	1658.1
138	1872.3	1760.3
172	1961.7	1840.2
207	2054.4	1935.1
241	2119.7	2015.7

Table 4.13. The isothermal and adiabatic bulk moduli, K and K_s, and the ratio of specific heats, c_p/c_V, of bis(2-ethylhexyl)sebacate.

p/MPa	0°C			38°C			99°C		
	K/GPa	K_s/GPa	c_p/c_V	K/GPa	K_s/GPa	c_p/c_V	K/GPa	K_s/GPa	c_p/c_V
0.1	1.65	2.08	1.26	1.35	1.65	1.22	0.99	1.14	1.15
34.5	2.02	2.55	1.26	1.72	2.14	1.25	136	1.59	1.17
69	2.38	2.94	1.23	2.08	2.52	1.21	1.72	2.00	1.16
103	2.74	3.32	1.21	2.43	2.91	1.20	2.07	2.33	1.13
138	3.09	3.69	1.19	2.78	3.23	1.16	2.41	2.67	1.11
172	3.44	4.00	1.16	3.13	3.58	1.15	2.75	2.98	1.09
207	3.79	4.33	1.14	3.47	3.91	1.13	3.09	3.32	1.08
241	4.13	4.65	1.13	3.81	4.23	1.11	3.42	3.64	1.07
276	4.47	5.08	1.14	4.14	4.60	1.11	3.74	3.94	1.05
310	4.80	5.40	1.12	4.47	4.93	1.10	4.07	4.29	1.05
345	5.14	5.64	1.10	4.80	5.27	1.10			

4.3 Empirical Equations of States

An equation relating the volume (or density) to temperature and pressure is known as an equation of state, EOS. An EOS is necessary for lubrication calculations to include compressibility and in Chapter 6 we will see that the most natural correlation of viscosity with temperature and pressure requires an EOS. Numerical solutions of the isothermal EHL problem for compressible liquid lubricants nearly always describe the pressure dependence of volume with the Dowson and Higginson [10] isothermal equation of state,

$$\frac{V}{V_0} = \frac{1 + \dfrac{K_0' - 1}{2K_0} p}{1 + \dfrac{K_0' + 1}{2K_0} p} \tag{4.4}$$

Here a standard form is used to write equation (4.5), where K_0 is the bulk modulus, K, at ambient pressure or $p = 0$ and K_0' is the pressure rate of change of K at ambient pressure or $p = 0$. Then the bulk modulus for the Dowson and Higginson equation of state depends on pressure as

$$K = K_0 \left(1 + \frac{K_0' + 1}{2K_0} p\right)\left(1 + \frac{K_0' - 1}{2K_0} p\right) \tag{4.5}$$

When used in EHL calculations, universal values of $K_0 = 1.67$ GPa and $K_0' = 6.67$ are employed. Dowson and Higginson [10] obtained this expression from curve fitting of data for a mineral oil at one temperature to about 350 MPa. This EOS shares a property of many early forms [3, p. 143] in that the volume approaches a limiting value, $V/V_0 \rightarrow 0.74$ for the universal parameters, as pressure increases. It is generally only accurate for low pressures and temperatures; however it has served well as a universal EOS for low-temperature EHL calculations. The limiting volume feature, while not realistic for liquids below the glass transition pressure, has probably compensated for the neglect of the glass transition on the compressibility of liquid lubricants.

Of the simple isothermal equations of state that can be written as $V = V_p(p)$, the Tait equation is often held to be the most accurate [11] and even accurate for extrapolation to very high pressures [12].

$$\frac{V}{V_0} = 1 - \frac{1}{1 + K_0'} \ln\left[1 + \frac{p}{K_0}\left(1 + K_0'\right)\right] \tag{4.6}$$

$$K = \left\{1 - \frac{1}{1 + K_0'} \ln\left[1 + \frac{p}{K_0}\left(1 + K_0'\right)\right]\right\}\left[K_0 + p\left(1 + K_0'\right)\right] \tag{4.7}$$

This EOS has been used very successfully by Cook et al. [13] in a free volume pressure–viscosity correlation.

Another useful isothermal EOS is the Murnaghan equation [14], derived from a linear theory of finite strain [15].

$$\frac{V}{V_0} = \left(1 + \frac{K_0'}{K_0} p\right)^{\left(-\frac{1}{K_0'}\right)} \tag{4.8}$$

$$K = K_0 + K_0' p \tag{4.9}$$

Of course, the application of the above equations of state assumes that the range of state variables does not include a transition; freezing, glass transition or polymorphic transition. MacDonald [15] has reviewed the uses of the Tait and Murnaghan equations for both liquids and solids.

The usefulness of the above isothermal equations of state would be improved by modifying them to include temperature, if only for a limited range of temperature. The usual approach taken for elastohydrodynamic calculations is to multiply the isothermal EOS by an expression for the temperature variation of volume or density at ambient pressure to arrive at a simple separable equation of the form, $V = V_0 \cdot F_T(T) \cdot F_p(p)$. While this can be reasonable for a very small range of temperature, it can result in large errors in compressibility at high temperature and also results in nonphysical sound velocity response to temperature. Note that in Tables 4.11 and 4.12 the sound velocity, c_L, decreases with increasing temperature. For the separable form above, dV/dp is independent of temperature and less than zero while V increases with temperature for organic liquids, leading to the incorrect conclusion that sound velocity increases with temperature or is relatively independent of temperature.

By far, the greater amount of PVT work has been done with the Tait equation (4.6), so the discussion of temperature modifications will begin here with the way that the Tait equation may be modified to accommodate changes in temperature. Many workers find that K_0' is independent of temperature. McDonald [15] gives a summary of work on the temperature dependence of K_0' which has been assigned a universal value of $K_0' = 10.2$ by Cutler et al. [16] based upon measurements for a wide variety of hydrocarbon liquids. For very light hydrocarbons, K_0' may have a value as low as nine and for high molecular weight lubricants, as high as 12.

If K_0' is taken to be independent of temperature, the effect of temperature on compressibility for the simple EOS is entirely described by the temperature dependence of K_0, the initial bulk modulus. Cutler et al. [16] and others have found that a quadratic in T can be useful.

$$K_0 = C_0 + C_1(T - T_R) + C_2(T - T_R)^2 \tag{4.10}$$

They found that for $T_R = 80°\text{C}$, the constants, C_0, C_1 and C_2, vary with chemical structure and are roughly equal to 1 GPa, -0.006 GPa/°C and 10^{-5} GPa/°C^2, respectively. A problem for the polynomial representation of $K_0(T)$ is that at some high temperature (~380°C) it begins to increase with temperature. The author [17] has used

$$K_0 = K_\infty + \frac{\dot{K}_0}{T} \tag{4.11}$$

for free volume correlations of low-shear viscosity, where K_∞ is approximately -0.8 GPa and $\dot{K}_0$ is approximately 600 GPa·K and T is absolute temperature. A problem for this representation of $K_0(T)$ is that above some high temperature (~500°C) K_0 becomes negative.

Fakhreddine and Zoller [18] and others have suggested that $K_0(T)$ be described by

$$K_0 = K_{00} \exp(-\beta_K T) \tag{4.12}$$

where K_{00} is approximately 9 GPa and β_K is approximately 0.006 K^{-1}. This relation seems to be well-behaved for all temperatures.

Now V_0 must also vary with temperature and it is generally held that a linear dependence of density on temperature,

$$\frac{V_0}{V_R} = \frac{\rho_R}{\rho_0} = \frac{1}{1 - a_\rho(T - T_R)} \tag{4.13}$$

is more accurate than a linear dependence of volume on temperature,

$$\frac{V_0}{V_R} = 1 + a_V(T - T_R) \tag{4.14}$$

however, in free volume formulations for viscosity, volumes are usually set out as linear functions of temperature as in (4.14). Subscript R denotes the reference state, $p = 0$ (or 0.1 MPa) and $T = T_R$.

The temperature-modified Tait and Murnaghan equations of state can produce good quality fits to the relative volume data in Table 4.7 for the model mineral oil, LVI260. This liquid has been the subject of many high-pressure measurements [19], traction studies [20] and numerical simulations [21]. The temperature dependence of V_0 was described by equation (4.14) and both equations (4.11) and (4.12) were used for $K_0(T)$. The parameters are listed in Table 4.14. There is little difference in the quality of the fit to the experimental data, the Tait equation with equation (4.11) having the least standard deviation. The Tait and Murnaghan equations are compared with the data in Fig. 4.10 using equation (4.11) for $K_0(T)$. In the next chapter, these results will be used to correlate the limiting low shear viscosity with temperature and pressure. The Tait parameters for some liquids are listed in Table 4.15.

For completeness, it must be mentioned that Hayward [22] has pointed out that the EOS generally attributed to Tait, equation (4.6), is not, in fact, the original form proposed by Tait. The original Tait equation is actually the form used by Dowson and Higginson, equation (4.5). To avoid confusion, however, the terminology in general use is retained here.

Now a general equation of state can be offered for numerical simulations of EHL. Employing equations (4.6), (4.12) and (4.14) with $K_0' = 11$, $a_V = 8 \times 10^{-4}\,\text{K}^{-1}$, $K_{00} = 9\,\text{GPa}$ and $\beta_K = 6.5 \times 10^{-3}\,\text{K}^{-1}$ should provide a description of the compressibility of liquid lubricants with sufficient accuracy for the calculation of film thickness over the usual range of temperature for which EHL lubricants are used. Exceptions are

Table 4.14. Parameters for the temperature-modified Tait and Murnaghan equations of state for LVI260.

EOS	$K_0(T)$	K_0'	$a_V/°\text{C}^{-1}$	K_∞/GPa	$\dot{K}_0/\text{GPa}\cdot\text{K}$	K_{00}/GPa	β_K/K^{-1}
Tait	(4.12)	12.38	7.158E-4			9.534	0.005728
Tait	(4.11)	11.36	6.727E-4	–0.7495	753.7		
Murnaghan	(4.12)	11.59	7.146E-4			9.504	0.005669
Murnaghan	(4.11)	10.57	6.673E-4	–0.6911	744.6		

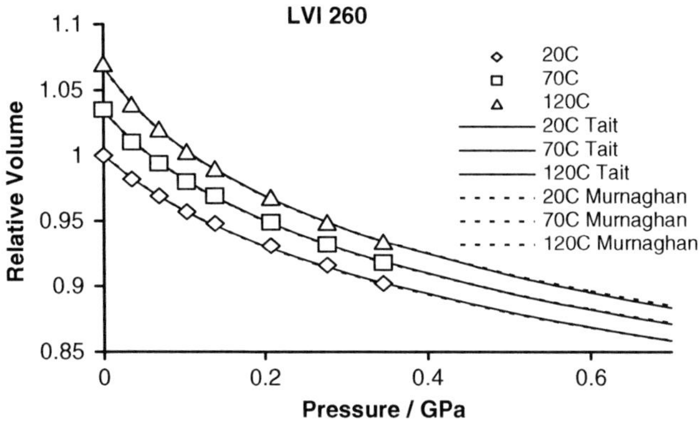

Fig. 4.10. The Tait and Murnaghan equations fitted to data for a model mineral oil.

Table 4.15. Parameters for the temperature-modified Tait equation using (4.12).

Liquid	K_0'	K_{00}/GPa	β_K/K^{-1}	a_V/°C^{-1}
Squalane	10.85	8.824	0.006321	8.35E-4
BCH	11.018	12.26	0.006479	7.744E-4
Z25	11.150	5.127	0.006119	1.0384E-3
DBEB	11.786	14.49	0.006444	6.912E-4
DOP	10.647	11.005	0.006327	8.33E-4

DOP is di(2-ethylhexyl)phthalate from the data of Ref [9].

silicone oils for which the compressibility is greater and polyphenyl ethers for which the compressibility is lower. This description of the lubricant compressibility will result in the reduction of the central film thickness for a Hertz pressure of 1.3 GPa by about 2% at 20°C and 11% at 180°C, compared to equation (4.5) that is generally employed in such simulations with its universal parameters. This general equation of state will be useful for generalized temperature–pressure–viscosity models offered in Chapter 6.

References

[1] Venner, C.H. and Bos, J., "Effects of Lubricant Compressibility on the Film Thickness in EHL Line and Circular Contacts", *Wear*, Vol. 173, 1994, pp. 151–165.

[2] Poulter, T.C., Ritchey, C. and Benz, C.A., "The Effect of Pressure on the Refractive Index of Paraffin Oil and Glycerine", *Phys. Rev.*, Vol. 41, 1932, pp. 366–367.

[3] Bridgman, P.W., *The Physics of High Pressure*, Dover, New York, 1970, first published 1931.

[4] Alssad, M., Bair, S., Sanborn, D.M. and Winer, W.O., "Glass Transitions in Lubricants: Its Relation to EHD Lubrication" *ASME J. Lubr. Techn.*, Vol. 100, No. 2, 1978, pp. 404–417.

[5] Burnham, W.C., Holloway, J.R. and Davis, N.F., *"Thermodynamic properties of water to 1,000°C and 10,000 bars"*, The Geological Society of America, Special Paper Number 132, 1969, pp. 1–96.

[6] Bair, S., "Recent Developments in High-Pressure Rheology of Lubricants", *Lubricants and Lubrication*, Dowson et al. (Ed.), Elsevier Science, Amsterdam, 1995, pp. 169–187.

[7] Back, P.J., Easteal, A.J., Hurle, R.L. and Woolf, L.A., "High-Precision Measurements with a Bellows Volumometer", *J. Phys. E: Sci. Instrum.*, Vol. 15, 1982, pp. 360–363.

[8] Bair, S., Jarzynski, J. and Winer, W.O., "The Temperature, Pressure, and Time Dependence of Lubricant Viscosity", *Tribol. Int.*, Vol. 34, No. 7, 2001, pp. 461–468.

[9] Kleinschmidt, R.V., Bradbury, D. and Mark, M., *Viscosity and Density of Over Forty Lubricating Fluids of Known Composition at Pressures to 150,000 psi and Temperatures to 425 F*, ASME, New York, 1953.

[10] Dowson, D. and Higginson, G.R., *Elastohydrodynamic Lubrication*, Pergamon Press, Oxford, 1966, p. 89.

[11] Hirschfelder, J.O., Curtiss, C.F. and Bird, R.B., *Molecular Theory of Gases and Liquids*, Wiley, New York, 1954, p. 261.

[12] Millat, J., Dymond, J.H. and de Castro, C.A.N., *Transport Properties of Fluids Their Correlation, Prediction and Estimation*, IUPAC, Cambridge, 1996, p. 172.

[13] Cook, R.L., King, H.E., Herbst, C.A. and Herschbach, D.R., "Pressure and Temperature Dependent Viscosity of Two Glass Forming Liquids: Glycerol and Dibutyl Phthalate", *J. Chem. Phys.*, Vol. 100, No. 7, 1994, pp. 5178–5189.

[14] Murnaghan, F.D., *Finite Deformation of an Elastic Solid*, Wiley, New York, 1951.

[15] McDonald, J.R., "Some Simple Isothermal Equations of State", *Reviews of Modern Physics*, Vol. 38, No. 4, 1966, pp. 699–679.

[16] Cutler, W.G., McMickle, R.H., Webb, W. and Schiessler, R.W., "Study of the Compressions of Several High Molecular Weight Hydrocarbons", *J. Chem. Phys.*, Vol. 29, No. 4, 1958, pp. 727–741.

[17] Bair, S., "The Pressure-Viscosity Coefficient of a Perfluorinated Polyether Over a Wide Temperature Range", *ASME J. Tribol.*, Vol. 123, 2001, pp. 50–53.

[18] Fakhreddine, Y.A. and Zoller, P., "The Equation of State of a Polydimethylsiloxane Fluid", *J. Appl. Polym. Sci.*, Vol. 41, 1990, pp. 1087–1093.

[19] Hutton, J.F., Phillips, M.C., Ellis, J.E., Powell, G. and Wyn-Jones, E., "Viscoelasticity studies of Lubricants and other Liquids in Shear and Bulk Deformation at Various Frequencies, Pressures and Temperatures", *Elastohydrodynamics and Related Topics*, Dowson et al. (Ed.), Mechanical Engineering Publications, London, 1979, pp. 173–187.

[20] Johnson, K.L. and Tevaarwerk, J.L., "Shear Behaviour of Elastohydrodynamic Oil Films", *Proc. Roy. Soc. London*, A. Vol. 356, 1977, pp. 215–236.

[21] Wang, S., Cusano, C. and Conry, T.F., "Thermal Analysis of Elastohydrodynamic Contacts Using the Ree-Eyring Fluid Model", *ASME J. Tribol.*, Vol. 113, No. 2, 1991, pp. 232–244.

[22] Hayward, A.T.J., "Compressibility Equations for Liquids: A Comparative Study", *Br. J. Appl. Phys.*, Vol. 18, 1967, pp. 965–977.

CHAPTER 5

The Pressure and Temperature Dependence of the Low-Shear Viscosity

5.1 Background

The full elastohydrodynamic lubrication (EHL) regime of concentrated contact lubrication relies upon the increase of viscosity with pressure to generate a sufficiently thick liquid film so that the roughness features of engineering surfaces can be effectively separated to reduce friction and avoid wear. The classic formulas, which were introduced in the first chapter for the calculation of film thickness, require a parameter to quantify the strength of the piezoviscous response of the liquid. This parameter, one of the pressure–viscosity coefficients to be described in a later section, is determined from measurements of the low-shear viscosity as a function of pressure to the moderate pressures (0.1–0.5 GPa) found in the contact inlet zone. Another coefficient, the local pressure–viscosity coefficient, may be defined as

$$\alpha(T, p) = \frac{\partial(\ln \mu)}{\partial p} \tag{5.1}$$

For a calculation of the friction (traction) at moderate pressure, the most important properties of the liquid seem to be both the low-shear viscosity and α at pressures near the maximum Hertz value.

The low-shear viscosity, μ, is an essential parameter in any description of non-Newtonian response as well, serving to locate the terminal plateau. It is also used in the method of reduced variables or time–temperature–pressure superposition (Section 2.5) where the horizontal shift factor is essentially the relative low-shear viscosity. It is easier to measure μ over a large range of temperature and pressure and shift the rate dependent viscosity data than it would be to measure the rate dependence over the same range of state variables.

A reader familiar with the literature regarding elastohydrodynamic lubrication, may already be aware of the substantial difference between the way in which the pressure dependence of viscosity is described inside and outside of the field of tribology. At sufficiently high pressure, α increases with pressure. The characteristic sigmoidal shape of a log(viscosity) versus pressure curve that can be seen in Fig. 1.1 has been conspicuously absent from most work dealing with elastohydrodynamics. This is perhaps best exemplified by the early classic paper by Crook [1] in which the value of the viscosity was determined which, if uniform throughout the Hertz region, would result in the observed traction. Obviously the technique can only be applied when the sliding velocity is small enough to avoid viscous heating and shear-thinning. See the curves in Fig. 5.1. Crook compared this effective viscosity, that was calculated from traction, with the viscosity from the ASME Pressure–Viscosity Report [2] of a mineral oil with similar pressure–viscosity coefficient to his oil. See the data represented by + symbols in Fig. 5.1. This, however, was an incomplete data set from the report, measurements having been obtained using only the first sinker. Another data set for an oil of similar low pressure behavior from the same report has been added to Fig. 5.1 as the • symbols. This is now a complete data set from the ASME Pressure–Viscosity Report [2] which clearly shows the sigmoidal shape that is absent from most of the EHL literature. Other oils of similar low-pressure behavior in the report show this type of high-pressure response.

The above description, of the pressure variation of viscosity, has been widely adopted and, in spite of lack of success over the years, it has been only recently that this experimental measurement technique was abandoned. There has been additional reinforcement for

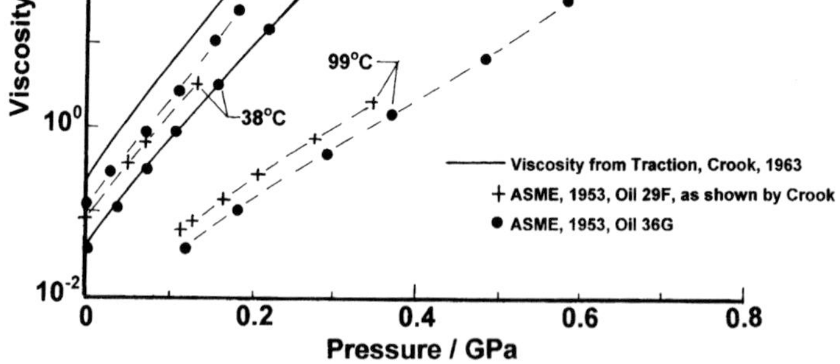

Fig. 5.1. The pressure dependence of the effective low shear viscosity of a mineral oil obtained from a measurement of elastohydrodynamic traction is shown as the curves. Points represent data from the ASME Pressure–Viscosity Report [2] for oils of similar low presure behavior. Adapted from Fig. 9 of Crook, A.W., "The Lubrication of Rollers IV. Measurements of Friction and Effective Viscosity", *Phil. Trans. Roy. Soc. Lond.*, Series A, Vol. 255, No. 1056, 1963, pp. 281–312, with permission of the Royal Society and the kind consent of the author.

underestimating the piezoviscous effect at Hertz zone pressures. While not often openly addressed, there seems to be a numerical stability problem with the usual EHL simulation that is associated with an increase of the log(viscosity) versus pressure slope, α, at high pressures. Then the piezoviscous effect, as quantified by α, has been understated at high pressures in order to obtain any numerical solution to the EHL problem [3] at all.

The temperature dependence of viscosity should be accurately described in an EHL analysis that includes the effects of temperature variations within the film. The local temperature–viscosity coefficient may be defined as

$$\beta(T, p) = -\frac{\partial(\ln \mu)}{\partial T} \tag{5.2}$$

Then the pressure rate of change of β is

$$\frac{\partial \beta}{\partial p} = -\frac{\partial \alpha}{\partial T} \tag{5.3}$$

At pressures for which α is increasing with pressure, β can become very much larger than its value at ambient pressure. Now if the value of α has been understated, it will be unlikely as well that the temperature–viscosity coefficient will be accurately described at the pressures of the Hertz zone. The sections that follow will provide a framework for accurate descriptions of the pressure and temperature dependence of the low-shear viscosity that may be employed in quantitative EHL calculations.

5.2 Viscometers for High Pressure

An instrument for the measurement of the limiting low-shear viscosity, μ, is known as a *viscometer*. There does not seem to be an exact division between the terms, viscometer and rheometer. If normal stress differences and other properties that may be related to viscoelastic behavior are measured then the instrument is known as a *rheometer*. For the measurement of the shear-dependent viscosity, η, a property of both a generalized Newtonian liquid and a viscoelastic liquid both terms have been used. An excellent resource for design and analysis of viscometers and rheometers of many types is Macosko [4].

The limiting low-shear viscosity is often very difficult to measure in high-molecular-weight polymers as the Newtonian limit may be less than the minimum rate or stress capability of common instruments. For the determination of the low-shear viscosity of high-molecular-weight polymers, it is common practice to measure the viscosity as a function of shear rate (or stress) and either demonstrate that, at the lowest possible rates, the viscosity has become independent of rate or fit the data to an empirical expression that contains μ as a parameter. For the liquids that are commonly used in EHL it should be sufficient to perform a viscosity measurement at a low shear stress, say $\tau < 100\,\text{Pa}$, to assume that the quantity being measured is the limiting low-shear viscosity. And for this low stress the viscous heating that must accompany any viscosity measurement should be negligible.

5.2.1 Capillary viscometers

Nearly every type of viscometer has been pressurized. One of the earliest to be pressurized was a capillary viscometer by Warburg and Sachs [5] who used the pressure head of a column of mercury to drive the flow through the capillary at system pressures to 15 MPa. They found that the viscosity of water decreased slightly with pressure and that the viscosity of benzene increased linearly to a greater extent than the loss of viscosity of water. A capillary viscometer measures the pressure change for a given flow across a length of a circular cross-section duct. The viscosity can then be calculated directly from the Hagen–Poiseuille equation,

$$\mu = \frac{\pi D_i^4}{128 Q} \frac{dp}{dx} \tag{5.4}$$

provided that the pressure change does not significantly affect the viscosity. At times, a kinetic energy correction and elastic energy effects must be applied to the pressure difference that is measured across the capillary. The shear stress at the capillary wall is

$$\tau_W = \frac{D_i}{4} \frac{dp}{dx} \tag{5.5}$$

Most other early capillary viscometers for the investigation of the pressure effect on viscosity had one end of the capillary open to ambient pressure. The large pressure variation along the length of the capillary then results in a variation of viscosity along the length of the capillary owing to the piezoviscous effect that is being measured. In order to extract the pressure dependence of viscosity from such measurements, a pressure–viscosity law must be assumed or measurements at various flow rates must be performed and the corrections that are applied reduce the accuracy. A simple solution to this problem was devised by Norton et al. [6]. They attached a long capillary to the exit of the short test capillary. The long capillary provided the elevated pressure at the exit to the test capillary so that the pressure change across the short capillary could be minimized.

The pressurized capillary viscometer was developed to its highest level by Novak and Winer [7] in the 1960s. Their instrument is shown in Fig. 5.2. It is capable of pressure to 0.6 GPa and temperature to 150°C. In an ingenious arrangement, the pressure within the capillary is maintained on the average at some high value with two intensifiers arranged in series. The translating piston shown in Fig. 5.2 is driven by a low-pressure hydraulic system. This piston is, at the same time, the high-pressure cylinder for both intensifiers. Then, with the rams fixed, motion of the moveable piston causes a known flow from one intensifier to the other which must pass through the capillary in the test section. Pressures are detected by strain gage based transducers and the electrical output from these transducers is subtracted and amplified by a special-built deviation amplifier. This viscometer was used to generate measurements of low-shear viscosity of lubricants over a period of about ten years. It was further developed to obtain shear stress to about 5 MPa and shear rate to 10^7 s^{-1} for the measurement of shear-dependent viscosity. Some limitations of this design are: a relatively low viscosity limit of 100 Pas, a great effort required to clean and replace the sample which occupies all of the high-pressure

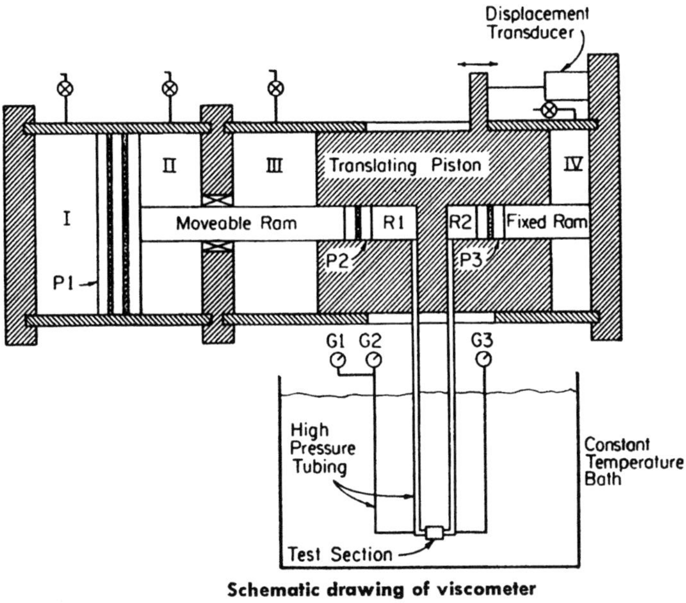

Fig. 5.2. The pressurized capillary viscometer of Novak and Winer, from Novak, J.D. and Winer, W.O., "Some Measurements of High Pressure Lubricant Rheology", *ASME J. Lubr. Techn.*, Vol. 90, Ser. F, No. 3, 1968, pp. 580–591, with permission of the American Society of Mechanical Engineers. The translating piston displaces liquid from reservoir R1 to R2 across the test section which contains the capillary. Piston P1 generates a force on both the moveable and fixed rams that determines the average capillary pressure. Pressure transducers are at G1–G3.

tubing, and viscous heating at relatively low values of viscous power, $\tau\dot{\gamma}$. Some shear dependent viscosity results from this pressurized capillary viscometer will be addressed in the following chapter.

5.2.2 Rotational Couette viscometers

The rotational couette viscometer shears the liquid sample in the gap between concentric cylinders with working diameters, D_o and D_i, and average diameter, D. The shear stress is obtained from measurement of the torque, T, to restrain the stationary cylinder or to drive the rotating cylinder and the shear rate is obtained from the rotation rate, Ω, and the radial gap, $(D_o - D_i)/2$. When the gap is small, $D_i/D_o > 0.99$, the curvature and end effects can be neglected [4] and the shear rate is

$$\dot{\gamma} = \frac{\Omega D}{D_o - D_i} \tag{5.6}$$

and the shear stress is

$$\tau = \frac{2T}{\pi D^2 L} \tag{5.7}$$

When the gap is large compared with the cylinder radius, corrections must be applied for the variation of shear stress across the gap and end effects [4]. This is a convenient and useful arrangement for the measurement of shear-dependent viscosity; however the great difficulty in providing a means for driving the rotating element and extracting a torque measurement in a high-pressure environment cannot generally be justified for the limiting low-shear viscosity measurement when simpler designs are available. The earliest pressurized Couette viscometer appears to have been constructed by Thomas et al. [8] who placed an electric motor within the pressure vessel to drive the rotation. They experienced many problems with this arrangement and the heat generated by the motor was significant [8]. Hutton and Phillips [9] overcame the problem of driving the rotation by passing a shaft through a high-pressure seal and situating the motor outside of the pressure vessel. The torque was measured by a strain gaged elastic element. The response time for the torque measurement was improved by immersing the torque sensor in a low viscosity liquid that was immiscible with the test sample. The innovations introduced by Hutton and Phillips made possible the measurement of shear-thinning in low molecular weight liquids that is discussed Chapter 8.

5.2.3 Dropping ball and rolling ball viscometers

In theory the velocity of a dropping ball can be employed for a measurement of Newtonian viscosity using Stoke's law. In practice, however, the interaction of walls with the flow around the ball makes this a difficult endeavor. In a high-pressure environment where the pressure containing vessel must be internally compact, the close proximity of the walls and the unpredictable trajectory of the ball usually makes the dropping ball a poor choice for a pressurized viscometer. The solution has been to allow the ball to roll down a close fitting inclined tube or along an inclined plane. The instrument is tilted in the opposite way to reset the ball. Many instruments may be utilized with the ball traveling in both directions. Typical inclination (from horizontal) angles are $\theta = 8\text{–}70°$. The time of fall, t, is measured for which the ball travels a fixed distance. The viscosity can be calculated from

$$\mu = C_{rb}\,(\rho_s - \rho)\,t \sin\theta \tag{5.8}$$

where ρ_s is the density of the ball. The density of the liquid sample, ρ, which varies with temperature and pressure, yields a buoyancy force that opposes gravity. The calibration factor, C_{rb}, must be obtained from measurements of a liquid of known viscosity. The creeping flow around a ball is a complex problem and it would be difficult to assign a characteristic shear stress to the measurement.

Most rolling ball viscometers have the ball roll along the inside diameter of a tube within the pressure vessel. With internal and external pressure acting on the tube, the inside diameter of the tube is affected by pressure only in as much as the compressibility of the tube material allows. Hersey [10], in 1916, analyzed the rolling ball configuration, constructed an instrument, and measured the viscosity of a lard oil to 50 MPa. He timed the "fall" of the ball with electrical contacts at the ends of the tube. The major distinction among the various rolling ball viscometers and all falling body viscometers, for that matter, has been the means for detecting the position of the ball or sinker. Sawamura et al. [11]

utilized a visible light detector to observe the light from a lamp through a pair of sapphire windows positioned across the tube in their pressure vessel. A glass ball refracted the light as it passed the detector position. The width of a characteristic waveform from the detector was used to establish a fall time, t. They worked at pressure to 400 MPa. Izuchi and Nishibata [12] utilized a pair of linear variable differential transformers, LVDTs, situated within the pressure vessel to inductively observe the time of fall of a steel ball at pressure to 1 GPa. Their instrument is shown in Fig. 5.3. The improved version on the right employs an integral bellows piezometer.

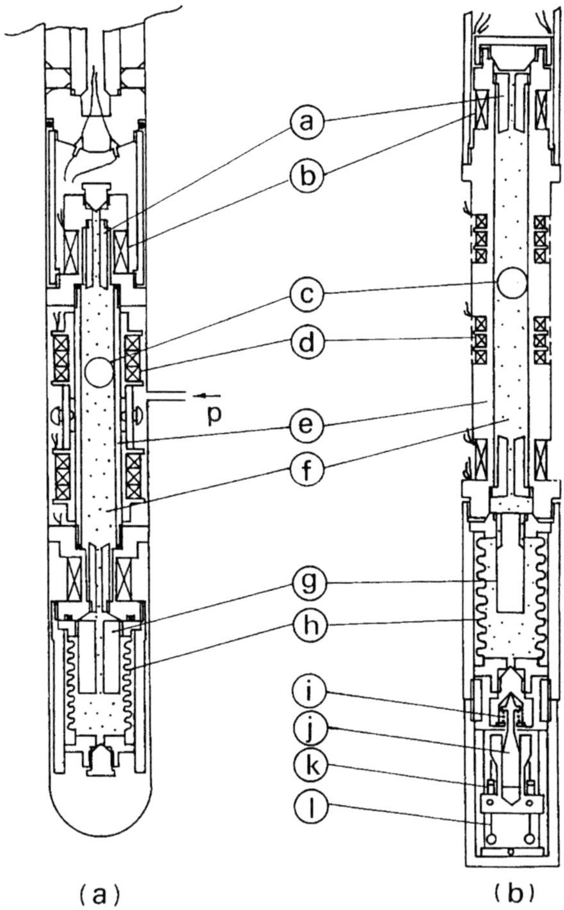

Viscometer cells of original design (a) and of improved design assembled with the potentiometer (b); a permalloy cylinder, b retaining coil, c ball, d differential transformer, e tube, f sample liquid, g spacer, h bellows, i coil spring, j connecting stem, k leaf spring contact, l manganin wire.

Fig. 5.3. A rolling ball viscometer from Izuchi, M. and Nishibata, K., "A High-Pressure Rolling-Ball Viscometer up to 1 GPa", *Japanese J. Appl. Phys.*, Vol. 25, No. 7, Part 1, 1986, pp. 1091–1096. Reproduced with permission of The Institute of Pure and Applied Physics.

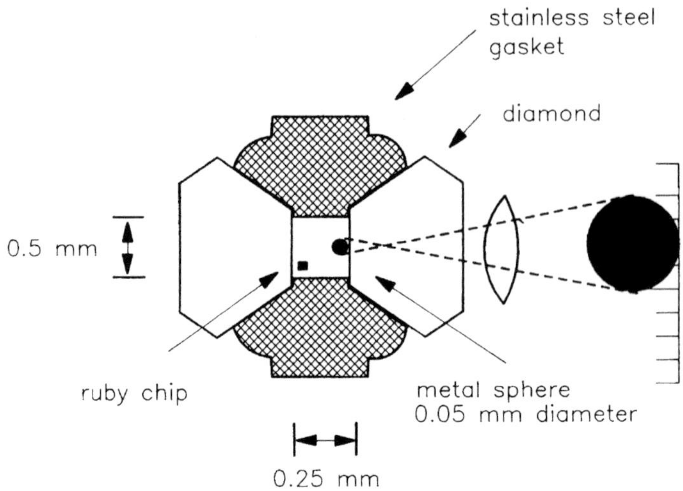

Fig. 5.4. The diamond anvil viscometer from King, H.E., Herbolzheimer, E. and Cook, R.L., "The Diamond-Anvil Cell as a Viscometer", *J. Appl. Phys.*, Vol. 71, No. 5, 1992, pp. 2071–2081, for presure to 10 GPa. This instrument is tilted to form a rolling ball viscometer. Reproduced with permission of American Institute of Physics.

The rolling ball viscometers above and the falling cylinder viscometers of the next section, which have been pressurized within cylindrical vessels, share the pressure limitations of their vessels. The pressure capability of a diamond anvil cell, however, exceeds the glass transition pressure for all of the organic materials that are liquids at ambient conditions. Piermarini et al. [13] placed a falling metal ball in a diamond anvil cell and measured the viscosity of a methanol–ethanol mixture to 7 GPa by visually determining the falling velocity through a microscope. They reported uncertainty due to the wall interaction of up to 50%. King et al. [14] converted their diamond anvil viscometer to the rolling ball configuration by utilizing one of the diamond faces as a plane surface for a rolling contact. The nickel ball was about 50 μm in diameter and their cell is shown in Fig. 5.4. The ruby chip provides the pressure measurement through the observation of a shift in fluorescence wavelength. The same group increased the viscosity limit of their device from 10^4 to 10^6 Pa s by substituting the acceleration of a centrifuge for the acceleration of gravity [15]. They measured the viscosity of methanol to a pressure of 8 GPa [15]. These diamond anvil viscometers have not been demonstrated to provide the same repeatability as the other types but they remain the only viscometers capable of pressures above 1.6 GPa excepting some unusual devices such as Bridgman's [16] pressure vessel within a pressure vessel and Barnet and Bosco's [17] capillary viscometer embedded within a hexahedral press.

5.2.4 Falling cylinder viscometers

The dropping and rolling ball viscometers have a rather limited range of viscosity unless the centrifuge technique of King and coworkers [15] is applied. In addition, due to the complexity of the flow, it is difficult to associate a characteristic stress with these stress-controlled measurements. In his study of the effects of pressure on viscosity, Bridgman [18] developed the falling cylinder configuration into an instrument capable of accurate relative viscosity measurements. His viscometer was utilized for the 1953 report of the ASME Research Committee on Lubrication, Advisory Board on Pressure–Viscosity [2] and is shown in Fig. 5.5. The cylindrical sinker falls within a cylindrical tube with the test pressure acting inside and outside of the tube so that the effect of pressure upon the tube dimensions is only that due to compressibility.

Irving and Barlow [19] made substantial improvements in the capability of the instrument. The falling velocity may be substantially increased and liquids of much greater viscosity studied, by providing a central hole within the sinker [19], shown as the lower sinker in the view of the Irving and Barlow viscometer in Fig. 5.6. The fall time is measured for the Bridgman viscometer in Fig. 5.5 by the making and breaking of electrical contact between the sinker, tube, and a pin. Then the fall time must always be large so that the time for the sinker to accelerate to the terminal velocity is a negligible part of the total. For the Irving and Barlow viscometer of Fig. 5.6, the position of the sinker is detected by monitoring the inductance of nine coils along the trajectory of the fall. Then the sinker is not required to traverse the entire length of the tube to obtain a viscosity measurement and the length of fall, l, can be any convenient value. Mclachlan [20] further improved upon the Irving and Barlow design by using a linear variable differential transformer, LVDT, placed within the pressure vessel to continuously monitor the sinker position within 2 μm. This raised the viscosity limit from about 300 to 10^6 Pa s. Dandridge and Jackson [21] used laser Doppler velocimetry to monitor the sinker velocity directly. The falling body viscometers employed in the author's laboratory use a commercial, large bore-to-core diameter ratio, LVDT placed around a nonmagnetic pressure vessel. An example is shown in Fig. 5.7 from [22].

An LVDT normally employs a central primary coil to which an AC excitation is applied and two secondary coils placed axially on either side of the primary and connected in a series opposing circuit. A magnetic core, the sinker in this case, centered between the secondary coils will produce equal induced voltages of opposite phase in the secondary coils which cancel in the series circuit. A small displacement of the sinker from center produces a voltage in the secondary circuit of a magnitude that is proportional to the displacement. The large bore-to-core diameter ratio LVDT has, instead of the usual one primary and two secondary coils, five primary and six secondary coils of varying number of turns distributed axially to give good linearity and signal strength.

Falling cylinder viscometers may be classified by the way in which the position of the sinker is detected, by the way in which the sinker is made to fall concentrically, and by the way in which the sinker is reset to the top of the tube to begin another measurement. Sinkers may be mechanically guided by protrusions from the sides or hydrodynamically guided by shaping the leading end of the sinker into a hemisphere. Irving [23] made a theoretical and experimental study of the hydrodynamically centered sinker and found that, when the viscometer axis was tilted from vertical, the falling velocity was increased due to the cylinder falling eccentrically in the tube. He also compared the advantages and disadvantages of each centering scheme. The viscometers that are shown in Figs. 5.5 and 5.7 use mechanically guided sinkers.

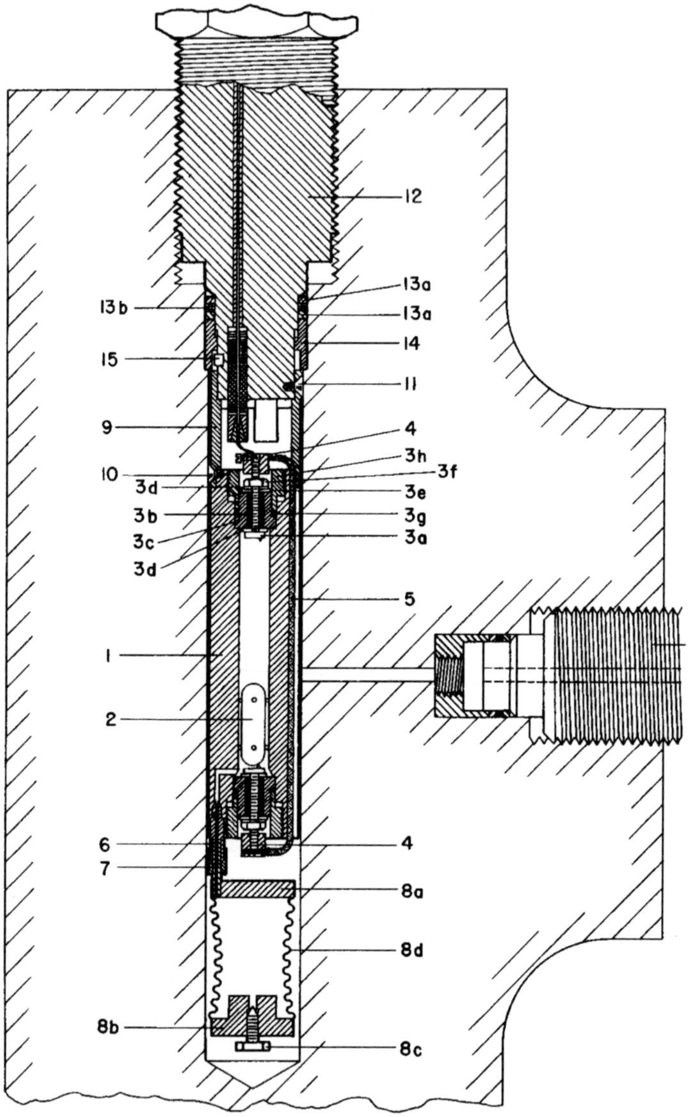

Fig. 5.5. The falling cylinder viscometer developed by Bridgman [18], from Kleinschmidt, Bradbury and Mark, *Viscosity and Density of Over Forty Lubricating Fluids of Known Composition at Pressures to 150,000 psi and Temperatures to 425 F,* ASME, New York, 1953, with permission of the American Society of Mechanical Engineers, for pressures to 1.2 GPa. The sinker is 2, the tube is 1, and the volume compensation bellows is 8. The viscometer is inverted for reset. Sinkers are guided by pins.

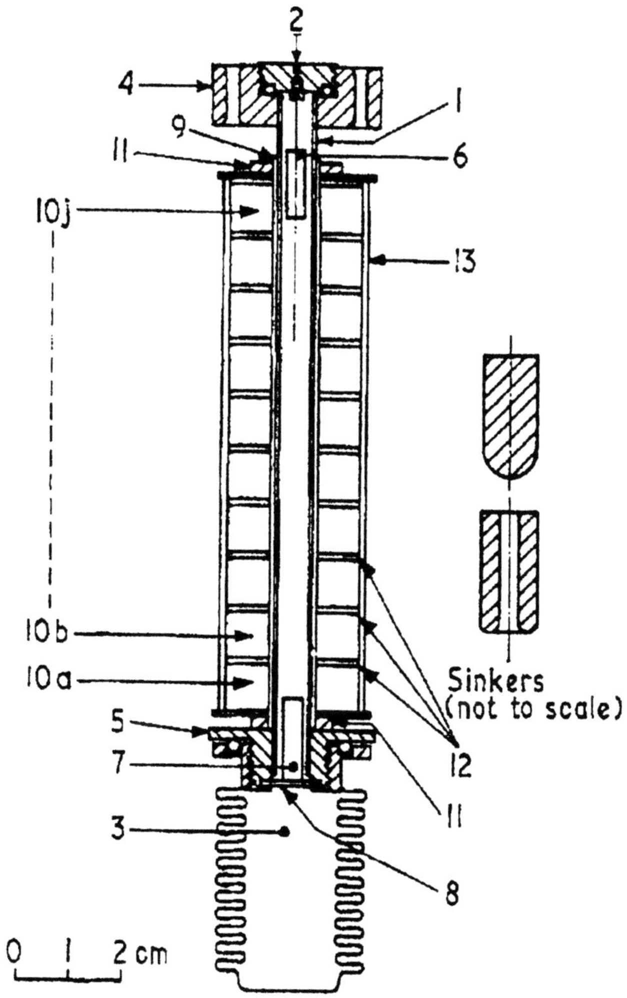

Fig. 5.6. The falling cylinder viscometer developed by Irving and Barlow, for pressures to 1.4 GPa, reproduced from Irving, J.B. and Barlow, A.J., "An Automatic High Pressure Viscometer", *J. Phys. E.*, Vol. 4, No. 3, 1971, pp. 232–236, with permission of Institute of Physics. Pressure vessel not shown. The tube is 1 and the sinker has been removed from the tube. Electrical coils, 10, both levitate the sinker for reset and detect its position during the fall. The permanent magnet, 6, holds the sinker while the heat from the levitating coils is dissipated. Sinkers are hydrodynamically centered.

The sinker may be reset by inverting the viscometer so that there is a "fall" in the opposite direction or by magnetically lifting the sinker with electrical coils. The viscometers shown in Figs. 5.5 and 5.7 are reset by rotation about the horizontal axis of the intensifier which remains attached to the vessel. The viscometer shown in Fig. 5.6 utilizes the same coils that are used to detect the sinker position to magnetically lift the sinker. There is a problem whenever a solenoid immersed in the high-pressure liquid is used either to lift

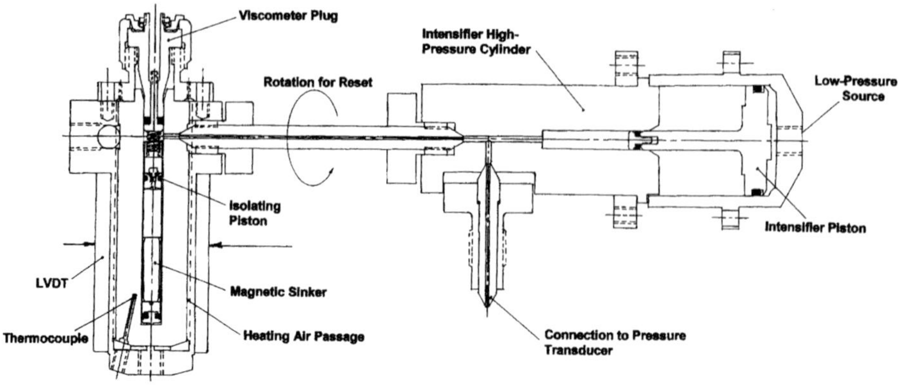

Fig. 5.7. A falling cylinder viscometer in the author's laboratory for 1.2 GPa, reproduced from Bair, S., "Routine High-Pressure Viscometer for Accurate Measurements to 1 GPa", *SILE Tribol. Trans.*, Vol. 47, No. 3, 2004, pp. 356–360, with permission of The Society of Tribologists and Lubrication Engineers. The viscometer is inverted by rotation about the horizontal axis of the intensifier for reset. An LVDT continuously detects the position of the sinker. The sinker is guided by pins and volume compensation is by isolating piston.

the sinker for reset or to apply the driving force to the sinker for a viscosity measurement. The solenoid resistively dissipates electrical energy so that the temperature of the liquid is increased in the same way as for the immersed electric motor of Thomas et al. [8] Irving and Barlow [19] overcame this limitation by holding the sinker with a permanent magnet until the temperature returned to the required test value and they seem to be the only group that was able to overcome this problem of the immersed solenoid.

The working equation for a falling cylinder viscometer is

$$\mu = C_{fc}(\rho_s - \rho)t \tag{5.9}$$

Irving and Barlow [19] derived the relationship between the calibration factor, C_{fc}, and the viscometer geometry and the distance of fall, L. They solved for the Newtonian flow through the annular clearance between the sinker outside diameter, D_o, and the tube inside diameter, D_i, and for the flow through a central circular hole of diameter, D_h, neglecting entrance and exit effects.

$$C_{fc} = \frac{g}{8L}\frac{\left(D_o^2 - D_h^2\right)}{\left(D_i^4 - D_o^4 + D_h^4\right)}\left[\left(D_i^4 - D_o^4 + D_h^4\right)\ln\left(\frac{D_i}{D_o}\right) - \left(D_i^2 - D_o^2\right)^2\right] \tag{5.10}$$

The above equation is useful in the design of a viscometer when the dimensionless calibration factor, $C_{cf}L/gD_i^2$, is plotted against relative cylinder diameter and relative hole diameter as in Fig. 5.8. In practice, the relative cylinder diameter is about $0.90 \le D_o/D_i \le 0.97$. It is extremely useful to be able to have large variations in C_{fc} so that the range of viscosities that may be conveniently measured is large. This might be accomplished by varying the tube diameter, D_i, but this would essentially require a

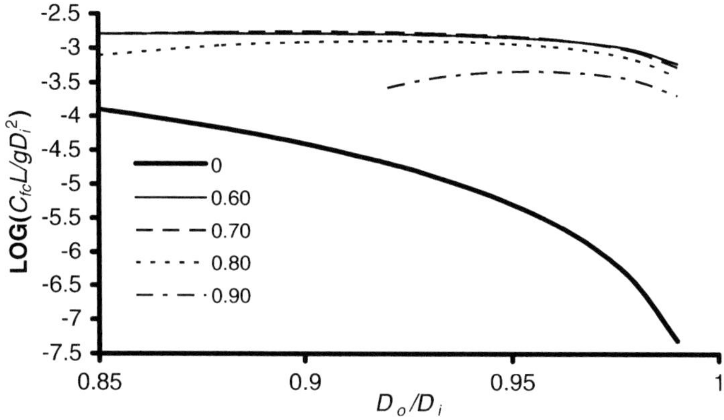

Fig. 5.8. The theoretical dimensionless calibration factor versus relative sinker diameter for falling cylinder viscometers. The individual curves represent different relative hole diameter, D_h/D_i, as indicated in the legend.

new viscometer. The length of fall, L, could be varied, but large values might include the displacement before terminal velocity is attained and small values are difficult to measure accurately. Figure 5.8 shows that the presence of a central hole can increase the calibration factor by two to three orders-of-magnitude and this technique has made the falling cylinder viscometer a good choice for measurements that approach the glass transition pressure. In practice, there is sufficient error in the application of equation (5.10) to real sinkers, generally about 20%, that it is used only as a guide for design and the calibration is accomplished by comparing equation (5.9) to the viscosity of a well-characterized standard.

For the usual dimensions, the operation of hollow sinkers and solid sinkers, sinkers with and without a central hole, are very different. In the case of the solid sinker at terminal velocity, the weight of the sinker is largely balanced by the force due to the pressure difference at opposing ends of the sinker. This pressure difference results in a Poiseuille dominated flow in the clearance space. The shear stress along the outside diameter of the sinker may be neglected in the force balance. The shear stress applied to the liquid can then be approximated by the shear stress distribution in a slit which would generate the value of the pressure difference, Δp, that would support the weight of the sinker. Ashare et al. [24] have analyzed the solid falling cylinder of length, l, in a closed tube for use as a viscometer to determine the shear-dependent viscosity. Neglecting the curvature and the velocity difference of the sinker and the tube in the boundary conditions, they found that the maximum shear stress, the wall shear stress, is approximately $\tau_w \approx \Delta p(D_i - D_o)\big/4l$. Neglecting the shear stress in the force balance, $\Delta p \approx (\rho_s - \rho)glD_o/D_i$. Then the characteristic shear stress for a solid sinker is approximately,

$$\tau_w \approx (\rho_s - \rho)(D_i - D_o)\frac{g}{4}\frac{D_o}{D_i} \tag{5.11}$$

Lohrenz et al. [25] experimentally determined the limit of laminar flow which determines the limit of usefulness of the linear working equation (5.9). They found for $l/D_i > 4$ that the critical Reynolds number, based on the diameter as the characteristic length, is equal to one and that the corrected calibration factor changes only slowly from Reynolds number of one to ten. This leads to the restriction that

$$\mu^2 > D_i l C_{fc} \rho(\rho_s - \rho) \text{ to } 10 D_i l C_{fc} \rho(\rho_s - \rho) \tag{5.12}$$

The situation is different for a hollow sinker that has a sufficiently large central hole to provide a significant increase in the calibration factor over that of the solid version. Here, the hole provides a flow path that reduces Δp and the weight of the sinker is essentially balanced by the shear stress on the outside diameter and the inside diameter of the sinker. An approximation for the maximum stress, the stress at the outside diameter of the hollow sinker, can be obtained from the analysis of Irving and Barlow [19] when the hole is large,

$$\tau_w \approx (\rho_s - \rho) \frac{g D_o}{4} \frac{\left(D_o^2 - D_h^2\right)}{\left(D_o^2 + D_h^2\right)} \tag{5.13}$$

The pressurized falling sinker viscometer is today perhaps the most important type. All of the viscometers mentioned above provide indirect measurements of viscosity, except possibly for the Couette type when the gap is large. The viscometer must be operated with a liquid of known properties and at least one of the data reduction parameters determined from that measurement on a reference liquid. A new instrument is briefly described next which may remove this restriction.

5.2.5 Vibrating wire viscometer and dielectric relaxation

A new pressurized viscometer was recently described by Caudwell et al. [26] that provides for the simultaneous measurement of density and viscosity for pressures up to 200 MPa. The steady-state vibrating wire technique drives the vibration of a thin wire with an electrical current of controlled amplitude and frequency close to the resonance of the tensioned wire. A magnetic field is provided by permanent magnets and the wire is tensioned by a weight. The real and imaginary components of the frequency-dependent voltage induced across the vibrating wire are analyzed using Navier–Stokes solutions of the vibrating wire problem [26] to yield both viscosity and density.

Results for toluene, n-dodecane and n-octadecane [26] indicate that this may be one of the most accurate techniques available and most of the parameters that are presently calibrated might be directly measured. The range of viscosity, about 100:1, of the present device is, however, restrictive for use with lubricants under pressure.

For polar liquids, for which there is a measurable change in dielectric permittivity as the dielectric relaxation transition is traversed, the relaxation time obtained from the reciprocal of the frequency at the midpoint of the transition is nearly proportional to the viscosity. In other words, the dielectric relaxation time is proportional to the mechanical relaxation time [27]. Once this constant of proportionality has been determined, measurements of viscosity can be obtained from a measurement of the dynamic permittivity under pressure [28].

The sections above are not meant as an exhaustive presentation of all of the successful high-pressure viscometers. All of the types of viscometers discussed above can be shown to give consistent values of low-shear viscosity as a function of temperature and pressure. Some results are presented in the next section.

5.3 General Pressure–Viscosity Response and Results for Pure Organic Liquids and Lubricants

The viscosity measurements of three laboratories using two different types of pressurized viscometers are shown in Fig. 5.9 for a low molecular weight dimethyl siloxane, octamethyltrisiloxane. Two laboratories have used mechanically guided falling cylinders [16,29] and one has employed a rolling ball diamond anvil viscometer [14]. A curve

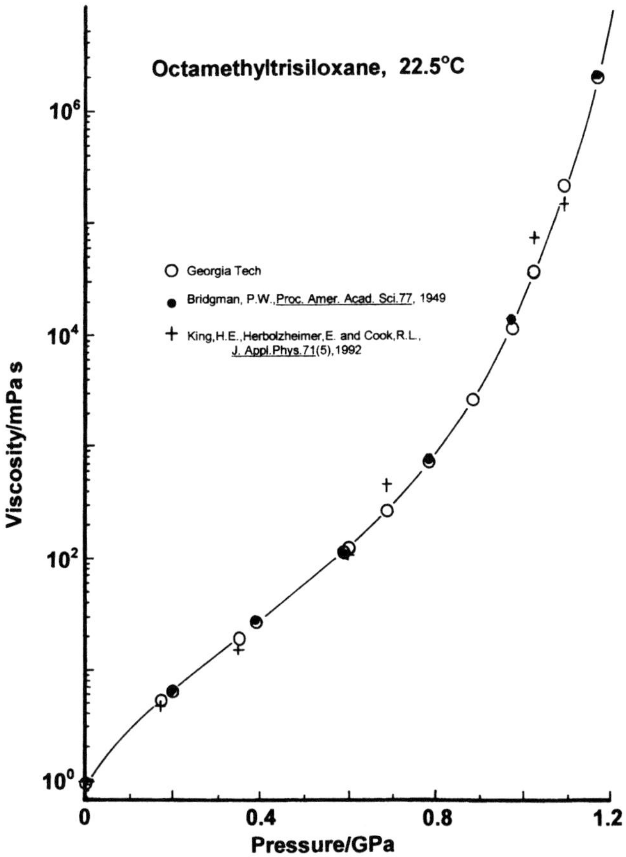

Fig. 5.9. The low-shear viscosity of a low molecular weight dimethyl silicone by three laboratories, one using a rolling-ball diamond anvil viscometer (King et al.) shown in Fig. 5.4 and the others using falling cylinder viscometers shown in Figs. 5.5 and 5.7, from [26]. Reprinted with permission of Elsevier.

has been drawn through the data for interpolation. The typical sigmoidal shape of the log viscosity–pressure curve is very evident in Fig. 5.9 and all of the instruments provide a consistent description of the piezoviscous response. There is, however, more scatter in the diamond anvil data. In comparing data in the literature from the diamond anvil with those from other falling body types for the two hydrocarbon liquids, dibutylphthalate and o-terphenyl, there seems to be a scatter in the diamond anvil results, which can amount to more than a factor of two in viscosity. These liquids have been used in the study of the dynamics of the glass transition. Some viscosity data for these liquids obtained in the author's laboratory are listed in Tables 5.1 and 5.2 for dibutylphthalate and o-terphenyl, respectively. The latter shows greater-than-exponential behavior at all investigated temperatures and pressures, while the former clearly shows an inflection in the log viscosity versus pressure response. The o-terphenyl is a crystalline solid at room temperature and solidifies at 100°C at a pressure of about 400 MPa. Some data are included for dibutylphthalate in Table 5.1 from Irving and Barlow [19] as well.

Table 5.1. The low-shear viscosity of dibutylphthalate by two laboratories.

	Low-Shear Viscosity, μ/mPa s		
p/MPa	30°C	30°C*	50°C
0.1	13.9	13.2	6.98
100	53.7	54.1	
115			28.0
200	191	192	
249			117
300	629	626	
400	1994	1952	
452			757
456	3850	3670	
573			2230
610	23,700	21,700	
733			9910
979			106,600
1248			2,280,000

*Irving and Barlow [19].

Table 5.2. The low-shear viscosity of o-terphenyl.

	Low-Shear Viscosity, μ/mPa s		
p/MPa	110°C	130°C	150°C
0.1	3.28	2.20	1.51
25	4.70		2.10
50	6.91	4.14	2.72
100	15.7	7.97	4.82
200	120	37.9	16.2
300	2050	289	69.7
400	solid	4140	508

Table 5.3. The low-shear viscosity of 1,1-(1,1,3-trimethyl-1,3-propanediyl)bis-cyclohexane by two laboratories and different methods.

	Low-Shear Viscosity, μ/mPa s							
p/MPa	20°C**	25°C*	40°C*	40°C**	60°C*	70°C**	80°C*	100°C**
0.1	43.6	38.6	17.5	18.1	8.27	5.96	4.72	2.98
50	297	236	73.3	75.5	26.2	17.2	12.2	6.96
100	2171	1580	336	347	90.9	50.4	33.5	16.0
200		81,400	9300	10,000	1427	548	306	87.9
300								564

*Rolling ball viscometer [30].
**Falling cylinder viscometer [31].

The viscosity of a pure hydrocarbon, 1,1-(1,1,3-trimethyl-1,3-propanediyl)bis-cyclohexane also known as 2,4-dicyclohexyl-2-methylpentane (see Table 4.2) has been measured by two laboratories using two different types of viscometer, a rolling ball [30] and a falling cylinder [31] viscometer. The data are listed in Table 5.3 where for the common temperature of 40°C the agreement is within 3% except at the highest pressure where it is 7%. The pressure response here is nearly exponential at the intermediate temperatures indicating that the temperature and pressure conditions of Table 5.3 are near the inflection in the log viscosity–pressure relationship.

It should not be inferred from the examples above that there is uniform quality of the pressure–viscosity data throughout the literature. Most laboratories claim good accuracy, 10% or less, but it is clear from comparisons using the same material that many have not done that well.

The viscosity of a jet oil meeting specification Mil-L23699 is shown as the logarithm of viscosity, a function of pressure for temperatures from 23 to 165°C in Fig. 5.10. These data are typical for EHL lubricants. The individual isotherms in the figure can be temperature shifted by simply rotating about a pole point located at $p_p = -110\,\text{MPa}$ and $\mu_p = 0.5$ mPa s except at the highest temperature where the resulting fit to the data is poor. Obviously, the shape of the isotherm for negative pressures is speculative, but the position of the pole has been obtained graphically and its position is well-defined by the data. In 1963, Roelands et al. [32] showed this form of temperature shifting of log viscosity plotted against p^m (with $0.75 \le m \le 1.10$) for mineral oil at low pressures where the response is strictly less-than-exponential. Apparently their shifting technique is useful for higher pressures as well for this particular liquid.

Several points can be made concerning the nature of the fanning out of the isotherms about the pole point in Fig. 5.10. The vertical separation of the isotherms which is proportional to the temperature–viscosity coefficient, β, increases strongly with increasing pressure at the highest values of viscosity. Because the pole point is at a negative pressure and the inflection point, marked by an $\times$ in Fig. 5.10, must remain a fixed distance along the isotherm from the pole, the pressure at which the inflection occurs, p_{inf}, moves toward the vertical axis as temperature decreases. For some low temperatures, the inflection pressure approaches zero and the initial response is exponential. For still lower temperatures, the inflection pressure is negative and the initial response must be greater-than-exponential.

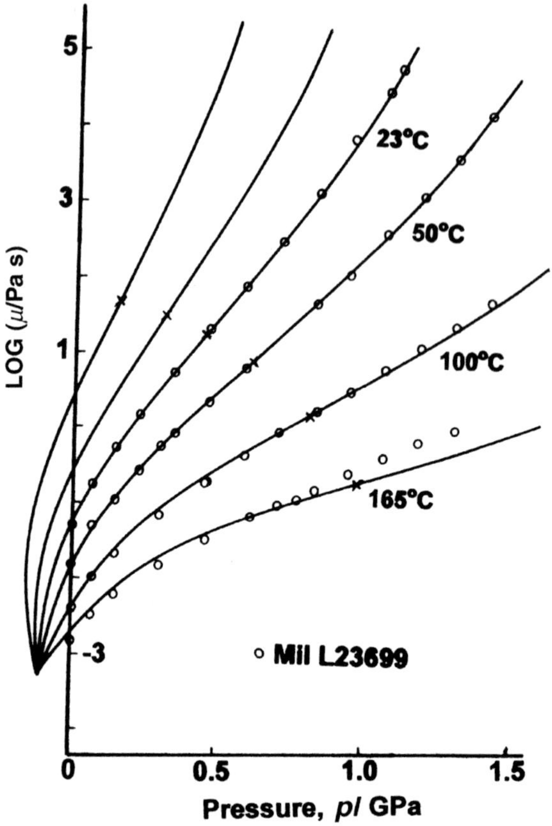

Fig. 5.10. The low-shear viscosity of a military specification jet oil. Pressure-log viscosity isotherms can, for a limited range of temperatures, be reproduced by rotation about a pole point located at a negative pressure, $p_p = -110$ MPa and viscosity of $\mu_p = 0.5$ mPa s.

The effect of chemical structure on viscosity is accentuated at high pressure. The data of Fig. 5.11 was generated in the author's laboratory during a study of the effect of pressure on the viscosity of isomers of dodecane, a 12-carbon alkane. Normal dodecane is not a good glass-former. It freezes (crystallizes) at a pressure of about 300 MPa at 40°C and 700 MPa at 100°C. Branching of the normal alkanes, however, promotes supercooling and allows the viscosity to be measured to higher pressures. The viscosities of the four structures are very similar at low pressures at the temperatures of 40 and 100°C, but diverge at the highest pressures. The isomers depicted in Fig. 5.11 are, in order of increasing viscosity at 1 GPa: 5-propylnonane, PN, a single three-carbon branch centered on a nine-carbon chain; 2,4-dimethyldecane, DMD, two single-carbon branches on a ten-carbon chain; 2-methyl 5-ethylnonane, MEN, a single-carbon and a double-carbon branch on a nine-carbon chain; and 2,2,4,6,6-pentamehtylheptane, PMH, five single-carbon branches on a seven-carbon chain.

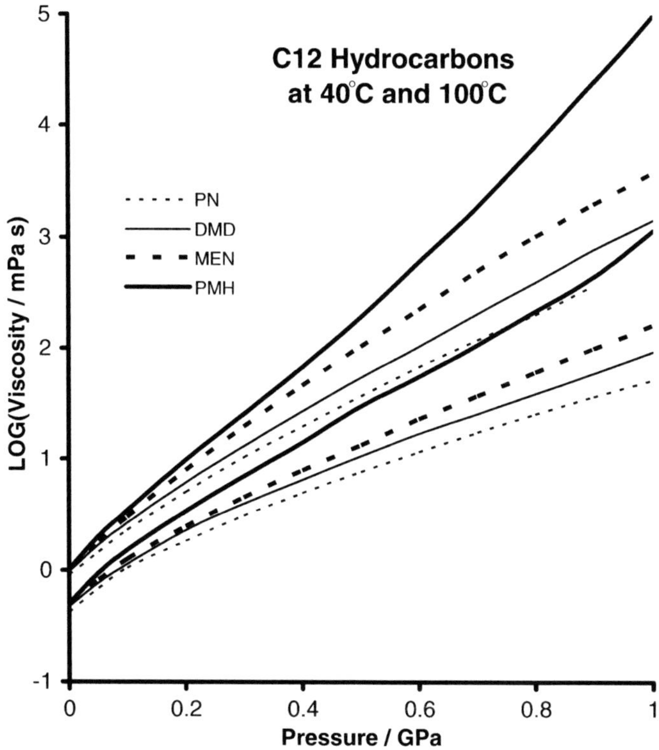

Fig. 5.11. The low shear viscosity of C12 isomers: 5-Propylnonane, PN; 2,4-Dimethyldecane, DMD; 2-methyl 5-ethylnonane, MEN; 2,2,4,6,6-Pentamethylheptane, PMH.

The trend is generally for the piezoviscous effect to be enhanced in going from a few long branches to many short branches – going from polyalphaolefin to polybutene, for example. The same is true in going from linear to cyclic and then to polycyclic structures.

Fatty acids and fatty acid esters have long been used as additives to reduce friction in gear oils for example. It is traditionally explained [33] that the acid reacts with the metal surfaces to form soaps that contribute to a lubricating surface layer. However, the friction reduction can extend to full films [34]. The explanation may be seen in Fig. 5.12 where the viscosities of 2% weight additive solutions relative to the viscosity of the mineral oil solvent are plotted against pressure. These additives are oleic acid and glycerol monooleate, a common friction modifier. The mineral oil is Shell turbine oil T9. For pressure less than about 0.6 GPa the effect of the additive on viscosity is negligible, however as the pressure increases to 1 GPa the additive clearly reduces the viscosity of the blend. The inflection pressure of the base oil at this temperature is about 0.37 GPa. One explanation for this behavior is that the additive functions as a plasticizer and effectively increases the glass transition pressure.

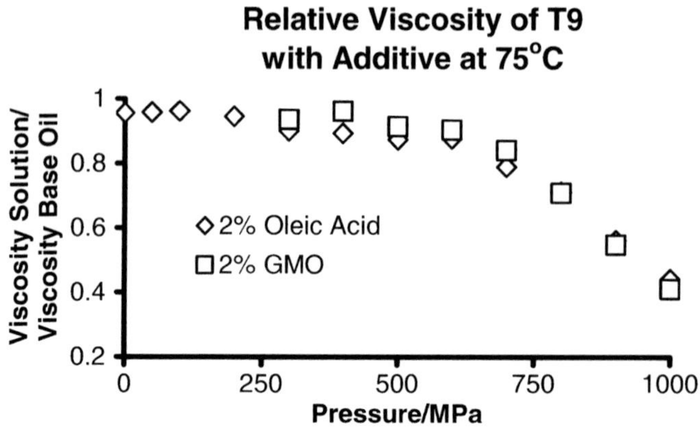

Fig. 5.12. Additives that reduce friction, friction modifiers, can act as plasticizers to reduce the viscosity at pressures greater than the inflection pressure.

Some viscosities are tabulated for lubricants used in aerospace applications in Tables 5.4 to 5.7. These applications see extremes of temperature and low ambient pressure. Several samples of the multiply alkylated cyclopentane of Table 5.4 have been examined and a large variation in viscosity among samples was observed. This is the lowest viscosity sample.

Table 5.4. The low-shear viscosity of a multiply-alkylated cyclopentane, X2001, used as a lubricant for space applications.

	Low-Shear Viscosity, μ/Pa s			
p/MPa	25°C	40°C	60°C	100°C
0.1	0.159	0.0820	0.0353	0.01055
25	0.260	0.132		
50	0.417	0.204	0.0820	0.0225
100	0.928	0.426	0.164	0.0404
200	3.76	1.54	0.536	0.112
230	5.93		0.699	
300			1.39	0.252
349	23.4		1.99	
467	83.7		5.37	
586	269		14.5	
705	864		34.1	
823	2740		84.4	
942	9900		199	
1060			443	
1179			1120	
α^*/GPa^{-1}	17.4	16.1	14.4	12.3

Table 5.5. The low-shear viscosity of a linear perfluoropoly-alkylether, type Z25, used as a lubricant for aerospace applications.

	Low-Shear Viscosity, μ/Pa s				
p/MPa	0°C	25°C	60°C	100°C	180°C
0.1	0.85	0.36	0.16	0.075	0.020
69	4.40	1.54	0.58	0.25	0.083
146	15.6	4.80	1.56	0.58	0.183
224	50				
230		14.7	3.62		
301	167	28		2.30	0.54
349		54.1	10.1		
456				6.40	1.24
467		175	25.7		
586		521	59.9		
705		1440	142		
823			356		
942			819		
1060			1690		
1179			3960		
α^*/GPa^{-1}	21.8	18.7	16.0	14.1	14.8

Table 5.6. The low-shear viscosity of a branched perfluoropolyalkylether, type 143AZ, used as a lubricant for aerospace applications.

	Low-Shear Viscosity, μ /Pa s		
p/MPa	65.6°C	98.9°C	148.9°C
0.1	0.0138	0.00606	0.00271
69	0.116	0.0374	0.0134
146	0.592	0.156	0.0469
224	2.98		
301		1.748	0.325
349	30.4		0.425
467	379	16.9	1.61
586	7150	110	6.26
705		1150	21.1
776		4380	
823			81.2
847		26,600	
942			425
1060			3160
1132			11,500
α^*/GPa^{-1}	29.3	24.5	20.8

Table 5.7. The low-shear viscosity of a military specification, Mil-L23699, jet oil used as a lubricant for aerospace applications.

	Low-Shear Viscosity, μ/Pa s				
p/MPa	23°C	50°C	100°C	165°C	220°C
0.1	0.0511	0.0159	0.00404	0.00152	0.000969
69	0.183	0.0493	0.00988	0.00333	0.00210
146	0.536	0.112	0.0203	0.00588	0.00357
230	1.50	0.267			
301		0.580	0.0702	0.0152	0.000831
349	5.57	0.826	0.097		
456			0.192	0.0327	0.01604
467	21.2	2.39	0.197		
586	83.8	6.70	0.438		
610				0.0642	
705	317		0.911	0.0956	
823	1450	49.8	1.85	0.168	
942	7520	133	3.35	0.276	
1060	33,120	423	6.44	0.432	
1096	63,000				
1179		1330	13.4	0.682	
1298		4200	24.2	1.054	
1416		16,200	53.7		
α^*/GPa^{-1}	16.6	13.8	10.6	8.2	7.6

Table 5.8. The low-shear viscosity of a traction fluid used in a commercial infinitely variable transmission for automotive applications.

	Low-Shear Viscosity, μ/Pa s				
p/MPa	20°C	70°C	100°C	140°C	200°C
0.1	0.079	0.00986	0.00472	0.00233	
25	0.179	0.0162			
50	0.416	0.0268	0.0109	0.00494	
100	2.39	0.0754	0.025	0.00961	
200	131	0.697	0.147	0.0362	
230		1.54	0.227		
250	1310				
300	12,800	8.44	0.953	0.138	
325	55,700				
349		29.5	2.20		
400				0.64	
467		1120	34.6	1.68	
527		7120	143	4.47	0.374
586			543	14.3	0.79
705			16,900	127	3.38
823				1560	20.8
883				4800	54.4
942				26,100	
α^*/GPa^{-1}	33.2	20.5	16.9	14.2	

The ordinary automobile transmission provides an adjustable ratio of engine rotational speed to tire rotational speed using a geared transmission with discrete gear ratios. Some benefits may be realized by replacing the toothed gears with smooth rollers connected by an EHL film. The input to output ratio can then be varied continuously by varying the working distance of the contact to the axis of rotation. The contact lubricant then has an unusual requirement for a lubricant; it must provide a high traction coefficient. A liquid for this purpose has the viscosities listed in Table 5.8. Notice that the viscosity reaches very high values at relatively low pressures.

Gear oils are examples of EHL lubricants that provide benefits from low traction coefficient through energy efficiency. For full-film contacts operating at high slide-to-roll ratio and moderate pressure, the liquid lubricant viscosity at the pressure of the Hertz zone is the primary property determining traction. The pressure variation of viscosity of two automotive gear oils is shown in Fig. 5.13. The mineral based oil has a greater viscosity at high-pressures than does the synthetic (polyalphaolefin) based oil, although its viscosity

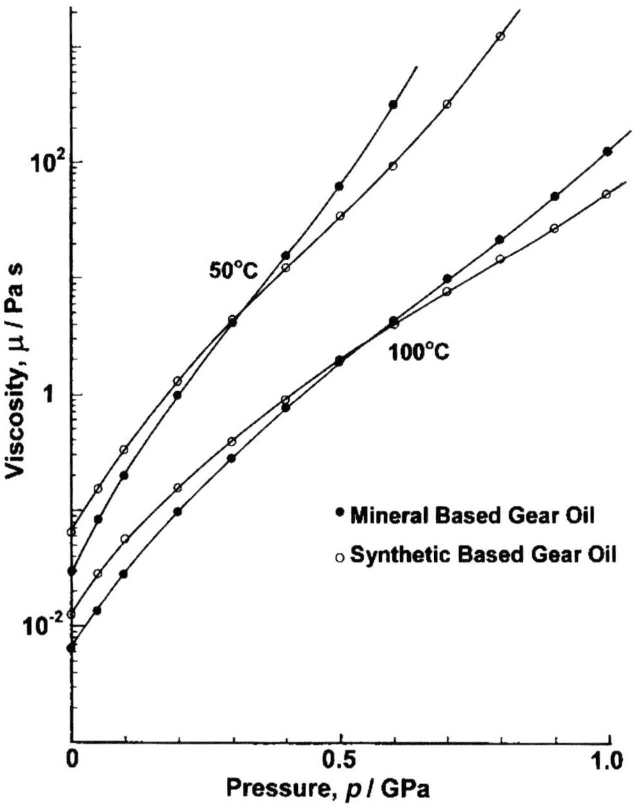

Fig. 5.13. The low-shear viscosity of two gear oils for automotive manual transmission applications, from Bair, S., "A Routine High-Pressure Viscometer for Accurate Measurements to 1 GPa", *STLE Tribol. Trans.*, Vol. 47, No. 3, 2004, pp. 356–360, with permission of the Society of Tribologists and Lubrication Engineers.

Table 5.9. The low-shear viscosity of a polyglycol gear oil, PG460, for worm gear applications.

	Low-Shear Viscosity, μ/Pa s		
p/MPa	40°C	65°C	100°C
0.1	0.437	0.178	0.0718
25	0.639		
50	0.870	0.333	0.125
100	1.68	0.561	0.200
200	5.63	1.462	0.435
230	7.72	1.82	0.571
300	16.8	3.69	0.898
349	27.7	4.78	1.23
467	96	13.5	2.46
586	322	34.9	5.22
705	1160	86.8	11.0
823	4200	227	21.8
942	17,900	591	42.8
1060		1680	85.5
1179		5850	166
1298			338
α^*/GPa^{-1}	13.1	11.0	9.1

at ambient pressure is less. These data clearly show the reason for reduced mechanical power loss and operating temperature with the synthetic gear oil. The temperature shifting rule of Roelands et al. [32] is not useful for these oils at pressures above the inflection. The viscosity of a polyglycol based gear oil for worm gears [35] is listed in Table 5.9.

The most common EHL lubricant is the paraffinic mineral oil. Unfortunately this material poses a special kind of problem for high-pressure viscometry. These oils contain, in varying amounts, linear alkanes or wax. This wax, even in small concentration will form a weak solid structure when cooled or compressed that imparts thixotropic behavior. Thixotropic materials have a yield stress below which flow does not occur, preventing a measurement of low-shear viscosity. Fortunately, the structure is weak and may be disrupted by shear, after which for some period of time the viscosity may be measured at low stress [36]. Then pressurized Couette viscometers can be used in place of the usual falling body type to obtain an estimate of the low-shear viscosity to the highest pressures. At ambient temperature the thixotropy appears at approximately 300 MPa increasing to about 600 MPa at 50°C and about 1000 MPa at 80°C.

In order to compare calculation with experiment, it would be extremely useful to have measurements of viscosity to high pressure of well-defined liquids with pressure and temperature dependence similar to EHL lubricants. Such data are provided in Tables 5.10–5.12 for some esters. Esters are often available in reasonably high purity from chemical supply houses, making them good choices as reference standards.

Table 5.10. The low-shear viscosity of pentaerythritol tetrahexanoate, a candidate standard reference material.

	Low-Shear Viscosity, μ/Pa s			
p/MPa	40°C	65°C	100°C	165°C
0.1	0.0202	0.00894	0.00398	0.00146
25	0.0296	0.01255	0.00542	0.00195
50	0.0418	0.0171	0.00717	0.00249
100	0.0779	0.0300	0.01122	0.00376
200	0.229	0.0774	0.0254	0.00703
300	0.605	0.172	0.0504	0.0122
400	1.51	0.386	0.0958	0.0198
500	3.68	0.828		
600	8.75	1.675		
700	20.6	3.36		
800	47.3	6.52		
900	122	12.8		
1000	286	25.4		
1100	656	49.8		
1200	1630	100.3		
α^*/GPa^{-1}	12.73	11.06	9.29	7.59

Table 5.11. The low-shear viscosity of 2-ethylhexylbenzoate, a candidate standard reference material.

	Low-Shear Viscosity, μ/Pa s		
p/MPa	20°C	60°C	100°C
0.1	0.00694	0.00236	0.00122
25	0.01052	0.00322	0.00157
50	0.01565	0.00427	0.00200
100	0.0326	0.00734	0.00312
200	0.144	0.0203	0.00667
300	0.614	0.0547	0.0135
400	2.54	0.144	0.0274
500	10.9	0.343	
600	45.2	0.956	
700	211	2.43	
800	1053	6.43	
900	6390	17.8	
1000		49.2	
1100		152	
1200		439	
α^*/GPa^{-1}	15.6	11.0	8.5

Table 5.12. The low-shear viscosity of di-n-nonyl phthalate, a candidate standard reference material.

	Low-Shear Viscosity, μ/Pa s		
p/MPa	40°C	65°C	100°C
0.1	0.0502	0.0165	0.00571
25	0.0926	0.0272	0.0086
50	0.167	0.0435	0.0126
100	0.489	0.106	0.0252
200	3.81	0.548	0.0933
300		2.48	0.294
400		11.25	0.919
500		53.3	2.89
600		275	8.91
700		1220	26.9
800		8030	94.7
900			296
1000			1014
α^*/GPa^{-1}	23.2	18.6	14.6

To incorporate the data presented in this chapter into an EHL analysis of almost any type, it is useful to have a correlation of viscosity with pressure. If thermal feedback is part of the analysis, the correlation must include temperature. The various models used for such correlations are investigated in the next chapter.

References

[1] Crook, A.W., "The Lubrication of Rollers IV. Measurements of Friction and Effective Viscosity", *Phil. Trans. Roy. Soc. Lond.*, Series A, Vol. 255, No. 1056, 1963, pp. 281–312.

[2] Kleinschmidt, R.V., Bradbury, D. and Mark, M., *Viscosity and Density of Over Forty Lubricating Fluids of Known Composition at Pressures to 150,000 psi and Temperatures to 425 F*, ASME, New York, 1953.

[3] Yang, P., Kaneta, M. and Masuda, S., Closure to "Discussion: 'Quantitative Comparisons between Measured and Solved Dimples in Point Contacts,' (Yang, P., Kaneta, M. and Masuda, S., 2003, *ASME J. Tribol.*, 125(1), pp. 210–214)", *ASME J. Tribol.*, Vol. 127, 2005, p. 457.

[4] Macosko, C.W., *Rheology Principles, Measurements, and Applications*, Wiley-VCH, New York, 1994, pp. 175–283.

[5] Warburg, S. and Sachs, W., "The Effect of Density on the Viscosity of Liquids", *Wiedmanns Annalen.*, Vol. 22, 1884, p. 518.

[6] Norton, A.E., Knott, M.J. and Muenger, J.R., "Flow Properties of Lubricants under High Pressure", *ASME Trans.*, Vol. 63, No. 7, 1941, pp. 631–643.

[7] Novak, J.D. and Winer, W.O., "Some Measurements of High Pressure Lubricant Rheology", *ASME J. Lubr. Techn.*, Vol. 90, Ser. F, No. 3, 1968, pp. 580–591.

[8] Thomas, B.W., Ham, W.R. and Dow, R.W., "Viscosity-Pressure Characteristics of Lubricating Oils", *Ind. Eng. Chem.*, Vol. 31, No. 10, 1939, pp. 1267–1270.

[9] Hutton, J.F. and Phillips M.C., "High Pressure Viscosity of a Polyphenyl Ether Measured with a New Couette Viscometer", *Nature Phys. Sci.*, Vol. 245, 1973, pp. 15–16.

[10] Hersey, M.D., "The Theory of the Torsion and the Rolling Ball Viscometers and their Use in Measuring the Effect of Pressure on Viscosity", *J. Washington Acad. Sci.*, Vol. VI, 1916, pp. 525–530.

[11] Sawamura, S., Takeuchi, N., Kitamura, K. and Tangiguchi, Y., "High Pressure Rolling-Ball Viscometer of a Corrosion-Resistant Type", *Rev. Sci. Instrum.*, Vol. 61, No. 2, 1990, pp. 871–873.

[12] Izuchi, M. and Nishibata, K., "A High Pressure Rolling-Ball Viscometer up to 1 GPa", *Japanese J. Appl. Phys.*, Vol. 25, No. 7, Part 1, 1986, pp. 1091–1096.

[13] Piermarini, G.J., Forman, R.A. and Block, S., "Viscosity Measurements in the Diamond Anvil Cell", *Rev. Sci. Instrum.*, Vol. 48, No. 8, 1978, pp. 1061–1066.

[14] King, H.E., Herbolzheimer, E. and Cook, R.L., "The Diamond-Anvil Cell as a Viscometer", *J. Appl. Phys.*, Vol. 71, No. 5, 1992, pp. 2071–2081.

[15] Cook, R.L., Herbst, C.A. and King, H.E., "High-Pressure Viscosity of Glass Forming Liquids Measured by the Centrifugal Force Diamond Anvil Cell Viscometer", *J. Phys. Chem.*, Vol. 97, No. 10, 1993, pp. 2355–2361.

[16] Bridgman, P.W., "Viscosities to 30,000kg/cm^3", *Proc. Amer. Acad. Sci.*, Vol. 77, 1949, pp. 117–128 and also in *P.W. Bridgman Collected Experimental Papers*, Harvard Univ. Press, Cambridge, 1964, p. 166.

[17] Barnett, J.D. and Bosco, C.D., "Viscosity Measurements on Liquids to Pressures of 60 kbar", *J. Appl. Phys.*, Vol. 40, No. 8, 1969, pp. 3144–3150.

[18] Bridgman. P.W., "The Effect of Pressure on the Viscosity of Forty-Three Pure Liquids", *Proc. Amer. Acad.*, Vol. 61, 1926, pp. 57–99.

[19] Irving, J.B. and Barlow, A.J., "An Automatic High Pressure Viscometer", *J. Phys. E.*, Vol. 4, No. 3, 1971, pp. 232–236.

[20] Mclachlan, R.J., "A New High Pressure Viscometer for Viscosity Range 10 to 10^6 Pa s", *J. Phys. E.*, Vol. 9, 1976, pp. 391–394.

[21] Dandridge, A. and Jackson, D.A., "Measurements of Viscosity under Pressure: A New Method", *J. Phys. D: Appl. Phys.*, Vol. 14, 1981, pp. 829–831.

[22] Bair, S. "A Routine High-Pressure Viscometer for Accurate Measurements to 1 GPa", *STLE Tribol. Trans.*, Vol. 47, No. 3, 2004, pp. 356–360.

[23] Irving, J.B., "The Effect of Nonvertical alignment on the Performance of a Falling-Cylinder Viscometer", *J. Phys. D: Appl. Phys.*, Vol. 5, 1972, pp. 214–224.

[24] Ashare, E., Bird, R.B. and Lescarboura, J.A., "Falling Cylinder Viscometer for non-Newtonian Fluids", *A.I.Ch. E. Journal*, Vol. 11, No. 5, 1965, pp. 910–916.

[25] Lohrenz, J., Swift, G.W. and Kurata, F., "An Experimentally Verified Study of the Falling Cylinder Viscometer", *A.I.Ch. E. Journal*, Vol. 6, No. 4, 1960, pp. 547–550.

[26] Caudwell, D.R., Trusler, J.P.M., Vesovic, V. and Wakeham, W.A., "The Viscosity and Density of n-Dodecane and n-Octane at pressures up to 200 MPa and Temperatures up to 473K", *Int. J. Thermophys.*, Vol. 25, No. 5, 2004, pp. 1339–1351.

[27] Harrison, G., *The Dynamic Properties of Supercooled Liquids*, Academic Press, London, 1976, p. 183.

[28] Bair, S. and Winer, W.O., "Some Observations on the Relationship between Lubricant Mechanical and Dielectric Transitions under Pressure", *ASME J. Lubr. Techn.*, Vol. 102, No. 2, 1980, pp. 229–235.

[29] Bair, S., Jarzynski, J. and Winer, W.O., "The Temperature, Pressure, and Time Dependence of Lubricant Viscosity", *Tribol. Int.*, Vol. 34, No. 7, 2001, pp. 461–468.

[30] Kuss, E. and Deymann, H., "Das Viskositats-Druckverhalten Einiger, Traction Fluids", *Schmiertechnik und Tribologie*, Vol. 27, No. 3, 1980, pp. 95–97.

[31] Bair, S., Qureshi, F. and Kotzalas, M., "The Low Shear Stress Rheology of a Traction Fluid and the Influence on Film Thickness", *Proc. Instn. Mech. Engrs.*, Vol. 218, Part J, 2004, pp. 95–98.

[32] Roelands, C.J.A., Vlugter, J.C. and Waterman, H.I., "The Viscosity–Temperature–Pressure Relationship of Lubricating Oils and Its Correlation with Chemical Composition", *ASME J. Basic Eng.*, Series D, Vol. 85, 1963, pp. 601–610.

[33] Boner, C.J., *Gear and Transmission Lubricants*, Reinhold Publishing, NewYork, 1964, p. 23.

[34] Costello, MT., "Effects of Basestock and Additive Chemistry on Traction Testing", *Tribol. Lett.*, Vol. 18, No. 1, 2005, pp. 91–97.

[35] Kong, S., Sharif, K., Evans, H.P. and Snidle, R.W., "Elastohydrodynamics of a Worm Gear Contact", *ASME J. Tribol.*, Vol. 123, 2001, pp. 268–275.

[36] Bair, S., "The Actual Eyring Models for Thixotropy and Shear-Thinning: Experimental Validation and Application to EHD", *ASME J. Tribol.*, Vol. 126, No. 4, 2004, pp. 728–732.

CHAPTER 6

Models for the Temperature and Pressure Dependence of Low-Shear Viscosity

6.1 Introduction

If the measurements of the low-shear viscosity presented in the previous chapter are to be useful in analyses of elastohydrodynamic lubrication, they should generally be replaced by equations that permit the calculation of the viscosity for arbitrary temperature and pressure. These equations then must accurately describe the behavior, preferably with a minimum number of parameters. It is always an advantage for some of the parameters to be physically meaningful and to be amenable to measurement by another means. There are many empirical models that can be very accurate but over a limited range of state variables. The free volume theory can provide a framework for correlations that describe the viscosity from the liquid/glass transition to close to the liquid/vapor transition with accuracy that should be sufficient for most EHL calculations and some of the parameters may be obtained from separate measurements.

It must also be recognized that the pressure and temperature dependence of the low-shear viscosity is so very complex that a correlation using as many as six fitting parameters is unlikely to be within the experimental accuracy over the entire measurable range of temperature and pressure. Correlations are often devised from an empirical expression that has been found to be useful for a small range of state variables, for example

$$\mu = \mu_R \exp\left[\alpha p - \beta\left(T - T_R\right)\right] \tag{6.1}$$

where the reference viscosity is $\mu_R = \mu(T_R, p = 0)$ and the pressure and temperature–viscosity coefficients are $\alpha = \partial \ln \mu / \partial p$ and $\beta = -\partial \ln \mu / \partial T$, as defined in equations (5.1) and (5.2). Then for a slightly greater range of temperature and pressure, if perhaps more data become available, equation (6.1) can be corrected by the introduction of additional terms inside the brackets in powers of $(T - T_R)$ and p and cross-products as well. The danger in this type of construction is that the correction terms may dominate at yet greater ranges of the independent variables and introduce behavior that was not intended in the correction process. Except for low temperatures, in general, α will first decrease with pressure, followed by an increase with pressure. An increase of temperature will generally cause a decrease in α except at low pressure combined with high temperature. In general, β will increase with pressure, approximately as the logarithm of viscosity, except at low pressure and high temperature.

When selecting a temperature–pressure–viscosity model some consideration of the ultimate use should be given. For a calculation of the central EHL film thickness, the viscosity at pressures greater than about $3/\alpha^*$ will have little influence on the determination of α^* and the film thickness and the selection of an appropriate model should consider mainly the accuracy at these low pressures. The regression of the parameters of the model might be weighted less heavily as pressure increases. Here, it is particularly important that the pressure rate of change be modeled accurately near ambient pressure. It is not uncommon that two models will provide similar average error for the values of viscosity but very different initial pressure–viscosity slopes.

On the other hand, for a calculation of a traction curve, once the central film thickness is known, the viscosity and pressure–viscosity slope at the high-pressures of the Hertz zone are the important parameters to be modeled and the model should be selected that most accurately describes this behavior. The regression of the parameters of the model might be weighted more heavily as pressure increases.

The temperature variation of viscosity is simpler than the pressure dependence and temperature–viscosity models will be introduced first. Some pressure–viscosity models for high pressure are analogs to the more successful temperature–viscosity models and these will be considered next. Low-pressure models will follow, ending with the free volume models that at present are most capable of a complete temperature–pressure–viscosity description.

6.2 Models for the Temperature–Viscosity Response

Consider the temperature dependence of the ambient-pressure viscosity of bis(phenoxyphenoxy)benzene, a mixture of isomers of a five-ring polyphenyl ether known as 5P4E, as shown in Fig. 6.1 in an Arrhenius plot, a plot of log viscosity against reciprocal absolute temperature. Measurements include capillary, falling cylinder and some published viscosities [1]. The measurements range in temperature from 5 to 425°C and in viscosity across six orders-of-magnitude. The Andrade equation [2] is plotted as the dashed line.

$$\mu = \mu_\infty \exp\left[\frac{E_a}{R_g T}\right] \tag{6.2}$$

with $\mu_\infty = 1.42 \times 10^{-5}$ Pa s and $E_a = 2.03 \times 10^7$ J kmol^{-1}. Here, E_a is a thermal activation energy and R_g is the gas constant. The temperature–viscosity coefficient for the Andrade equation is $\beta = E_a/R_g T^2$. Liquids display Arrhenius behavior at temperatures near the boiling temperature and far from the glass transition temperature.

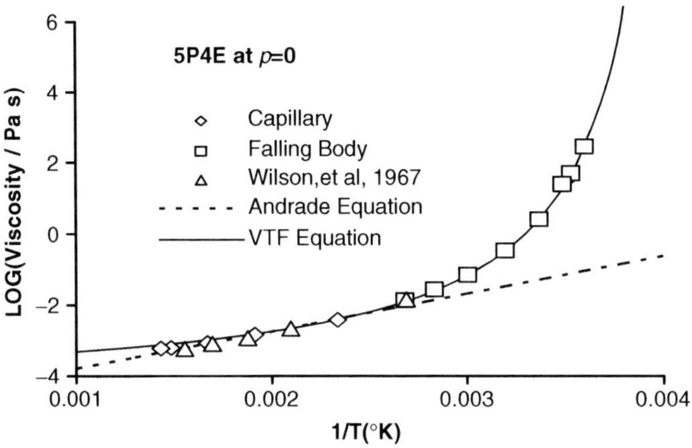

Fig. 6.1. An Arrhenius plot of the viscosity of the polyphenyl ether, 5P4E, comparing the Andrade and VTF equations. The measurements range in temperature from 5 to 425°C

The Vogel, Tammann and Fulcher (VTF) equation [3] is plotted in Fig. 6.1 as the solid curve.

$$\mu = \mu_\infty \exp\left[\frac{D_F T_\infty}{T - T_\infty}\right] \tag{6.3}$$

with $\mu_\infty = 2.455 \times 10^{-4}$ Pa s, $D_F = 2.184$ and $T_\infty = 240.5$ K. The temperature–viscosity coefficient for the VTF equation is $\beta = D_F T_\infty/(T - T_\infty)^2$. Here, D_F is the fragility parameter and T_∞, the Vogel temperature, is the temperature at which the viscosity diverges. It has been found that T_∞ is equal to the Kauzmann temperature [4], the temperature at which the crystal and the extrapolated liquid would have the same entropy. The parameter, T_∞, has also been referred to as the *idealized glass transition temperature* and is roughly 15–40°C less than T_g, the ordinary glass transition temperature.

The behavior described by the Andrade equation, Arrhenius behavior, which results in the straight line in Fig. 6.1 is an accurate description of the viscosity of 5P4E only for $1/T$ less than about 0.0027 K^{-1} or for $T > 100$°C. For lower temperatures, greater $1/T$, the non-Arrhenius behavior, the upturn in the curve, is a characteristic of glass-forming liquids that are supercooled [3]. A useful modification of the VTF equation is the Williams, Landel and Ferry (WLF) equation [5], written in standard form as

$$\mu = \mu_g \exp\left[\frac{-2.303 C_1 (T - T_g)}{C_2 + (T - T_g)}\right] \tag{6.4}$$

Here, the glass transition is employed as a reference state with viscosity equal to μ_g, the viscosity at the glass transition, and C_1 and C_2 are the well-known WLF constants. The WLF and VTF equations are related by $D_F T_\infty = C_2 \ln(\mu_g/\mu_\infty)$, $\ln(\mu_g/\mu_\infty) = 2.303 C_1$ and $T_\infty = T_g - C_2$. From dilatometry, for the 5P4E, $T_g = -18$°C and then the WLF parameters are $\mu_g = 1.32 \times 10^{12}$ Pa s, $C_2 = 14.5$ K and $C_1 = 15.73$.

There exist many additional empirical models for temperature–viscosity behavior. For example, the form $\mu = \mu_\infty \exp(C_3/T^q)$ has been used by Litovitz [6] with $q = 3$ and Barlow and Lamb [5] with $q = 4$ for a limited range of temperature.

6.3 Fragility and Empirical Models for High Pressure Behavior

The term *fragility* was popularized by Angell [3,4] to describe differences in the behavior of supercooled glass-forming liquids. Fragile liquids experience relatively greater changes in their properties as the glass transition is approached from low viscosity states than do *strong* liquids. The D_F parameter from the VTF equation (6.3) can be used to quantify the fragility. The graph shown in Fig. 6.2 is an *Angell plot*, a plot of log viscosity against reciprocal absolute temperature scaled by the glass transition temperature. The glass transition can be measured by dilatometry or calorimetry or other direct means or it may be specified by a characteristic relaxation time or viscosity. For liquids with T_g less than room temperature (at ambient pressure), the glass transition viscosity is roughly $\mu_g = 10^7$ to 10^{10} Pa s [7,8], the greater value being used in constructing Fig. 6.2. It has been assumed in Fig. 6.2 that $\mu_\infty = 10^{-5}$ Pa s. Strong liquids for which $D_F > 100$ appear on an Angell plot to be almost indistinguishable from the Arrhenius response of equation (6.2). Highly fragile liquids, for which $D_F < 20$, experience rapid changes in viscosity with temperature near the glass transition.

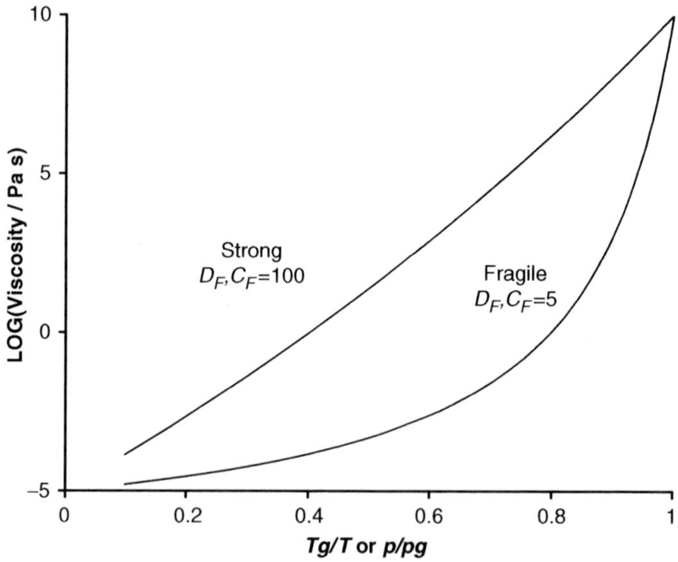

Fig. 6.2. An Angell plot comparing the temperature and pressure response of strong and fragile liquids.

A temperature-fragility index [9] has been defined as

$$m_T = \frac{1}{2.303}\left|\frac{d\ln\mu}{d\left(T_g \backslash T\right)}\right|_{T=T_g} = \frac{T_g \beta_g}{2.303} \tag{6.5}$$

The dimensionless temperature–fragility index, m_T, varies linearly with the temperature–viscosity coefficient at the glass transition, β_g, and also with the reciprocal of D_F. The number, 2.303, appears in equation (6.5) because m_T was defined in terms of the logarithm base ten. For organic liquids, m_T can assume values from about 40 to 200 at atmospheric pressure [9]. Similarly, a pressure–fragility index has been defined by Paluch et al. [8] as

$$m_p = \frac{1}{2.303}\left|\frac{d\ln\mu}{dp}\right|_{p=p_g} = \frac{\alpha_g}{2.303} \tag{6.6}$$

It is not helpful here to scale the pressure in equation (6.6) by dividing by the glass transition pressure, p_g, in the same way that reciprocal temperature was scaled by multiplying by T_g in equation (6.5) since it is possible for p_g to be zero. Therefore m_p is dimensional with units of reciprocal pressure and it is proportional to the local pressure–viscosity coefficient at the glass transition, α_g.

The analog of the VTF equation (6.3) for the pressure dependence of viscosity is suggested by the definition of m_p. Temperatures in equation (6.3) can be replaced by the reciprocal of pressure [10] to arrive at

$$\mu = \mu_0 \exp\left(\frac{C_F p}{p_\infty - p}\right) \tag{6.7}$$

a form suggested by Paluch et al. [8]. The parameter, p_∞, has also been referred to as the *idealized glass transition pressure* and is the pressure at which the viscosity diverges. Here, μ_0 must not be thought of as the zero pressure viscosity but as an extrapolated zero pressure viscosity, instead, since equation (6.7) will not be accurate in general for low pressures. This relation plots as the same curve as the VTF equation in Fig. 6.2 if the fragility parameters are equal, $C_F = D_F$, assuming quite arbitrarily that $\mu_0 = 10^{-5}$ Pa s. Equation (6.7) is equivalent to an earlier expression by Johari and Whalley [11],

$$\mu = \mu_\infty \exp\left(\frac{C_F p_\infty}{p_\infty - p}\right) \tag{6.8}$$

with μ_∞ related to μ_0 in equation (6.7) by $\mu_\infty = \mu_0 \exp\left(-C_F\right)$ and the pressure fragility is proportional to $\alpha_g = \left(\ln\left(\mu_g/\mu_\infty\right)\right)^2 / C_F p_\infty$. Equations (6.7) and (6.8) have a property that has been observed in the high-pressure behavior of fragile glass-forming liquids; the derivative Stickel function, $1/\sqrt{\alpha}$, decreases linearly with increasing pressure [12].

$$\frac{1}{\sqrt{\alpha}} = \frac{p_\infty - p}{\sqrt{C_F p_\infty}} \tag{6.9}$$

The pressure fragility or alternatively, α_g, can be related to EHL traction through the representative stress, τ_E, that was introduced in equation (1.19).

$$\bar{\tau} = \tau_E \sinh^{-1}(C_E \Sigma) \tag{1.19}$$

This traction property, τ_E, is the slope of the linear, increasing, portion of a traction curve for a lubricated contact that has been plotted as average shear stress against the natural logarithm of any measure of the sliding velocity. The parameters of the Johari and Whalley model have been obtained from viscosity measurements for four liquid lubricants and listed in Table 6.1. The pressure fragility, m_p, and the pressure–viscosity coefficient at the glass transition, α_g, have been calculated from the Johari and Whalley parameters and are listed as well. Notice that for the bright stock, two temperatures have been employed and pressure fragility decreases with increasing temperature. The representative stresses were obtained from Fig. 1.7, measurements in the author's laboratory, and the literature [13] for the same liquids and are compared with fragility in Table 6.1. The trend is that more fragile liquids tend to have lower values of the representative stress, τ_E. Relatively stronger liquids have higher representative stress. Then the decrease in fragility with temperature is consistent with the observed [13] increase in τ_E with temperature. The reason for this will be made clear in Chapter 12.

There has been an additional model proposed for the high-pressure, faster-than-exponential response. The Munro, Block and Piermarini [14] model (MBP model) is

$$\mu = \mu_0 \left(\frac{p_\infty}{p_\infty - p} \right)^q \tag{6.10}$$

This model can be useful at low pressures as well when the less-than-exponential behavior does not appear.

Table 6.1. The relationship of pressure fragility and representative stress.

Liquid	SHF 401	Bright Stock, BS 98		LVI260	5P4E
T/°C	50	60	100	80	83
Viscosity Parameters					
C_F	30.8	16.7	21.9	21.7	39.3
p_∞/GPa	5.20	2.10	3.60	1.61	1.63
μ_∞/Pa s	3.3×10^{-13}	7.2×10^{-8}	9.5×10^{-11}	6.8×10^{-11}	5.3×10^{-20}
Measured Pressures/GPa	0.8–1.2	0.6–0.8	0.7–1.15	0.35–0.6	0.2–0.35
Fragility Parameters					
α_g/GPa^{-1}	16.7	44.5	27.0	61.7	71.1
m_p/GPa^{-1}	7.3	19.3	11.7	26.8	30.9
Traction Parameters					
τ_E/MPa	11.6	–	6.0	4	2
p_H/GPa	1.93	–	1.29	1.0	1.0
Traction Source	Fig. 1.7		Twin Disc	Ref. [13]	Ref. [13]

6.4 The Pressure–Viscosity Coefficient and Empirical Models for Low Pressure Behavior

It will be helpful to introduce low-pressure models in the context of the pressure–viscosity coefficient since within EHL the most important use of low-pressure models is for the calculation of film thickness which is determined by the viscosity in the inlet zone. Traction, on the other hand, is for the greatest part generated in the high-pressure Hertz zone. We saw in Chapter 1 that the Newtonian formulas for EHL film thickness require two parameters to describe the rheology of the lubricant; μ_0 and a measure of the strength of the piezoviscous response, known as the pressure–viscosity coefficient. There have been many definitions proposed for this parameter, some have been based upon a particular pressure–viscosity model. In addition to the local pressure–viscosity coefficient, α, there is the conventional pressure–viscosity coefficient,

$$\alpha_0 = \left[\frac{d\,(\ln \mu)}{dp}\right]_{p=0} \tag{6.11}$$

the secant pressure–viscosity coefficient,

$$\alpha_B(p) = \frac{\ln(\mu(p)) - \ln(\mu(0))}{p} \tag{6.12}$$

and the previously defined, reciprocal asymptotic isoviscous pressure coefficient,

$$\alpha^* = \frac{1}{p_{ai}} = \left[\int_0^\infty \frac{\mu(p=0)dp}{\mu(p)}\right]^{-1} \tag{1.3}$$

among some others [15]. The latter was proposed by Blok [16] to provide a measure of the piezoviscous response for general forms of $\mu(p)$. The values of these various coefficients diverge from one another as temperature increases and may differ by as much as 2:1 at high temperature [17].

At about the same time as the writing of this book, another approach has been taken to the definition of the appropriate pressure–viscosity coefficient for EHL film-forming with Newtonian liquids of general piezoviscous response [18]. The approach was to use numerical simulation to find the piezoviscous parameter, α_{film}, that together with the ambient viscosity, μ_0, would completely quantify the viscosity throughout the inlet zone so that the film thickness could be written as $h = h(\mu_0, \alpha_{film}, \ldots)$ regardless of the particular details of the pressure–viscosity response. The definition obtained [18] is

$$\alpha_{film} = \frac{1 - \exp(-3)}{p_{iv}(3/\alpha^*)} \tag{6.13}$$

where the isoviscous pressure is

$$p_{iv}(p) = \int_0^p \frac{\mu(p=0)dp^*}{\mu(p^*)} \tag{6.14}$$

In the case of purely exponential behavior, equation (6.1), $\alpha = \alpha_0 = \alpha_B = \alpha^* = \alpha_{film}$. Notice that $p_{ai} = p_{iv}(\infty) = 1/\alpha^*$.

Having measured viscosity at discrete pressures to obtain a table of data as in the previous chapter, some formulas are needed to calculate the coefficients above. First the local pressure viscosity coefficient is approximated by a finite difference as

$$\alpha_i = \frac{\ln\left(\mu_i/\mu_{i-1}\right)}{(p_i - p_{i-1})} \tag{6.15}$$

where $\mu_i = \mu(p_i)$ are the measured data and $p_0 = 0$ or 0.1 MPa and $1 \le i \le N$ where p_N is the highest experimental pressure. Then,

$$\alpha^* \approx \left[\frac{\mu_0}{\alpha_N \mu_N} + \sum_{i=1}^{N} \frac{\mu_0}{\alpha_i} \frac{\mu_i - \mu_{i-1}}{\mu_i \mu_{i-1}}\right]^{-1} \tag{6.16}$$

and

$$\alpha_{film} \approx [1-\exp(-3)]\left\{\frac{\mu_0}{\alpha_J \mu_{J-1}}\left[1-\exp\left(-\alpha_J(3/\alpha^* - p_{J-1})\right)\right] + \sum_{i=1}^{J-1} \frac{\mu_0}{\alpha_i} \frac{\mu_i - \mu_{i-1}}{\mu_i \mu_{i-1}}\right\}^{-1} \tag{6.17}$$

where J is defined by $p_{J-1} < 3/\alpha* \le p_J$. In addition, α_0 can be approximated as

$$\alpha_0 \approx \frac{\dfrac{p_2}{p_1}\ln\left(\dfrac{\mu_1}{\mu_0}\right) - \dfrac{p_1}{p_2}\ln\left(\dfrac{\mu_2}{\mu_0}\right)}{p_2 - p_1} \tag{6.18}$$

Equation (6.18) is based on the assumption that α_B varies linearly with pressure near ambient pressure and equation (6.16) assumes that the pressure–viscosity response is exponential for pressures greater than p_N with $\alpha = \alpha_N$. Assuming the usual repeatability of a falling body viscometer ($\approx$1%) and pressure–viscosity behavior typical for commercial lubricating oils, in order to accurately apply equations (6.16) and (6.18), experimental pressures of 0.1, 50, 100, 200, 300 MPa, ...$3/\alpha^*$ seem to be optimum. For equation (6.17) for α_{film}, slightly more data are required; 0.1, 25, 50, 100, 150, 200, 300 MPa, ...$3/\alpha^*$ being more appropriate.

Pressure–viscosity models useful for low pressures generally have α and α_B decreasing with increasing pressure with a rate of decrease that is reduced with increasing pressure. It must be repeated here that α_0 is experimentally very difficult to measure with good accuracy. The differential in equation (6.11) can be approximated by a finite difference of two measured viscosities at two experimental pressures as in equation (6.15) with $i = 1$. Since the local coefficient, α, may be changing rapidly with pressure, the pressures must be close to ambient or to zero or one pressure may even be negative [19] (in tension). This in turn means that small errors in viscosity and pressure can result in large errors in α_0. Fortunately, the conventional pressure–viscosity coefficient is not the best predictor of film thickness; both the reciprocal asymptotic isoviscous pressure coefficient [20] and α_{film} are better suited for that purpose.

There is an even greater variety of empirical models for low-pressure behavior than for high-pressure, attesting to the wide variety of pressure–viscosity response under this condition. More than one hundred years ago, measurements of the effect of pressure on viscosity were made by a number of researchers. In 1892 Cohen [21] measured the viscosity of turpentine to 60 MPa and found a linear increase to twice the ambient value. Barus, in 1892–1893, published a series of papers [22] describing the extrusion of a semi-solid marine glue from an open-ended capillary. Barus reported viscosity as a function of the average pressure using the linear model,

$$\mu = \mu_0(1 + \alpha_0 p) \tag{6.19}$$

Of course, α^* and α_{film} are not defined for this behavior. This model has been found to be accurate for normal octane in the author's laboratory at 115°C to a pressure of 175 MPa using $\mu_0 = 0.228$ mPa s and $\alpha_0 = 14.36$ GPa^{-1}. Hersey [23], in 1916, measured the viscosity of two lubricating oils to $p = 20$ MPa and utilized the linear Barus equation (6.19) to analyze the results. In this paper, Hersey may have originated the term "pressure coefficient of viscosity", α_0, for linear behavior but in subsequent measurements to 50 MPa found a faster-than-linear increase in viscosity.

Bridgman [24], in 1926 was apparently the first to generate sufficient pressure, 1.2 GPa, in a viscosity measurement to observe exponential response following less-than-exponential behavior. The pure exponential form has been employed for simplicity and because it often provides an analytical solution to idealized problems.

$$\mu = \mu_0 \exp(\alpha p) \tag{6.20}$$

For the exponential law, $\alpha = \alpha_0 = \alpha_B = \alpha^* = \alpha_{film}$ and

$$\frac{1}{p_{iv}(p)} = \alpha_0 \left[\frac{1}{1 - \exp(-\alpha p)} \right] \tag{6.21}$$

This two parameter model is accurate for 2,4-dicyclohexyl-2-methylpentane (Table 5.3) at 40°C to a pressure of 200 MPa using $\mu_0 = 18.1$ mPa s and $\alpha = 31.56$ GPa^{-1}.

In 1938, Skinner [25] reviewed the available measurements on lubricating oils and proposed the Skinner model,

$$\mu = \mu_0 \left[1 - \frac{\alpha_0}{\alpha_s} + \frac{\alpha_0}{\alpha_s} \exp(\alpha_s p) \right] \tag{6.22}$$

See also equation (1.12). This three parameter model has been found to be accurate for pentaerythritol tetrahexanoate (Table 5.10) to a pressure of 1.2 GPa at 40°C using $\mu_0 = 20.2$ mPa s, $\alpha_0 = 19.15$ GPa^{-1}, and $\alpha_s = 8.77$ GPa^{-1} and at 65°C using $\mu_0 = 8.94$ mPa s, $\alpha_0 = 18.13$ GPa^{-1} and $\alpha_s = 7.02$ GPa^{-1}. The Skinner model yields exponential response in the limit of high-pressure and liquids for which this model is accurate in the exponential regime must be classified as strong as regards the pressure fragility. For the Skinner model, equation (6.22),

$$\frac{1}{p_{iv}(p)} = \frac{\alpha_0 - \alpha_s}{\ln\left[1 - \frac{\alpha_0}{\alpha_s} + \frac{\alpha_0}{\alpha_s}\exp(\alpha_s p)\right] - \alpha_s p} \tag{6.23}$$

and

$$\alpha^* = \frac{\alpha_0 - \alpha_s}{\ln\left(\frac{\alpha_0}{\alpha_s}\right)} \tag{6.24}$$

In 1952, McEewen [26] proposed the following model:

$$\mu = \mu_0\left(1 + \frac{\alpha_0}{q}p\right)^q \tag{6.25}$$

This three-parameter equation is accurate for dipentaerythritol hexaisostearate at 100°C to a pressure of 300 MPa using $\mu_0 = 25.1$ mPa s, $\alpha_0 = 17.03$ GPa^{-1} and $q = 3.77$. For the McEwen model,

$$\frac{1}{p_{iv}(p)} = \alpha_0 \frac{\left(1 - \frac{1}{q}\right)}{1 - \left(1 + \frac{\alpha_0}{q}p\right)^{1-q}} \tag{6.26}$$

and

$$\alpha^* = \alpha_0\left(1 - \frac{1}{q}\right), \quad q > 1 \tag{6.27}$$

The Munro, Block and Piermarini [14] model (MBP model) was introduced above in equation (6.10) for high-pressure, greater-than-exponential behavior. At low temperatures or for liquids for which the glass transition state is close to ambient conditions, this three parameter model can be useful for low pressures as well. For this application it is useful to write equation (6.10) with $p_\infty = q/\alpha_0$ as

$$\mu = \mu_0\left(\frac{q}{\alpha_0}\right)^q\left(\frac{q}{\alpha_0} - p\right)^{-q} \tag{6.28}$$

Then, for the MBP model,

$$\frac{1}{p_{iv}(p)} = \frac{(q+1)\left(\frac{q}{\alpha_0}\right)^q}{\left(\frac{q}{\alpha_0}\right)^{q+1} - \left(\frac{q}{\alpha_0} - p\right)^{q+1}} \tag{6.29}$$

and

$$\alpha^* = \alpha_0 \left(1 + \frac{1}{q}\right) \tag{6.30}$$

This model is accurate for bis(phenoxyphenoxy)benzene at 83°C to 350 MPa using $\mu_0 = 25.1$ mPa s, $\alpha_0 = 17.0$ GPa^{-1} and $q = 9.4$.

From 1963 to 1966, Roelands [27,28] proposed three low-pressure models. The first was a modification of the exponential law and is not useful for inlet zone calculations since the pressure–viscosity coefficient behaves badly.

$$\mu = \mu_0 \exp\left(C_R p^q\right), \quad \alpha_0 = 0 \quad \text{if } q > 1, \quad \alpha_0 \to \infty \quad \text{if } q < 1 \tag{6.31}$$

The second Roelands model is similar to equation (6.7) except that the pressure for which the argument of the exponential function is unbounded is less than zero, $p_{-\infty} < 0$.

$$\mu = \mu_0 \exp\left(\frac{\alpha_0 p_{-\infty} p}{p_{-\infty} - p}\right) \tag{6.32}$$

This three parameter model is accurate for an L7808 jet oil at 220°C to a pressure of 300 MPa using $\mu_0 = 0.709$ mPa s, $\alpha_0 = 14$ GPa^{-1} and $p_{-\infty} = -0.286$ GPa. Here, the viscosity approaches a limiting value of $\mu_0 \exp\left(-\alpha_0 p_{-\infty}\right)$ at high-pressure and therefore α^* and α_{film} cannot be defined.

The third Roelands equation (6.33) is interesting because, for low temperatures and less-than-exponential response, the initial rate of change with pressure of the local pressure viscosity coefficient, α, varies realistically with the magnitude of α_0.

$$\mu = \mu_p \left(\frac{\mu_0}{\mu_p}\right)^{\left(\frac{p_p - p}{p_p}\right)^Z} \tag{6.33}$$

$$\alpha_0 = \frac{Z \ln\left(\dfrac{\mu_0}{\mu_p}\right)}{-p_p} \tag{6.34}$$

Here Roelands made use of the viscosity pole concept [27] that was introduced in Fig. 5.10 and Section 5.3. The viscosity pole is located at the intersection of pressure-viscosity isotherms when extrapolated to negative pressures. The intersection occurs at $p = p_p$ and $\mu = \mu_p$ and isotherms can be constructed about the pole point. This concept fails above some temperature and pressure so that equation (6.33) should be considered only for relatively low pressures and temperatures. At high temperatures the decrease of α with pressure becomes more rapid than can be described by the third Roelands model. Roelands found that reasonably accurate results could be obtained by setting the pole parameters to universal values [28] of $\mu_p = 6.31 \times 10^{-5}$ Pa s and $p_p = -0.196$ GPa. Then the third model with the universal pole parameters is a two parameter model and, for a model with just two adjustable parameters, it is remarkable in the range of temperature and pressure for which it is accurate with high-viscosity paraffinic mineral oils and high-viscosity polyalphaolefins and some esters. It is often found that the *pressure–viscosity index*, Z, is independent of temperature up to about 100°C.

The isoviscous pressure for the third Roelands model is

$$p_{iv}(p) = \frac{\mu_0}{\mu_p} \int_0^p \left(\frac{\mu_0}{\mu_p}\right)^{-\left(1-\frac{p}{p_p}\right)^Z} dp \tag{6.35}$$

Roelands [28] found that a change in the integration variable to $s = \left(1 - \left(p/p_p\right)\right)^Z \times \ln\left(\mu_0/\mu_p\right)$ would allow the isoviscous pressure to be evaluated from

$$p_{iv}(p) = \frac{-p_p}{Z}\frac{\mu_0}{\mu_p}\left[\ln\left(\frac{\mu_0}{\mu_p}\right)\right]^{-\frac{1}{Z}} \int_{\ln\left(\frac{\mu_0}{\mu_p}\right)}^{\left(1-\frac{p}{p_p}\right)^Z \ln\left(\frac{\mu_0}{\mu_p}\right)} \exp(-s) s^{\frac{1}{Z}-1} ds \tag{6.36}$$

where the integral may be represented in terms of the incomplete gamma functions. It should then be possible to obtain α^* and α_{film}.

In recent years, numerical simulations of EHL have almost exclusively utilized equation (6.33) and these simulations have generally assumed pressures well beyond the useful upper pressure limit which was said by Roelands [28] to be from 0.15 to 0.5 GPa. However, as an idealization of typical inlet zone behavior with few parameters, this is a very reasonable model to use. This author has found one liquid for which equation (6.33) does accurately describe the behavior to high pressure. For trimethylolpropane trioleate at 50 and 100°C, with $Z = 0.461$ and $\mu_0 = 31.3$ and 8.73 mPa s respectively, the pressure–viscosity behavior is accurately modeled up to the pressure at which the liquid crystallizes which is about 0.9 GPa and 1.3 GPa respectively. For this liquid, then, the inflection pressure must be greater than these pressures. It is not surprising then that this is a liquid for which the traction coefficient is remarkably low.

The pressure–viscosity models described above do not include the effect of variable temperature. Roelands, however, incorporated variable temperature forms into his first and third models [27,28]. For Roelands third model, including temperature,

$$\mu = \mu_p \left(\frac{\mu_R}{\mu_p}\right)^{\left[\left(\frac{p_p - p}{p_p}\right)^Z \left(\frac{T_R - T_\infty}{T - T_\infty}\right)^S\right]} \tag{6.37}$$

where S is a temperature–viscosity parameter called the *slope index*, and as before, $\mu_R = \mu(T_R, p = 0)$ and T_∞ is the divergence temperature where the viscosity is unbounded. Roelands specified a universal value of $T_\infty = -135$°C so that equation (6.37) with the universal parameters is a model with only three adjustable parameters. The previous two sections showed that accurate models for temperature–viscosity behavior at low temperatures and models for the pressure–viscosity behavior at high pressures require a parameter that specifies the state at which the viscosity would be unbounded.

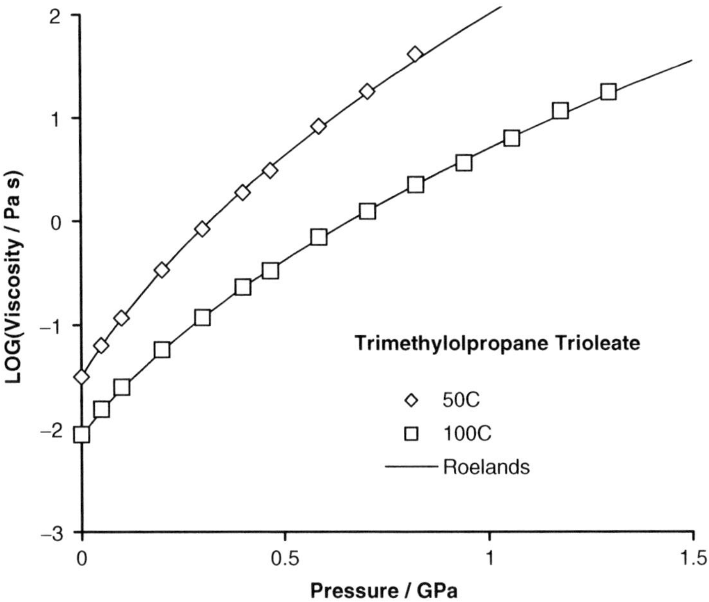

Fig. 6.3. The viscosity of a polyol ester compared with the Roelands temperature–pressure–viscosity model. This liquid generates remarkably low traction in EHL.

For equation (6.37), this state is $T = -135°C$ regardless of pressure. In other words the singularity in viscosity can only be reached by a decrease in temperature and cannot be reached by an increase in pressure. This problem reinforces the limitation of this model to low pressures. When $S = 1$ and pressure is constant, equation (6.37) reduces to a form similar to the VTF equation (6.3).

When the inflection pressure is very high, equation (6.37) may be accurate for some liquids to high pressures, as it is for trimethylolpropane trioleate at 50 and 100°C. With $Z = 0.461$, $S = 0.98$, $\mu_R = 31.3$ mPa s and $T_R = 50°C$, the pressure–viscosity behavior is accurately modeled as shown in Fig. 6.3 up to the pressure at which the liquid crystallizes. When the crystallization of pure liquids is observed under pressure, it is generally associated with the combination of relatively low viscosity and high pressure (low value of Z) such as in Fig. 6.3.

6.5 Empirical Models for Large Pressure Intervals

Since elastohydrodynamic numerical calculations generally cover pressures from ambient to the maximum contact pressure, the Hertz pressure, p_H, and this is generally larger than the inflection pressure, p_{inf}, for the lubricant, a useful model should describe the pressure-log viscosity inflection. Apparently, Irving and Barlow [29] in 1971 first proposed a double-exponential model incorporating an inflection to describe the pressure–viscosity response of diesters. Their four parameter model can be written as

$$\mu = \mu_0 \exp\left(\frac{\alpha_n + \alpha_p - \alpha_0}{\alpha_p}\right) \cdot \exp\left[\frac{\alpha_0 - \alpha_n}{\alpha_p} \exp\left(\alpha_p p\right) - \exp\left(-\alpha_n p\right)\right] \tag{6.38}$$

where the inflection pressure is

$$p_{inf} = \frac{\ln\left(\dfrac{\alpha_n^2}{\alpha_p\left(\alpha_0 - \alpha_n\right)}\right)}{\alpha_p + \alpha_n} \tag{6.39}$$

This model applied to the data of Table 5.7 for the L23699 jet oil at 50°C results in $\alpha_0 = 20.42\ \mathrm{GPa}^{-1}$, $\alpha_p = 0.1978\ \mathrm{GPa}^{-1}$ and $\alpha_n = 12.60\ \mathrm{GPa}^{-1}$ with $p_{inf} = 0.362$ GPa.

Recently Bair and Kottke [30] proposed a five-parameter empirical model with an inflection pressure that features the singularity of the second Roelands model (6.32) at negative pressure, $p_{-\infty}$, and the singularity of the Johari and Whalley model (6.8) at positive, p_∞. They used the model in successful comparative calculations of EHL traction for a polyalphaolefin and a polyolester to be revisited in Chapter 12. The model can be reduced to a four parameter equation, giving up some flexibility.

$$\mu = \mu_0 \exp\left[1 - p_\infty\left(\alpha_0 + {}^1/_{p_{-\infty}}\right)\right] \cdot \exp\left[\frac{\left(\alpha_0 + {}^1/_{p_{-\infty}}\right) p_\infty^2}{p_\infty - p} + \frac{p_{-\infty}}{p - p_{-\infty}}\right] \tag{6.40}$$

The inflection pressure is

$$p_{inf} = \frac{p_\infty\left(-p_{-\infty}\right)^{\frac{1}{3}} + p_{-\infty}\left[\left(\alpha_0 + {}^1/_{p_{-\infty}}\right) p_\infty^2\right]^{\frac{1}{3}}}{\left(-p_{-\infty}\right)^{\frac{1}{3}} + \left[\left(\alpha_0 + {}^1/_{p_{-\infty}}\right) p_\infty^2\right]^{\frac{1}{3}}} \tag{6.41}$$

For the data of Table 5.7 for the L23699 jet oil at 50°C, $\alpha_0 = 22.97\ \mathrm{GPa}^{-1}$, $p_{+\infty} = 16.62$ GPa and $p_{-\infty} = -0.0679$ GPa with $p_{inf} = 0.434$ GPa. This model is compared to the Irving and Barlow equation (6.38) in Fig. 6.4 where there is little difference between the models in the quality of the fit to the data.

The VTF equation (6.3) is often used to describe the temperature dependence of viscosity at elevated pressures. It is has been noted [12] that the idealized glass transition temperature, T_∞, varies with pressure with the same functional relationship as the glass transition temperature, T_g. (The same relationship has been found between p_∞ and p_g as well [12].) Ole and Rehage [31] discovered that, for polystyrene, the rate of change of (glass transition) pressure with respect to T_g increases linearly with pressure.

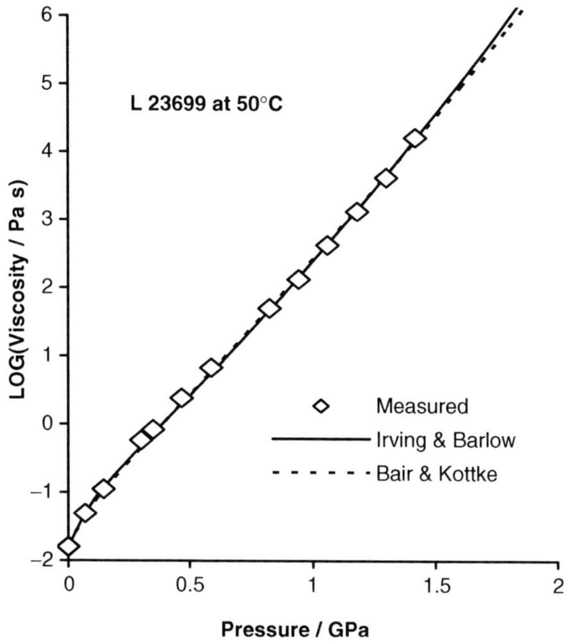

Fig. 6.4. The viscosity of a specification L 23699 jet oil compared with the Irving and Barlow and the Bair and Kottke empirical models.

If T_∞ follows the same rule,

$$T_\infty = T_0 - \frac{T_0'^2}{T_0''} \ln\left(1 - \frac{T_0''}{T_0'} p\right) \tag{6.42}$$

where $T_0' = |dT_\infty/dp|_{p=0}$ and $T_0'' = |d^2T_\infty/dp^2|_{p=0}$. Then it should be possible to describe the variation of viscosity with both temperature and pressure using the VTF equation,

$$\mu = \mu_\infty \exp\left[\frac{D_F T_\infty}{T - T_\infty}\right] \tag{6.3}$$

if suitable functions can be found for the pressure dependence of μ_∞ and D_F. A reasonably accurate representation of the viscosity of diisodecyl phthalate can be obtained by setting μ_∞ and D_F constant and applying equation (6.42) for T_∞. This liquid has been proposed as a viscosity standard [32]. The results for $\mu_\infty = 3.655 \times 10^{-6}$ Pa s, $D_F = 16.30$, $T_0 = -158.8$°C, $T_0' = 146.0$°C/GPa and $T_0'' = -331.1$°C/GPa^2 are plotted in Fig. 6.5 for four temperatures and compared with measurements. This relatively simple model reproduces the character of the data, including the pressure inflection, with only five adjustable parameters; however the accuracy at low pressure suffers.

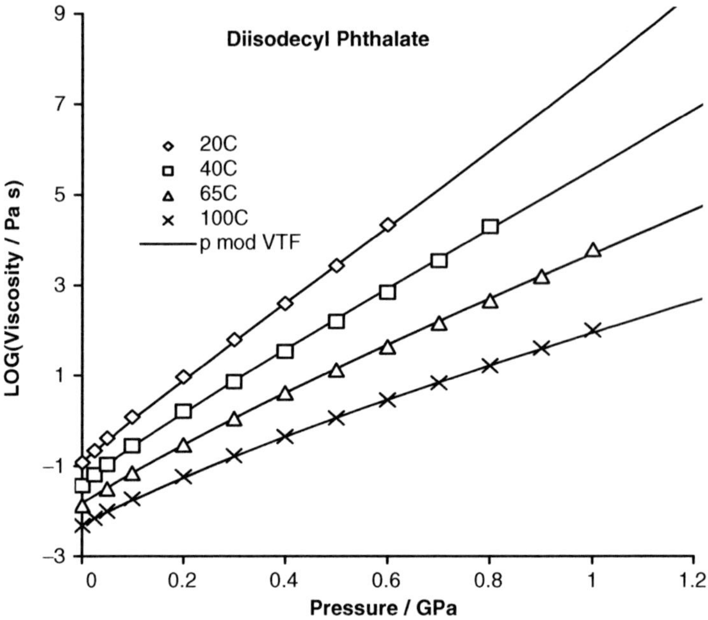

Fig. 6.5. The viscosity of diisodecyl phthalate, a proposed viscosity standard, compared with the pressure-modified VTF equation.

6.6 Models Based on Free Volume Theory

We have seen that the pressure–viscosity behavior of liquids at low pressure cannot be described accurately for all liquids by any single simple empirical model. Models for the entire pressure range of measurements become more complex with more parameters. It is not generally clear how to accurately implement the temperature dependence of viscosity into these empirical forms for a large range of pressures. The free volume theory of viscous flow can provide a framework for a natural description of the temperature and pressure dependence of viscosity with the additional benefit that at least some of the parameters have physical meaning outside of the application to viscosity and may be independently measured.

The basic equation of free volume theory is the Doolittle [33] equation,

$$\mu \propto \exp\left(B\frac{V_\infty}{V - V_\infty}\right) \tag{6.43}$$

for total volume, V, and occupied volume, V_∞. The *free volume* is $V - V_\infty$. Doolittle proposed this originally as an empirical expression to correlate the viscosity of normal alkanes with temperature at ambient pressure for temperatures of about 100°C above the freezing temperature up to the boiling point. The viscosities of the liquids were less than 3 mPa s under these conditions. Doolittle [33] defined the occupied volume as a constant equal to the volume of the liquid extrapolated to a temperature of 0 K. Williams, Landel and Ferry [5] recognized that the Doolittle equation, with its abrupt increase in viscosity when the free volume becomes small, can describe the behavior near the glass transition where the viscosity is very large. They defined the occupied volume to be a function of temperature as $V_\infty = V_g\left[0.975 + a_g\left(T - T_g\right)\right]$, where V_g and a_g are the volume and expansivity of the glass, respectively, at the glass transition. Cohen and Turnbull [34] gave physical meaning to the free volume concept. They hypothesized that molecular transport occurs when a fluctuation in the free volume opens up a void greater than some critical size required to allow movement of a molecule into that void. The probability of finding a void exceeding the critical volume was found to be equal to $\exp\left[-BV_\infty/(V - V_\infty)\right]$.

Making use of the viscosity at the reference state, $\mu_R = \mu\left(T_R, p = 0\right)$, where the total volume and occupied volume are V_R and $V_{\infty R}$ yields a more useful form of the Doolittle equation,

$$\mu = \mu_R \exp\left[BR_0\left(\frac{\dfrac{V_\infty}{V_{\infty R}}}{\dfrac{V}{V_R} - R_0\dfrac{V_\infty}{V_{\infty R}}} - \frac{1}{1 - R_0}\right)\right] \tag{6.44}$$

Notice that temperature and pressure are, at this stage, absent from the governing equation for viscosity and the occupied volume fraction at the reference state is $R_0 = V_{\infty R}/V_R$. With viscosity and PVT data available over a large range of temperature and pressure, it becomes possible to investigate the behavior of the Doolittle equation for a much greater range of occupied volume fraction, V_∞/V.

First consider the proposition that V_∞ and B are independent of temperature and that the volume at ambient pressure follows equation (4.14),

$$\frac{V_0}{V_R} = 1 + a_V\left(T - T_R\right) \tag{4.14}$$

where for ambient pressure $V = V_0$. Then the Doolittle equation becomes

$$\mu = \mu_R \exp\left[\frac{-BR_0}{1 - R_0}\right] \exp\left[\frac{BR_0}{1 - R_0 + a_V\left(T - T_R\right)}\right] \tag{6.45}$$

which is equivalent to the VTF equation (6.3),

$$\mu = \mu_\infty \exp\left[\frac{D_F T_\infty}{T - T_\infty}\right] \tag{6.3}$$

with the Doolittle parameters related to the VTF parameters by $\mu_\infty = \mu_R \exp\left[-BR_0/(1 - R_0)\right]$, $D_F = \left[BR_0/(a_V T_R + R_0 - 1)\right]$ and $T_\infty = T_R - (1 - R_0)/a_V$. If we examine the behavior of the 5P4E, bis(phenoxyphenoxy)benzene, that was used an example in Sections 6.2, 6.3 and 6.4, it will be helpful to choose $T_R = 83°\text{C}$ since we have elevated-pressure viscosity measurements for this temperature. Then for the VTF parameters that were extracted from the ambient-pressure, variable-temperature data of Fig. 6.1 in Section 6.2, we have $R_0 = 0.8572$, $B = 0.7572$ and $a_V = 1.236 \times 10^{-3}\ \mathrm{K}^{-1}$. This value for the expansivity, a_V, is larger than the value, $0.7 \times 10^{-3}\ \mathrm{K}^{-1}$, that can be obtained from measurements [1] of density at ambient pressure.

Returning to equation (6.44), consider that the occupied volume may vary linearly with temperature as

$$\frac{V_\infty}{V_{\infty R}} = 1 + \varepsilon(T - T_R) \tag{6.46}$$

Let us retain the values of $R_0 = 0.8572$ and $B = 0.7572$ that were obtained from the VTF fit to the ambient pressure viscosity data and set the expansivity equal to the value from the ambient-pressure density measurements, $a_V = 0.7 \times 10^{-3}\ \mathrm{K}^{-1}$. Then value of the occupied volume expansivity required to fit the temperature–viscosity data in Fig. 6.1 becomes $\varepsilon = -0.589 \times 10^{-3}\ \mathrm{K}^{-1}$ and the quality of the fit is comparable to that of the original VTF equation only now the expansivity, a_V, is more realistic. Now it should be possible to test the applicability of the Doolittle equation (6.44) to pressure–viscosity response with an appropriate equation of state (EOS).

The parameters of the Tait isothermal EOS (4.6),

$$\frac{V}{V_0} = 1 - \frac{1}{1 + K_0'} \ln\left[1 + \frac{p}{K_0}\left(1 + K_0'\right)\right] \tag{4.6}$$

have been estimated [18] from PVT data to be $K_0 = 1.70$ GPa and $K_0' = 13$ for 5P4E at 83°C. Now if the relative volume, V/V_R, is calculated from the product of equations (4.6) and (4.14) and substituted into the Doolittle equation (6.44) using the values of $R_0 = 0.8572$ and $B = 0.7572$ that were obtained from the VTF fit to the ambient pressure viscosity data, the resulting viscosity can be successfully compared with measurements as shown in Fig. 6.6. The ability to extrapolate from temperature–viscosity effects to pressure–viscosity effects makes a powerful statement for the usefulness of the free volume theory as a framework for the correlation of viscosity with temperature and pressure. Notice that the occupied volume, V_∞, is a function of temperature but is independent of pressure and that the Doolittle parameter, B, has been assumed to be a constant for a specific liquid.

In Chapter 4, a rather complete description was obtained of the PVT response of the model mineral oil, LVI260. There, the Tait EOS (4.6) was made temperature dependent through equation (4.14), repeated above, and equation (4.11) repeated below.

$$K_0 = K_\infty + \frac{\dot{K}_0}{T} \tag{4.11}$$

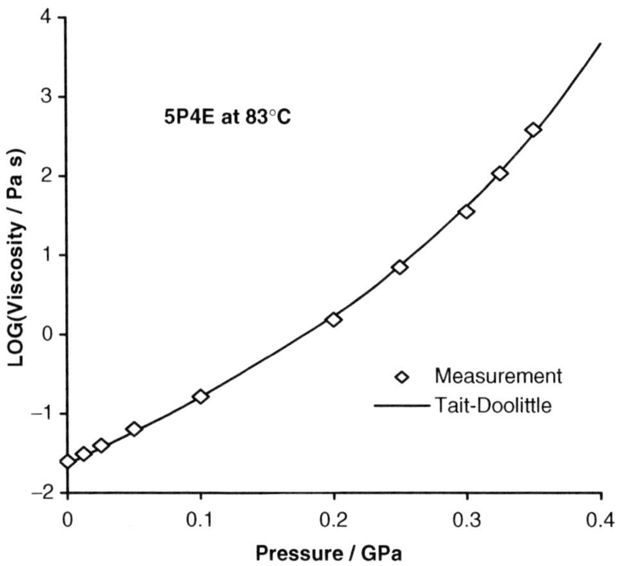

Fig. 6.6. The viscosity of the polyphenyl ether, 5P4E, as a function of pressure at one temperature. The curve represents the Doolittle model with compressibility described by the Tait EOS. The Doolittle parameters, ε, B and R_0 were obtained from the temperature dependence of viscosity shown in Fig. 6.1 and the Tait (compressibility) parameters, K_0 and K_0', were estimated from the data of Fig. 4.6.

As before, K_0', is assumed to be independent of temperature. In addition, the parameters for another EOS, the Murnaghan equation,

$$\frac{V}{V_0} = \left(1 + \frac{K_0'}{K_0} p\right)^{\left(-\frac{1}{K'}\right)} \tag{4.8}$$

were obtained for the mineral oil in Chapter 4. All of the required EOS parameters are listed in Table 4.14. The additional parameters of the Doolittle equation were obtained from a regression using viscosity data from a falling body viscometer and are listed in Table 6.2. The calculated viscosity is compared with the data in Fig. 6.7 where very good agreement is shown and the Tait and Murnaghan equations yield similar results.

Table 6.2. The Doolittle Parameters for LVI260. $T_R = 20°C$.

Parameter	Using the Tait EOS	Murnaghan EOS
μ_R/Pa s	1.733	1.659
B	2.547	2.513
R_0	0.8296	0.8306
ε/K^{-1}	-0.6559×10^{-3}	-0.6460×10^{-3}

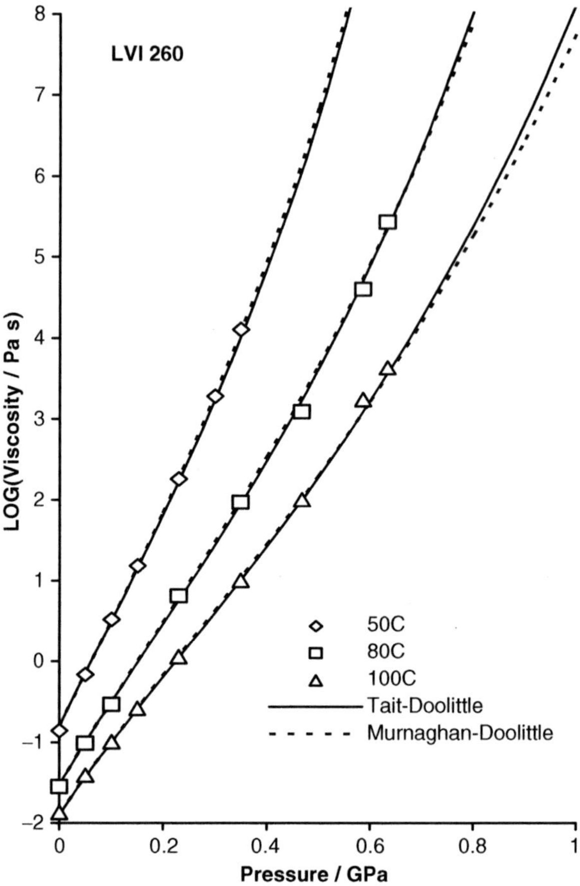

Fig. 6.7. The viscosity of the mineral oil LVI260 compared with the free volume model using the Tait and the Murnaghan equations of state.

In the event that PVT data are not available for a particular liquid, the EOS parameters may be regressed from the viscosity data along with the free-volume parameters and these are often found to be an accurate representation of the liquid compressibility. To do so the viscosity data should include a substantial number of points for pressures greater than the inflection pressure, p_{inf}. This is because the parameters are not uniquely defined from a regression unless the pressure inflection is clearly evident. Complete sets of free-volume and EOS parameters for eight liquids have been regressed from viscosity data and listed in Table 6.3. When the expansivity, a_V, is known for the particular liquid it is entered in the table, otherwise the general value of $8 \times 10^{-4}{}^{\circ}\mathrm{C}^{-1}$ is entered. Note that any error in a_V is simply absorbed into the value of ε. One problem for this framework of temperature–pressure–viscosity correlation is that for very high temperatures and pressures the viscosity is predicted to increase with temperature, a physically unacceptable behavior. The temperature at which this happens is usually much greater than the typical application for EHL.

Table 6.3. The free volume parameters for selected lubricants for $K_0 = K_\infty + \frac{\dot{K}_0}{T}$.

Fluid	Doolittle Equation Parameters					Tait Equation Parameters			
	μ_R/Pa s	T_R/°C	B	R_0	ε/°C^{-1}	K_0'	K_∞/GPa	$\dot{K}_0$/GPa · K	a_V/°C^{-1}
Nissan traction fluid	0.0538	20	1.900	0.7795	-8.12×10^{-4}	10.391	−0.8427	502.1	8×10^{-4}
S50 traction fluid	0.0821	20	2.741	0.7426	-9.65×10^{-4}	10.754	−0.9225	512.5	7.96×10^{-4}
Mineral oil SAE 20	0.1089	20	3.520	0.6980	-10.34×10^{-4}	10.4	−0.9282	580.7	8×10^{-4}
Mineral base for gear oil	0.0707	70	3.431	0.6502	-13.22×10^{-4}	10.82	−1.014	571.4	8×10^{-4}
PAO-100	0.1803	70	3.811	0.6193	-13.44×10^{-4}	11.74	−1.051	568.5	8×10^{-4}
PAO-ISO 68	0.0819	20	3.966	0.6622	-10.35×10^{-4}	11.38	−0.9881	580.8	8×10^{-4}
Automatic transmission fluid	0.0780	20	3.877	0.6774	-10.27×10^{-4}	10.52	−0.9662	585.4	8×10^{-4}
L23699	0.0467	20	3.382	0.6641	-12.80×10^{-4}	10.741	−1.0149	570.8	7.42×10^{-4}

It was mentioned in Chapter 4 that equation (4.11) behaves badly at high temperature. Since $K_\infty < 0$, the physically unacceptable result that $K_0 < 0$ will occur at some high temperature and even before this temperature is attained the prediction should become inaccurate with the viscosity even increasing with temperature. Another expression for the temperature dependence of the initial bulk modulus that is better behaved at high temperature was introduced in Chapter 4 and is repeated below.

$$K_0 = K_{00} \exp(-\beta_K T) \tag{4.12}$$

The jet turbine oil L23699 is used for rolling bearing lubrication at extremely high temperature. Bulk oil temperatures in this application can exceed 240°C [35]. The free volume model was applied to viscosity data, Table 5.7, for this liquid with equation (4.12) in place of equation (4.11) and the parameters are listed in Table 6.4. The two expressions for K_0 are similar for $20 < T < 150$°C but diverge at higher temperatures. The previous parameters given in Table 6.3 for equation (4.11) do more faithfully reproduce the data but the high-temperature viscosity response is more realistic for the parameters of Table 6.4.

6.7 Generalized Temperature–Pressure–Viscosity Models

In the closing paragraph of Chapter 4, an equation of state to be used in the EHL simulation of general liquid behavior was suggested. The last two entries in Table 6.4 supply suggested parameters for general pressure–temperature–viscosity behavior of liquids to be used in EHL simulations. The model liquids use the same general equation of state parameters suggested in Chapter 4. The temperature dependence of K_0 is described by the exponential equation (4.12) to avoid the unrealistic temperature–viscosity response at high temperature. Since very diverse pressure response can be had for the viscosity of ordinary lubricants, two different model liquids are suggested in Table 6.4, models of strong and fragile liquids, with the expectation that numerical simulations can be performed with both model liquids to investigate the effects of fragility upon the calculated contact behavior.

The properties of the model liquids are investigated in Figs. 6.8–6.11. Of course, the fragile liquid is more pressure sensitive than the strong liquid in Fig. 6.8, with isotherms turning sharply upward as pressure increases. This is behavior typical of naphthenic mineral oils and some traction fluids and polybutenes. The strong liquid is typical of polyalphaolefins and some polyglycols and esters. Fragility may be quantified by the pressure–viscosity coefficient of the liquid at the glass transition, α_g. Assuming that $\mu_R = 1$ Pa s and $\mu_g = 10^{10}$ Pa s, then α_g is about 70 and 50 GPa^{-1} for the model fragile liquid at 60 and 100°C, respectively, and is about 13 and 8 GPa^{-1} for the model strong liquid at 60 and 100°C, respectively.

The local pressure–viscosity coefficient, $\alpha = \partial \ln \mu / \partial p$, is plotted in Fig. 6.9 where it can be seen to have a minimum value that occurs at a pressure that increases with temperature. As of the writing of this book this behavior has seldom been modeled in EHL simulations. The local temperature–viscosity coefficient, $\beta = -\partial \ln \mu / \partial T$, is plotted in Fig. 6.10 where it can be seen to increase with temperature, faster for the

Table 6.4. The free volume parameters for selected lubricants for $K_0 = K_{00} \exp(-\beta_K T)$.

Fluid	Doolittle Equation Parameters					Tait Equation Parameters			
	μ_R/Pa s	T_R/°C	B	R_0	ε/°C^{-1}	K_0'	K_{00}/GPa	β_K/K^{-1}	a_V/°C^{-1}
L23699	0.05004	20	3.161	0.6723	-12.64×10^{-4}	10.835	10.142	8.021×10^{-3}	7.42×10^{-4}
Model fragile liquid	–	20	1.8	0.82	-10×10^{-4}	11	9	6.5×10^{-3}	8×10^{-4}
Model strong liquid	–	20	4	0.7	-15×10^{-4}	11	9	6.5×10^{-3}	8×10^{-4}

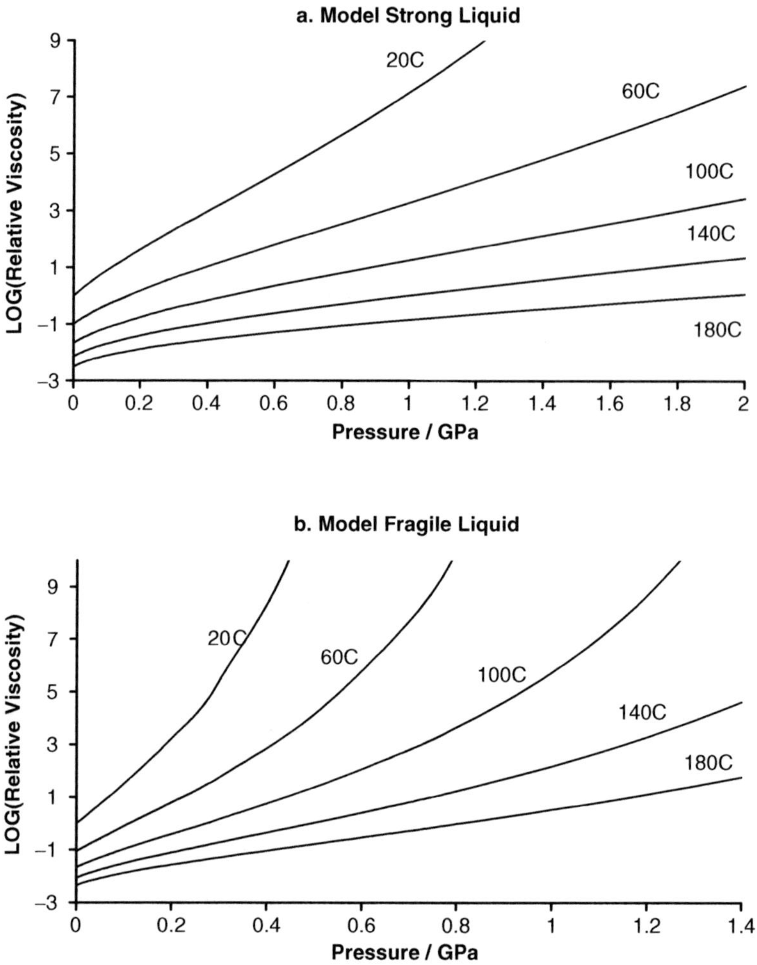

Fig. 6.8. The logarithm of relative viscosity, $\log \mu/\mu_R$, of the model liquids.

fragile liquid, but in both cases reaching large values at high pressure. As of the writing of this book, these large values of β have not been a part of thermal EHL simulations. It has been observed that the derivative Stickel function [36], $1/\sqrt{\alpha}$, decreases linearly with increasing pressure for high pressure. This behavior is approximated by the model liquids in Fig. 6.11 at low temperatures for some range of high pressures.

6.8 Multi-Component Systems

Figure 5.12 showed that by the addition of a small weight fraction, c, of an additive component to a lubricant base oil the viscosity at high pressure could be substantially reduced with little effect for low pressures. Such additives may provide significant

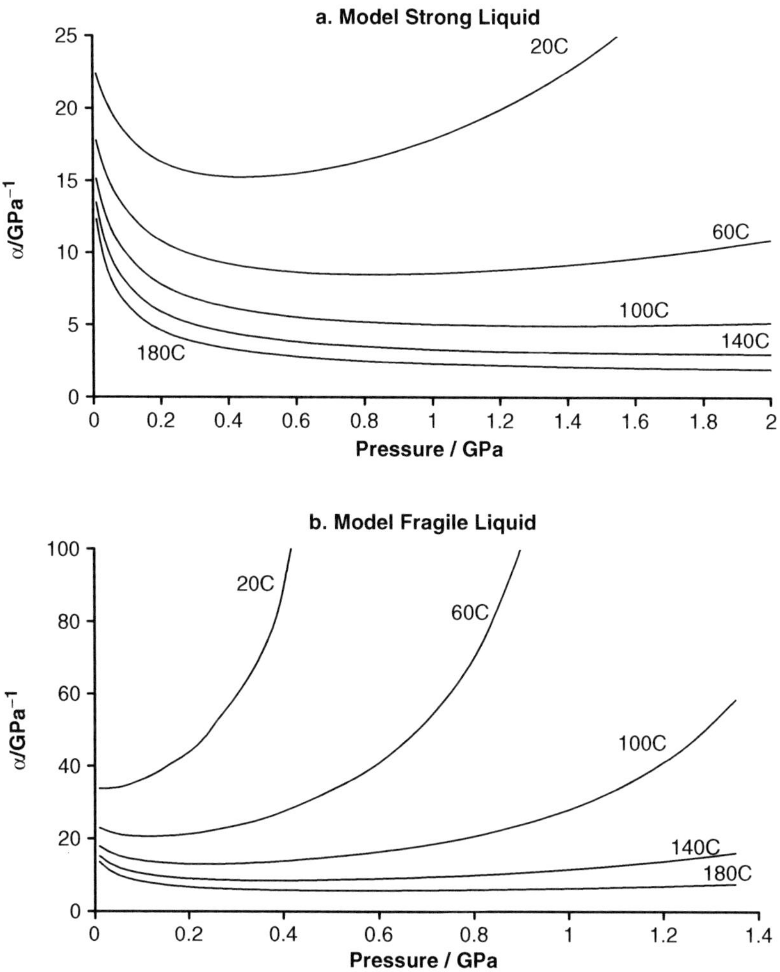

Fig. 6.9. The local pressure–viscosity coefficient, $\alpha = \partial \ln \mu / \partial p$, of the model liquids.

friction reduction in full film lubricated contacts. See Section 5.3. Liquid lubricants are seldom single component materials. Different "cuts" of base stocks are blended to arrive at a specification for the ambient pressure viscosity. Additives are also blended into the mixture to provide a beneficial chemical activity or to increase the viscosity as a thickener or to reduce the temperature dependence as a viscosity index (VI) improver.

The Grunberg and Nissan mixing rule [37], assuming that the densities of the components are similar, is

$$\ln \mu = (c - 1) \ln \mu_b + c \ln \mu_a + c (c - 1) d \tag{6.47}$$

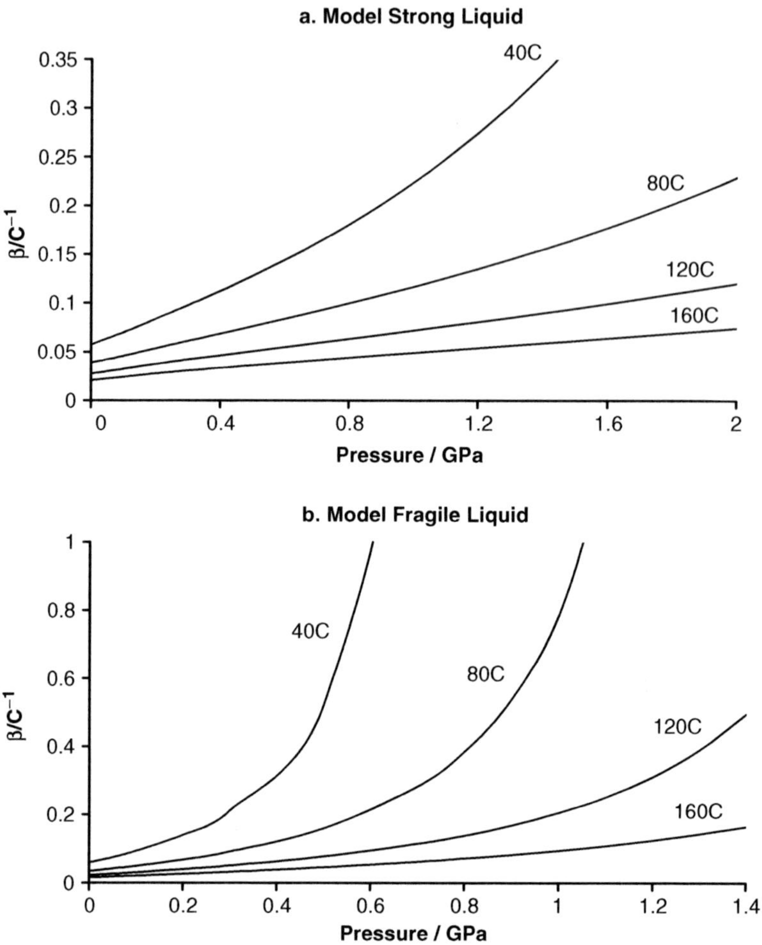

Fig. 6.10. The temperature–viscosity coefficient, $\beta = -\partial \ln \mu / \partial T$, of the model liquids.

With the interaction parameter, d, set to zero, equation (6.47) is the idealized or Arrhenius equation [38]. Mamagakis and Panayiotou [38] found a correlation of d with the excess volume that results from mixing, a correlation that is consistent with free volume theory. When the excess volume is positive, then $d < 0$ and the solution viscosity is less than the ideal. An example of this less than ideal mixture viscosity is shown in Fig. 6.12 for a blend of a 4cS grade of polyalphaolefin, PAO, with a diester, di(2-ethylhexyl)sebacate, DOS. The straight curve is the idealized, Arrhenius mixing rule with $d = 0$. The mixing rule for the two component system in Fig. 6.12 is pressure independent. This is not the case for the two component system of Fig. 6.13, where a polyolester, trimethylolpropane trioleate, has been blended with an aromatic hydrocarbon, dibenzylethylbenzene [39]. The mixing response here is strongly pressure-dependent, with the low-pressure mixture viscosity greater than the ideal and the high-pressure mixture viscosity less than

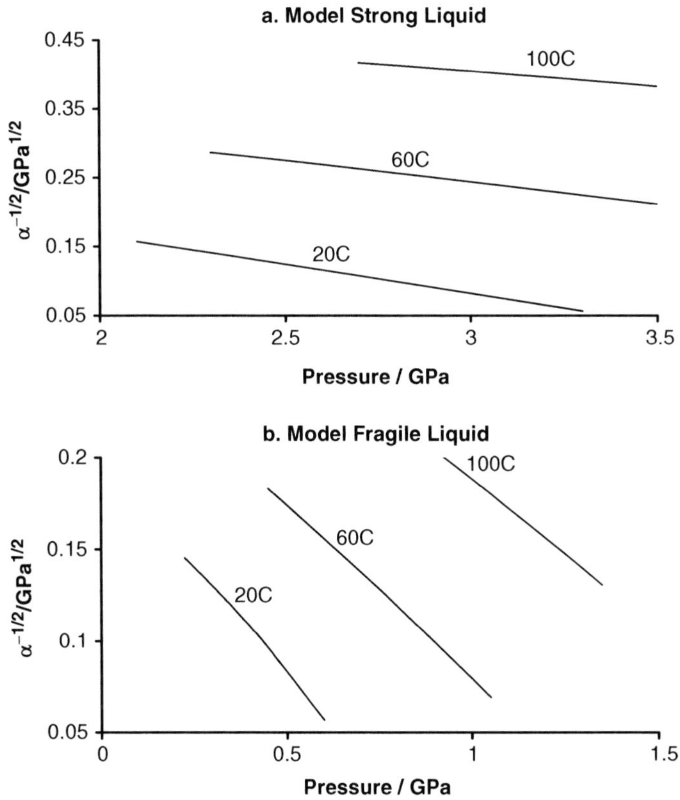

Fig. 6.11. The derivative Stickel function, $1/\sqrt{\alpha}$, of the model liquids.

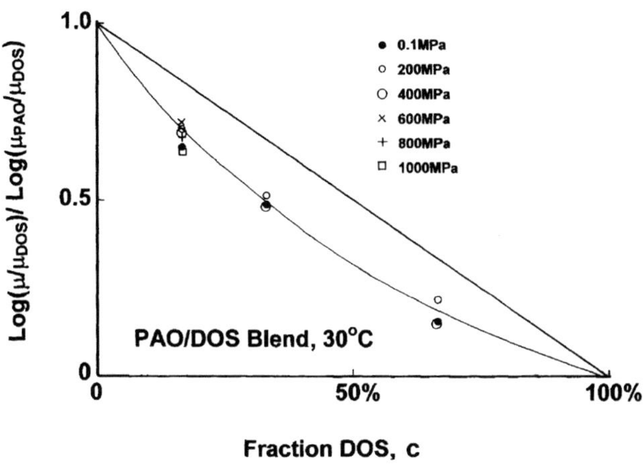

Fig. 6.12. The viscosity of a blend of poly alphaolefin and diester showing less than ideal mixture viscosity at all measured pressures. Reproduced from Bair, S., "The High-Pressure Rheology of Mixtures", *ASME J. Tribol.*, Vol. 126, No. 4, 2004, pp. 697–702, by permission of the American Society of Mechanical Engineers.

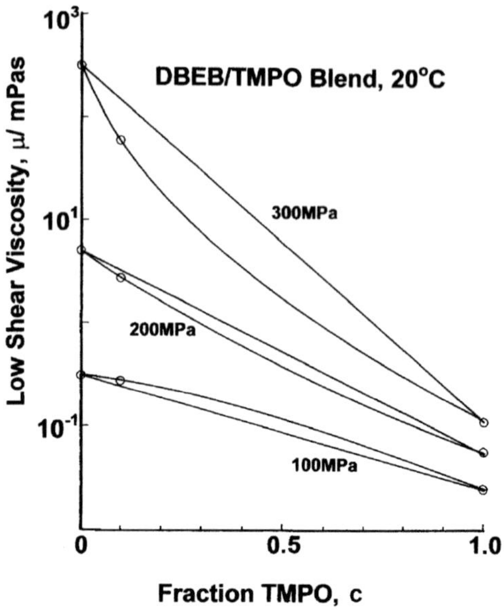

Fig. 6.13. The mixing response of a polyolester, trimethylolpropane trioleate (TMPO) in an aromatic hydrocarbon, dibenzylethylbenzene (DBEB). The mixing response is strongly pressure-dependent. Reproduced from Bair, S., "The High-Pressure Rheology of Mixtures", *ASME J. Tribol.*, Vol. 126, No. 4, 2004, pp. 697–702, by permission of the American Society of Mechanical Engineers.

the ideal. This pressure-dependent nonideal mixing response is believed [40] to result from the polyol ester in Fig. 6.13 behaving like the ester and the fatty acid in Fig. 5.12 to delay the glass transition of the base oil to higher pressure, that is, behaving as a plasticizer.

A similarly interesting and potentially useful alteration of the pressure–viscosity response can come about from blending of different molecular weight cuts of the same chemical structure as illustrated by Fig. 6.14. The pressure–viscosity response at 40, 100 and 150°C of a "straight cut" of polybutene is shown in the figure as the solid curves.

In addition, two other cuts of polybutene, one of higher and one of lower molecular weight were blended to arrive at the same ambient pressure viscosity at 40°C. The blend is shown to have a reduced temperature dependence of the ambient pressure viscosity. Perhaps even more interesting in Fig. 6.14 is the effect of blending on the viscosity at high pressure. The pressure dependence of viscosity is markedly reduced for the blend only at the higher pressures.

Two chapters have been devoted to the low-shear viscosity, but it has long been appreciated that a description of the viscosity relevant to EHL is incomplete without addressing the shear dependence. The next two chapters will describe measurements and the nature of shear dependence.

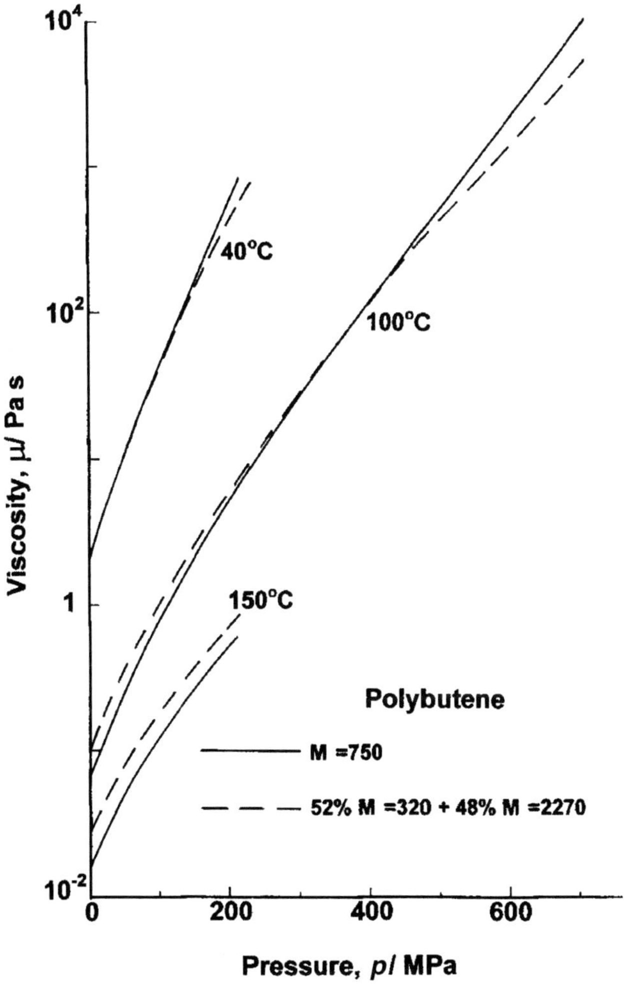

Fig. 6.14. The viscosity of polybutene base oils of the same ambient-pressure viscosity at 40°C. Solid curve is a straight cut. Broken curve is a blend of two different molecular weight cuts. Reproduced from Bair, S., " The High-Pressure Rheology of Mixtures", *ASME J. Tribol.*, Vol. 126, No. 4, 2004, pp. 697–702 by permission of the American Society of Mechanical Engineers.

References

[1] Wilson, D.R., Marshall, W.A., Dolle, R.E. and Benzing, R.J., "Physical, Chemical, and Mechanical Characteristics of a Polyphenyl Ether", *I & EC Product Research and Development*, Vol. 6, No. 2, 1967, pp. 81–88.

[2] Andrade, E.N. and da, C., "The Viscosity of Liquids", *Nature*, Vol. 125, 1930, pp. 309–310.

[3] Angell, C.A., "Formation of Glasses from Liquids and Biopolymers", *Science*, Vol. 276, 1995, pp. 1924–1935.

[4] Angell, C.A., "Relaxation in Liquids, Polymers and Plastic Crystals-Strong/Fragile Patterns and Problems", *J. Non-Cryst. Solids*, Vol. 131–133, 1991, pp. 13–31.

[5] Williams, M.L., Landel, R.F. and Ferry, J.D., "The Temperature Dependence of Relaxation Mechanisms in Amorphous Polymers and Other Glass-Forming Liquids", *J. Amer. Chem. Soc.*, Vol. 77, No. 14, 1955, pp. 3701–3707.

[6] Harrison, G., *The Dynamic Properties of Supercooled Liquids*, Academic Press, London, 1976, p. 21.

[7] Xu, W., Cooper, E.I. and Angell, C.A., "Ionic Liquids: Ion Mobilities, Glass Temperatures, and Fragilities", *J. Phys. Chem. B.*, Vol. 107, No. 25, 2003, pp. 6170–6178.

[8] Paluch, M., Denzik, Z. and Rzoska, S.J., "Scaling of High-Pressure Viscosity Data in Low-Molecular-Weight Glass-Forming Liquids", *Phys. Rev. B.*, Vol. 60, No. 5, 1999, pp. 2979–2982.

[9] Bohmer, R., Ngai, K.L., Angell, C.A. and Plazek, D.J., "Non-Exponential Relaxations in Strong and Fragile Glass Formers", *J. Chem. Phys.*, Vol. 99, No. 5, 1993, pp. 4201–4209.

[10] Reiser, A., Kasper, G. and Hunklinger, S., "Effect of Pressure on the Secondary Relaxation in a Simple Glass Former", *Phys. Rev. Lett.*, Vol. 24, No. 12, 2004, pp. 125701-1–125701-4.

[11] Johari, G.P. and Whalley, E., "Dielectric Properties of Glycerol in the Range 0.1–10^5 Hz, 218–357K, 0–53kb", *Faraday Symp. of Chem. Soc.*, Vol. 6, 1972, pp. 23–41.

[12] Roland, C.M., Hensel-Bielowka, S., Paluch, M. and Casalini, R., "Supercooled Dynamics of Glass-Forming Liquids and Polymers Under Hydrostatic Pressure", *Reports on Progress in Physics*, Vol. 68, 2005, pp. 1405–1478.

[13] Johnson, K.L. and Tevaarwerk, J.L., "Shear Behaviour of Elastohydrodynamic Oil Films", *Proc. Roy. Soc. London A*, Vol. 356, 1977, pp. 215–236.

[14] Munro, R.G., Block, S. and Piermarini, G.J., "Correlation of the Glass Transition and the Pressure Dependence of Viscosity in Liquids", *J. Appl. Phys.*, Vol. 50, No. 11, 1979, pp. 6779–6783.

[15] Bair, S. and Qureshi, F., "Accurate Measurements of Pressure-Viscosity Behavior in Lubricants", *STLE Tribol. Trans.*, Vol. 45, 2002, pp. 390–396.

[16] Blok, H., "Inverse Problems in Hydrodynamic Lubrication and Design Directives for Lubricated Flexible Surfaces", *Proceedings International Symposium on Lubrication and Wear*, Houston, Texas; Muster, D. and Sternlicht, B. (Eds.), McCutchan Publishing, Berkeley, 1963, pp. 7–79.

[17] Bair, S. and Gordon, P., "Rheological Challenges and Opportunities for EHL", *Proc. IUTAM Symp. Elastohydrodynamics and Micro Elastohydrodynamics*, Springer, 2006, pp. 23–43.

[18] Bair, S., Liu, Y. and Wang, Q.J., "The Pressure-Viscosity Coefficient for Newtonian EHL Film Thickness with General Piezoviscous Response", *ASME J. Tribol.*, Vol. 123, No. 3, 2006, pp. 624–631.

[19] Kottke, P., Bair, S. and Winer, W.O., "The Measurement of the Viscosity of Liquids Under Tension", *ASME J. Tribol.*, Vol. 125, 2003, pp. 260–266.

[20] Bair, S., "An Experimental Verification of the Significance of the Reciprocal Asymptotic Isoviscous Pressure", *STLE Tribol. Trans.*, Vol. 36, 1993, pp. 153–162.

[21] Cohen, R., "The Effect of Pressure on the Viscosity of Liquids", *Wiedmanns Annalen.*, Vol. 44, 1892, pp. 666.

[22] Barus, C., "Isothermals, Isopiestics and Isometrics Relative to Viscosity", *Am. J. Sci.*, Third Series, Vol. XLV, No. 266, 1893, pp. 87–96.

[23] Hersey, M.D., "The Theory of the Torsion and the Rolling Ball Viscometers and their Use in Measuring the Effect of Pressure on Viscosity", *J. Washington Academy Sci.*, Vol. VI, 1916, pp. 525–531.

[24] Bridgman, P.W., "The Effect of Pressure on the Viscosity of Forty-Three Pure Liquids", *Proc. Amer. Acad.*, Vol. 61, 1926, pp. 57–99.

[25] Skinner, S.M., "Film Lubrication of Finite Curved Surfaces", *J. Appl. Phys.*, Vol. 9, 1938, pp. 409–421.

[26] McEwen, E., "The Effect of Variation of Viscosity with Pressure on the Load-Carrying Capacity of the Oil Film Between Gear-Teeth", *J. Inst. Petroleum*, Vol. 38, No. 344–345, 1952, pp. 646–672.

[27] Roelands, C.J.A., Vlugter, J.C. and Waterman, H.I., "The Viscosity-Temperature-Pressure Relationship of Lubricating Oils and Its Correlation with Chemical Constitution", *ASME J. Basic Engr.*, Vol. 11, 1963, pp. 601–611.

[28] Roelands, C.J.A., *Correlational Aspects of the Viscosity-Temperature-Pressure Relationship of Lubricating Oils*, Ph.D. Thesis, University of Technology, Delft, 1966, pp. 94–95.

[29] Irving, J.B. and Barlow, A.J., "An Automatic High Pressure Viscometer", *J. Phys. E.*, Vol. 4, No. 3, 1971, pp. 232–236.

[30] Bair, S. and Kottke, P., "Pressure-Viscosity Relationships for Elastohydrodynamics", *STLE Tribol. Trans.*, Vol. 46, 2003, pp. 289–295.

[31] Oels, H.J. and Rehage, G., "Pressure-Volume-Temperature Measurements on Atactic Polystyrene, A Thermodynamic View", *Macromolecules*, Vol. 10, No. 5, 1977, pp. 1036–1043.

[32] Caetano, F.J.P., Fareleira, J.M.N.A., Oliveira, C.M.B.P. and Wakeham, W.A., New Measurements of the Viscosity of Diisodecyl Phthalate Using a Vibrating Wire Technique", *J. Chem. Engr. Data*, Vol. 50, 2005, pp. 1875–1878.

[33] Doolittle, A.K., "Studies in Newtonian Flow. II. The Dependence of the Viscosity of Liquids on Free-Space", *J. Appl. Phys.*, Vol. 22, No. 12, 1951, pp. 1471–1475.

[34] Cohen, M.H. and Turnbull, D. "Molecular Transport in Liquids and Gases", *J. Chem. Phys.*, Vol. 31, No. 5, 1959, pp. 1164–1169.

[35] Loomis, R.L. and Fusaro, R.L., "Liquid Lubricants for Advanced Aircraft Engines", NASA TM-104531, National Aeronautics and Space Administration, Lewis Research Center, 1992.

[36] Casalini, R. and Roland, C.M., "Viscosity at the Dynamic Crossover in o-Terphenyl and Salol under Pressure", *Phys. Rev. Lett.*, Vol. 92, No. 24, 2004, pp. 245702-1–245702-4.

[37] Et-Tahir, A., Boned, C., Lagourette, B. and Xans, P., Determination of the Viscosity of Various Hydrocarbons and Mixtures of Hydrocarbons versus Temperature and Pressure", *Int. J. Thermophys.*, Vol. 16, No. 6, 1995, pp. 1309–1334.

[38] Mamagakis, N. and Panayiotou, C., "Excess Volume and Dynamic Viscosity of Ternary Liquid Mixtures", *Zeitschrift fur Physikalische Chemie Neue Folge*, Vol. 162, 1989, pp. 57–72.

[39] Hirst, W. and Moore, A.J., "The Effect of Temperature on Traction in Elastohydrodynamic Lubrication", *Phil. Trans. Roy. Soc. London Series A*, Vol. 298, No. 1438, 1980, pp. 183–208.

[40] Bair, S., "The High-Pressure Rheology of Mixtures", *ASME J. Tribol.*, Vol. 126, No. 4, 2004, pp. 697–702.

CHAPTER 7

Measurement Techniques for the Shear Dependence of Viscosity at Elevated Pressure

7.1 Introduction

Elastohydrodynamics is apparently a field that is unique for the necessity of assuming shear-dependent viscosity even for low molecular weight liquids. When the pressure dependence of the low-shear viscosity is accurately accounted for, traction and, at times, film thickness cannot be explained by the Newtonian assumption. For many working in EHL, the effect of shear-thinning on traction may be found to be disappointingly weak and on film thickness to be surprisingly strong. It is only at relatively low Hertz pressures that shear-thinning can fully explain the nonlinearity in the relation between traction and sliding velocity for isothermal response. At higher pressures, in point contact, where the logarithmic traction behavior described by equation (1.19) presents itself,

$$\bar{\tau} = \tau_E \sinh^{-1}(C_E \Sigma) \tag{1.19}$$

shear-thinning has the surprising effect of reducing the value of C_E in a way that would be similar to a reduction in the pressure dependence of the viscosity. For the highest pressure contacts, where the small-slip behavior is dominated by the roller compliance, shear-thinning has little effect on the traction.

Shear-thinning can be said to occur for a given liquid when the Weissenberg number, $Wi = \lambda\dot{\gamma}$, becomes greater than unity. Many experimental measurements on polymeric liquids and microstructual theories [1] put the value of the characteristic time, λ, at approximately

$$\lambda = \frac{\mu}{G} \approx \frac{\mu M}{\rho R_g T} \tag{7.1}$$

where M is molecular weight, ρ is mass density and R_g is the universal gas constant. Nonequilibrium molecular dynamics simulations for low-molecular-weight liquids have shown that λ is the rotational relaxation time for the molecule [2]. Equation (7.1) is the

Einstein–Debye relation for the rotational relaxation time for a molecule. For the material to be shear-thinning, $\lambda\dot{\gamma} > 1$, we must have the shear stress, $\tau > G \approx \rho R_g T/M$ in a simple shear experiment. As an example, for the standard reference material, SRM2491, of Fig. 2.4 the average molecular weight is $M = 308{,}000$ kg/kmol [3] and $\rho = 980$ kg/m^3. Then for a temperature of 300 K, equation (7.1) yields $G \approx 7900$ Pa which is intermediate to the value of 4100 Pa that was used by Schultheisz et al. [3] to fit the Carreau equation (Table 2.1) and the value of $G = 9000$ Pa that was used to fit the corotational Maxwell model in Fig. 2.4 to experimental measurements of shear-thinning.

Now, for a mineral oil of similar density having an average molecular weight of $M = 600$ kg/kmol, $G \approx 4$ MPa and shear-thinning will not be observed until the shear stress exceeds about 4 MPa. This is an extremely high shear stress relative to the capability of conventional viscometers and rheometers, but this is a relatively low shear stress for lubricated concentrated contacts. A typical traction coefficient of $\bar{\tau}/\bar{p} = 0.03$ with average pressure of 1 GPa yields an average stress of 30 MPa in the load-bearing Hertz zone and the viscosity of the mineral will clearly be shear-dependent in this region of the contact.

The effect of molecular weight on the stress at the onset of shear-thinning is demonstrated by the flow curves obtained with a pressurized Couette viscometer [4] for various base oils of well-defined chemical structure and shown in Fig. 7.1. DMS is a polydimethylsiloxane, BCH is 2,4-dicyclohexyl-2-methylpentane (see Table 4.2), and DOP is di(2-ethylhexyl)phthalate. This instrument is capable of applying a shear stress of as much as 20 MPa without significant thermal softening owing to the very thin films used (of the order of 1 μm) and the very large value of the low-shear viscosity. The shear stress, τ,

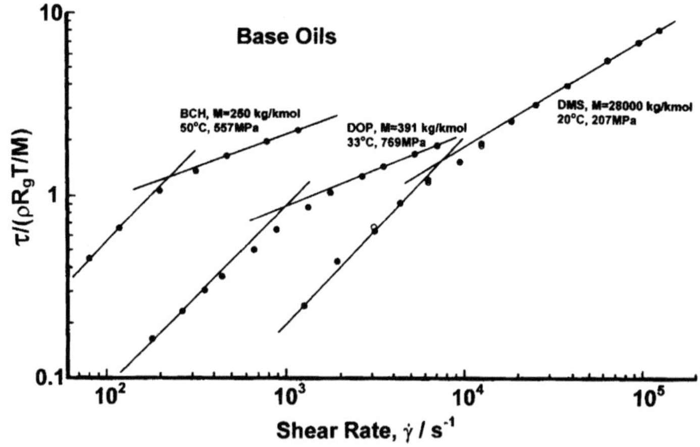

Fig. 7.1. Flow curves generated with a high-pressure Couette viscometer for three lubricant base oils of very different molecular weights. DMS is a polydimethylsiloxane, a very high molecular weight oil that is well-known to generate films much thinner than the Newtonian prediction. BCH is a traction fluid base oil and DOP is a diester. Adapted from [4]. Reprinted with permission of Elsevier.

on the vertical axis of Fig. 7.1 has been normalized by the theoretical value of G. The molecular weights vary by about 100–1, but the normalization of shear stress by $\rho R_g T/M$ brings the onset of shear-thinning to approximately the same value of normalized stress equal to unity. Clearly, the shear-thinning behavior of the low-molecular weight liquids, BCH and DOP, is no different, except for the magnitude of the stress, from that of the liquid polymer, DMS.

For polymer solutions the relation is slightly different,

$$\lambda = \frac{\mu}{G} \approx \frac{\mu M}{c\rho R_g T} \tag{7.2}$$

where M is now the molecular weight of the polymer and c is the weight fraction concentration of the polymer. Now for $\lambda\dot{\gamma} > 1$, we have the shear stress, $\tau > G \approx c\rho R_g T/M$. For example, with the standard reference material [5], SRM 2490, of Fig. 2.5 the number average molecular weight of the polymer additive is $M = 600{,}000$ kg/kmol, $\rho = 800$ kg/m^3 and the concentration is $c = 0.114$. Then for a temperature of 300 K, equation (7.2) yields $G \approx 380$ Pa, which is very close to the value of $G = 428$ Pa used in fitting the cross equation in Fig. 2.5. As concentration, c, goes to zero the stress at which the nonlinearity begins, G, goes to zero which does not seem sensible except that the power-law exponent, n, then goes to one.

Multigrade lubricants are made less temperature dependent by blending a base oil, the solvent in this case, with a small concentration of a high molecular weight polymer. For a multigrade gear oil the polymer molecular weight may be about 30,000 kg/kmol with a concentration of about 15%. Here $G \approx 10{,}000$ Pa. For a multigrade motor oil the molecular weight is generally higher and the concentration lower so that G may be less than 1000 Pa. Motor oils often show a second Newtonian while gear oils will generally reach the stress at which the base oil shear thins without displaying a second plateau.

Greenwood and Kauzlarich [6] found that the maximum inlet zone shear rate in a rolling line contact for an incompressible power-law fluid with exponent, n, is

$$\dot{\gamma}_{\max} = \left(\frac{1}{n} + 2\right)\frac{\bar{u}}{2h_c} \tag{7.3}$$

With a Newtonian liquid, $n = 1$, and the maximum shear rate occurs in the inlet at the position where $h = 2h_c$ and $\alpha p \approx 1$ [6]. Then the shear stress in the inlet zone where the shear rate is greatest is about $3\mu(p = 1/\alpha)\bar{u}/(2h_c) \approx 4\mu_0\bar{u}/h_c$, which works out to be about 2×10^5–2×10^6 Pa. This is not, however, the maximum shear stress occurring in the inlet zone of a rolling contact because the pressure and, therefore, the viscosity are increasing rapidly downstream of this location. Numerical solutions of the inlet zone problem show that the maximum shear stress occurs much closer to the Hertz boundary and reaches to about 5×10^5–5×10^6 Pa. Therefore polymer blended lubricants and high molecular weight base oils with G much smaller than this maximum stress will shear thin through a large portion of the inlet. The film will then be thinner than the Newtonian prediction.

7.2 Phenomena Producing Behavior Similar to Shear-Thinning

It should be helpful before going further to define specifically what we mean by shear-dependent viscosity or shear-thinning as there has been some confusion within the EHL literature on this subject. A discussion of similar phenomena that are not shear-thinning will be useful to this end. In a measurement of the viscosity of a liquid, a flow is produced that is, in at least the vicinity of the measurement, free of extensional kinematics and on a microscopic scale can be visualized as surfaces within the liquid sliding with respect to each other. The sliding velocity divided by the separation between the surfaces is the local shear rate. The apparent viscosity is reported as the shear stress on the material surfaces divided by the shear rate.

At low shear rate (or stress), the viscosity is found to be independent of the shear rate (or stress). At high shear rate (or stress), the apparent viscosity may be observed to decrease with shear rate (or stress). The reduction in apparent viscosity with shear may be shear-thinning. For the case of shear-thinning, the apparent viscosity measured at a particular stress or rate is immediately repeatable, even after a measurement at a higher stress. Then the term, shear-thinning, does not include either thixotropy [7] or molecular degradation. Shear-thinning is constitutive behavior. In isothermal plane Couette shear between parallel plates, the shear rate is uniform across the gap between the plates and equal to the difference in velocities of the plates divided by the gap. Therefore shear-thinning does not include the thermal softening that results from viscous heating. It also does not include the effect of slip, whether within the liquid or at the interface between the liquid and a solid boundary. Then shear-thinning also does not include effects of cavitation.

7.2.1 Thixotropy

Thixotropic materials display a viscosity that decreases with shear rate and time of shear as well. They possess a solid structure that requires time to form under high-pressure and that structure is disrupted by shear stress. For a truly thixotropic material (in contrast to false body) this structure is completely removed at a sufficiently large shear stress so that, for a time, Newtonian response can be obtained. The structure recovers again given sufficient time with the stress removed. Thixotropy is often observed in falling body viscometers during measurements on paraffinic mineral oils where the stress applied by the sinker is too small to disrupt the structure of crystallizing wax. Couette viscometers may apply sufficiently high stress that the thixotropy in paraffinic mineral oils goes unnoticed. An example of the thixotropic behavior of a solution of 5% by weight of normal tetracosane, a waxy solid at ambient conditions, in squalane is shown in Fig. 7.2. The pressurized Couette viscometer is described below and a complete description of the experiment can be found in Ref. [8]. The solid structure seems to be completely formed in about one hour at the pressure of 207 MPa and can be seen in Fig. 7.2 to be disrupted

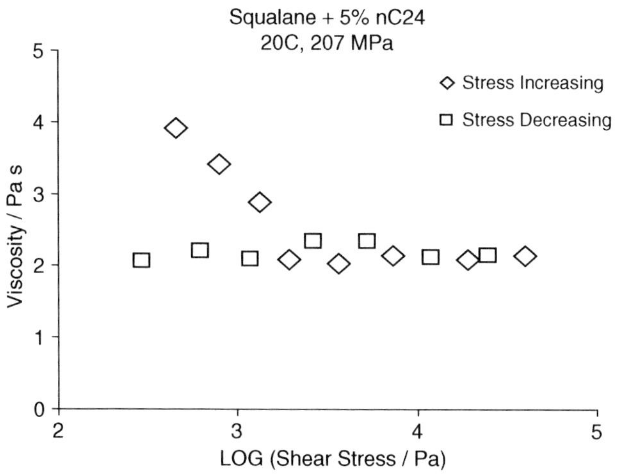

Fig. 7.2. Flow curves generated with a high-pressure Couette viscometer for squalane and wax solution showing viscosity dependent upon stress history, that is, thixotropy.

for shear stress greater than about 2000 Pa. Then for lower stress, the viscosity is history dependent. This response is not shear-thinning.

7.2.2 Molecular degradation

Molecular degradation describes a chemical change in the material brought about by shear. Chemical bonds are broken, resulting in a lower molecular weight and a lower viscosity, often called *permanent viscosity loss* as opposed to the temporary viscosity loss of shear-thinning. The original viscosity cannot be recovered, even after a long time without shear.

An example of what is believed to be molecular degradation in a pressurized Couette viscometer is shown in Fig. 7.3 using the NIST non-Newtonian standard reference material [5], SRM2490, a solution of high-molecular-weight polyisobutylene in pristane, a low-molecular-weight branched alkane. The first set of measurements was performed at $p = 5$ MPa and agrees reasonably well with the shear-thinning viscosity values certified by NIST at ambient pressure. Following the highest stress measurement of 1.4 kPa at this low pressure two of the lower stress measurements were repeated within about 2%. Then flow curves were generated in order of increasing pressure and increasing maximum shear stress. At a pressure of 100 MPa the maximum stress reached 2.0 kPa and, from the shape of the flow curve, it appears that the viscosity is approaching a second Newtonian plateau. The next flow curve at $p = 200$ MPa also shows the development of a second Newtonian plateau that is abruptly terminated at a stress of 4.8 kPa where the viscosity decreases again up to the maximum stress of 7.1 kPa. The next flow curve at $p = 300$ MPa shows a similar process occurring where the plateau terminates at a stress of 6.1 MPa and the maximum stress reaches to 9.9 kPa.

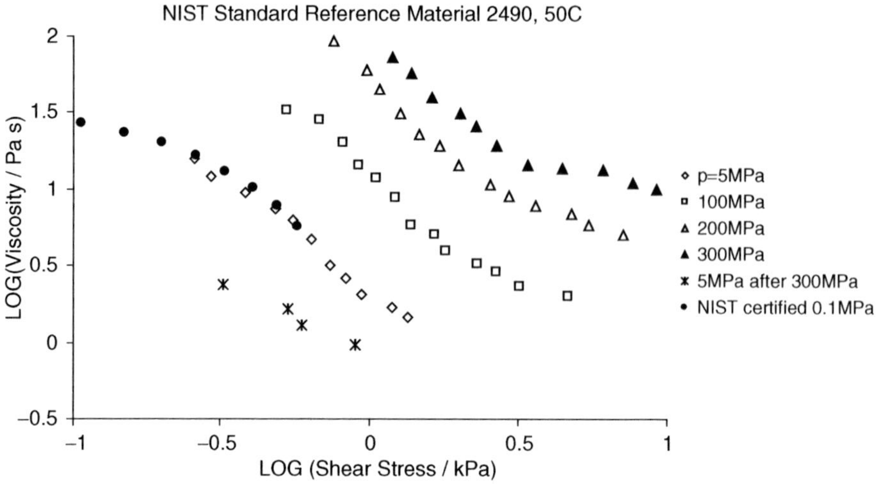

Fig. 7.3. Flow curves generated with a high-pressure Couette viscometer for a non-Newtonian Standard Reference Material, SRM2490, that is supplied by and the viscosities are certified by the NIST. Measurements were taken in order of increasing pressure. Then the $p = 5$ MPa measurement was repeated. There is evidence of molecular degradation in the permanent change in viscosity at 5 MPa following shear at high shear stress and in the departure from the second Newtonian plateau in the 300 MPa data.

After the high pressure measurements, new measurements were taken at the pressure of 5 MPa and are shown in Fig. 7.3. Now the flow curve is very different from the one supplied by NIST and the first one generated in the pressurized Couette viscometer. The new viscosity is both lower than the original and less sensitive to shear stress than the original viscosity. Both observations are consistent with a reduction in molecular weight of the polymer in solution. Then the increased shear dependence of the viscosity shown in Fig. 7.3 in the 200 and 300 MPa curves at the highest stresses is not shear-thinning.

7.2.3 Thermal-softening resulting from viscous heating

Any conventional measurement of the shear viscosity requires that a shear flow be generated. In a shear flow there is dissipation and therefore a temperature increase of the liquid. When the shear stress is low, as in a measurement of the limiting-low-shear viscosity, μ, at stress less than say about 100 Pa, the resulting temperature increase is sufficiently small compared with the reciprocal of the temperature–viscosity coefficient that the thermal-softening may be ignored without checking. For the measurement of shear-thinning in typical lubricants where the behavior of interest only begins at stresses from 10^3 to 10^7 Pa, the challenge for the experimenter is to provide a measurement that is sufficiently close to the isothermal ideal that the shear-thinning is not obscured by the thermal-softening that, on first look, itself appears to be a shear dependence of viscosity. Unfortunately, the EHL literature provides multiple examples of thermal-softening that has been mistaken for shear-thinning. See Ref. [9] for an example in a short-time split Hopkinson torsion bar experiment.

The effect of thermal softening on a capillary viscosity measurement is shown in Fig. 7.4 where the Winer and Novak [10] viscometer has been employed in the author's laboratory for a measurement under elevated pressure on a polyalphaolefin. For capillary flow the wall shear stress is given by equation (5.5), repeated here.

$$\tau_W = \frac{D_i}{4}\frac{dp}{dx} \tag{5.5}$$

The apparent shear rate at the wall is obtained from the assumption of a uniform viscosity along the capillary radius, isothermal and Newtonian flow, and is

$$\dot{\gamma}_{app} = 32Q/\pi {D_i}^3 \tag{7.4}$$

There are techniques for correcting the wall shear rate in capillary flow for shear-thinning. (See Macosko [7], pp. 240–244.) The nonlinearity in the flow curves of Fig. 7.4 is, however, not the result of nonlinear constitutive behavior but viscous heating.

Two different capillaries have been employed in Fig. 7.4. The effectiveness of thermal conduction in removing the heat of viscous dissipation varies inversely with the square of the capillary inside diameter, $1/D_i^2$. The Nahme–Griffith number, which is for

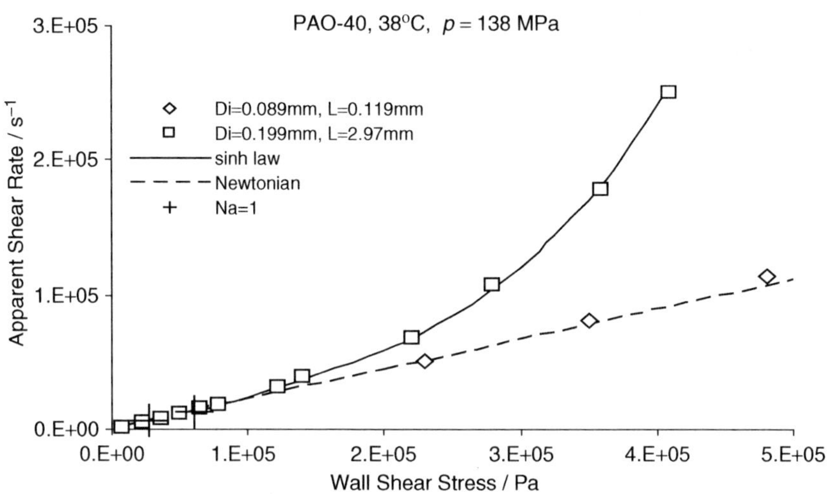

Fig. 7.4. The apparent shear rate plotted against the wall shear stress for a polyalphaolefin. Data generated with a high-pressure capillary viscometer using two different diameter capillaries. The large capillary is more severely affected by viscous heating than the smaller capillary resulting in apparent sinh-law response.

capillary flow [7]

$$Na = \frac{\beta \tau_W \dot{\gamma}_{app} D_i^2}{16k} \tag{7.5}$$

determines how much the viscous heating will affect the viscosity that is measured for a liquid with thermal conductivity, k. For long capillaries, viscous heating significantly affects the measurement when $Na > 1$. For short capillaries there is insufficient time for development of a steady temperature distribution and the effect of viscous heating is delayed to higher values of Na. The disadvantage of the short capillary is the greater need to correct for end effects. The position within the data for the two capillaries where $Na = 1$ is marked by "+" in Fig. 7.4 for each capillary. The capillary with the larger diameter, having $Na = 1$ closest to the origin, shows the greatest departure from linearity. The Newtonian (broken) curve in Fig. 7.4 represents

$$\dot{\gamma}_{app} = \frac{\tau_W}{\mu} \tag{7.6}$$

where $\mu = 4.5$ Pa s is the viscosity that has been measured by a falling body viscometer. The smaller capillary shows agreement with the linear equation (7.6) to a stress of about 0.3 MPa. The solid curve, fitted to the larger capillary data in the figure, represents a sinh-law,

$$\dot{\gamma}_{app} = \frac{\tau_E}{\mu} \sinh\left(\frac{\tau_W}{\tau_E}\right) \tag{7.7}$$

where the representative stress is $\tau_E = 0.15$ MPa, about fives times the stress for which $Na = 1$. Now this equation (7.7) accurately describes the shear stress dependence of the apparent shear rate for a capillary operating in a regime for which the temperature distribution across the capillary radius severely affects the flow but it does not describe constitutive behavior in this case.

Figure 7.4 and the explanations given above for the observed viscous heating in capillary viscometry have a special significance to a series of misunderstandings that have had a profound effect upon the EHL field. In 1941, Norton et al. [11] generated flow curves for liquid lubricants using a capillary viscometer which maintained both the inlet and outlet to the capillary at elevated pressures. These capillaries had an inside diameter of $D_i = 0.456$ mm and very large lengths of $L = 161$–790 mm. It should not be surprising, then, from the example above that nonlinear flow response was observed by Norton [11] for $Na > 1$. This nonlinearity was, however, mistakenly attributed to thixotropy of the oils thought to be due to the formation of a solid structure. Whorlow [12] has found that there have been many published reports of thixotropic behavior which were, in reality, the result of temperature variations. In a 1958 paper, Hahn, Eyring et al. [13] found that equation (7.7) reproduced the Norton data with $3 < \tau_E < 30$ kPa depending on the viscosity and proposed that the stress-increasing regime of thixotropic flow should be accurately modeled with equation (7.7). This idea was introduced to EHL with a paper by Bell et al. who in a 1964 paper [14] mistakenly cited the 1958 Eyring paper [13] on thixotropy to support the use of a sinh-law description of shear-thinning. Motivated by the success of the sinh-law in the form of equation (1.19),

$$\bar{\tau} = \tau_E \sinh^{-1}(C_E \Sigma) \tag{1.19}$$

which is similar in appearance to equation (7.7), for describing the area averaged shear stress in a regime of traction measurement, many workers have continued to use this thixotropy model to represent shear-thinning in EHL films.

Apparent sinh-law behavior can also be found when the maximum shear stress is reported for highly transient applications of high shear rate to a thick liquid film in a split Hopkinson bar experiment [9]. The liquid responds as in adiabatic homogeneous shear with the stress reaching a sharp peak and then rapidly decreasing with time. A plot of the peak stress against shear rate can be described by an equation of the form of (7.7).

The effect of viscous heating on a continuous circular Couette shear experiment is somewhat different in appearance. If the temperature of the material of the cylinders does not change significantly with time, then the stress will be constant with time on the time scale of typical measurements. The temperature distribution within the film reaches steady state quickly. However, if the apparent shear rate is increased from measurement to measurement until viscous heating is significant, the imposed stress will show a maximum value and eventually decrease with increasing apparent shear rate so that a double-value of apparent shear rate is obtained for a given stress [15]. The apparent shear rate for a Couette shear experiment is

$$\dot{\gamma}_{app} = \frac{|u_1 - u_2|}{h} \tag{7.8}$$

7.2.4 Shear cavitation

Published data for the shear rate dependence of dimethylsilicone oils at atmospheric pressure often show viscosity for a number of different viscosity grades converging along a curve that represents approximately $\tau = 10^5$ Pa or $\tau = p = 1$ atmosphere. This behavior has the appearance of the limiting shear stress that has been observed in EHL traction where the shear stress averaged over the contact area becomes independent of sliding velocity and generally does not exceed some material specific fraction of the pressure averaged over the contact. The origins of these behaviors are, however, very different. If the viscosity measurement of the silicone oil is performed at variable pressure, an unusual sensitivity to pressure is observed as shown in Fig. 7.5 for a 60,000 cS grade oil. These data were obtained using pressurized Couette viscometers of the type described by Winer and Bair [16] who were apparently first to set forth the principal normal stress cavitation criterion; if a liquid has zero tensile strength, then cavitation is predicted to occur when the largest principal stress becomes tensile. Therefore, for a generalized Newtonian liquid, cavitation will occur when $|\tau| > p$. For a non-Newtonian liquid,

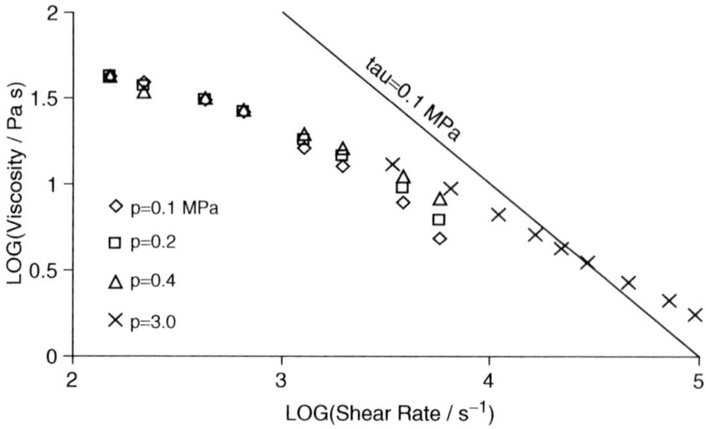

Fig. 7.5. The apparent viscosity plotted against apparent shear rate of a silicone oil measured with a pressurized Couette viscometer. Viscosities measured at atmospheric pressure for this liquid polymer generally approach a line representing the shear stress equated to the atmospheric pressure as shown in the figure.

the cavitation criterion is

$$\left(N_1 + \sqrt{N_1^2 + 4\tau^2}\right)\Big/2 > p \tag{7.9}$$

Silicone polymers can generate large first normal stress difference, N_1, as shown by Fig. 2.4. In Fig. 7.5 the apparent viscosity diverges for pressures less than four times atmospheric as shear stress increases toward the value equal to atmospheric pressure. Measurements at lower pressure diverge at lower stress. The measurements at $p = 3$ MPa continue to show ordinary power-law response to stress well above 10^5 Pa.

Kottke [17] performed a similar measurement as that shown in Fig. 7.5 in a pressurized Couette viscometer with the outer cylinder fabricated from transparent glass so that the cavitation could be visualized. His results are shown in Fig. 7.6 for a polybutene which is Newtonian at these low values of shear stress. This remarkable visualization clearly shows the formation of cavitation voids when the liquid is subjected to a shear stress that exceeds the pressure. Analysis suggests that the cavitation is gaseous and originates from a preexisting nucleus [18]. Whatever the origin of the cavitation, the criterion (7.9) provides a useful test to determine whether a measurement suffers from this departure from constitutive behavior.

7.3 Rheometers for High Pressure

The rheology of polymer melts at elevated pressures has always been of some interest for polymer processing. Experimental investigations of the rheology of polymer melts have utilized rotating Couette and capillary instruments. Semjonow [19] measured the shear

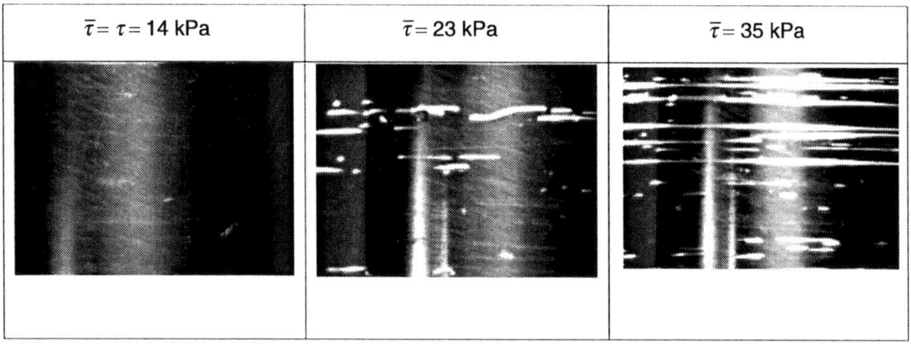

Fig. 7.6. Cavitation occurs in the viscometer gap when the average shear stress exceeds the pressure. Photographs taken through a transparent glass outer cylinder. The pressure was maintained at a subambient 19 kPa. From Kottke, P.A., Bair, S. and Winer, W.O., "Cavitation in Creeping Shear Flows", *AICHE Journal*, Vol. 51, No. 8, 2005, pp. 2150–2170. Reproduced with permission of John Wiley & Sons.

rheology of polyethylene and polystyrene melts to pressures of 100 MPa. Cogswell [20] pressurized a Couette rheometer to 180 MPa and generated flow curves for polypropylene and polyethylene melts. Early pressurized Couette rheometers used the outer cylinder as a pressure vessel. The gap then varies with pressure and if the gap is large this variation is relatively small. Large gaps limit the viscous power, $\tau\dot{\gamma}$, for which the measurement may be considered to be isothermal.

The high-pressure capillary viscometer of Novak and Winer [10] was described in Chapter 5 and shown in Fig. 5.2. This instrument, developed in the 1960s, has provided measurements of the shear dependence of viscosity for shear stress from about 10 to 10^6 Pa and to pressures of about 550 MPa. Flow curves generated from the Novak and Winer instrument are shown in Fig. 7.7 where, for a polymer blended mineral oil typical of multigrade lubricants, two plateaus can be seen to be developing. Binding et al. [21] have developed a high-pressure capillary rheometer for the measurement of both shear and extensional viscosity, η and η_e, and have published flow curves for multigrade oils to pressures of 80 MPa showing power-law response to shear over about two decades of shear rate for both viscosities.

7.3.1 The pressurized Couette rheometer

The pressurized Couette viscometer as used in the author's laboratory has become the standard instrument for the observation of shear-dependent viscosity in liquids of the molecular weight of liquid lubricant base stocks at elevated pressure. The data of Fig. 7.1 were generated with this type of instrument. In the conventional rotational Couette viscometer for ambient pressure, the liquid sample only occupies the space between the concentric cylinders and possibly within the bottom of the outer cylinder called the cup. An air/liquid meniscus exists along the top end of the outer cylinder. For a pressurized instrument, obviously there can be no air/liquid meniscus. At a relatively low pressure

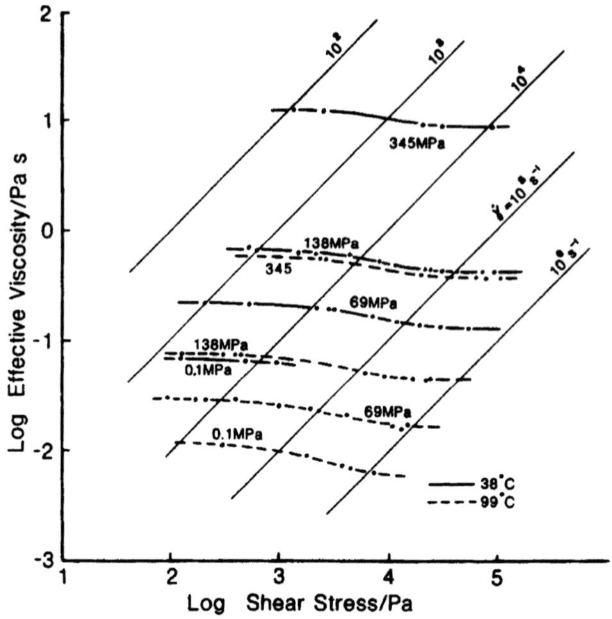

Fig. 7.7. Flow curves generated with a high-pressure capillary viscometer for a mineral oil blended with 4% polyalkylmethacrylate. Adapted from Novak, J.D. and Winer, W.O., "Some Measurements of High Pressure Lubricant Rheology", *ASME J. Lubr. Techn.*, Vol. 90, Ser. F, No. 3, 1968, pp. 580–591, by permission of the American Society of Mechanical Engineers.

air bubbles are collapsed to the point of disappearance and probably enter into solution in the oil. Therefore the concentric cylinder pair is contained within a vessel which is filled with liquid as shown in Fig. 7.8.

For the pressurized instrument, the cylinder pair is completely immersed in liquid. Two types of cylinder pairs are shown in Fig. 7.9; a set with a long working surface in the axial direction and large working diameters for low stress and a set with short working surface and small working diameters for high stress. The inner cylinders in this case are rotated by a drive shaft from above that is flexibly coupled to the inner cylinder at the hexagonal drive socket shown in the figure. The drive shaft is attached by similar means to the connection to the stepper motor shown in Fig. 7.8. The connection to the motor passes through a rotary seal of the spring energized type described in Section 3.4.1. The outer cylinders in Fig. 7.9 are fastened to a splined spring assembly shown below them in Fig. 7.8. The spring maintains the relative axial position of the two cylinders and the spline provides the torque transfer to the transducer below. The torque is measured by strain gages bonded to a thin aluminum cylinder. The electrical connections between the strain gages within the vessel and a signal conditioner outside pass through an electrical feed-through of the swaged wire cable type described in Section 3.4.2.

In the design shown in Fig. 7.8, there are two liquids occupying the pressure vessel. The sample under study resides in the upper portion beginning at the level marked in

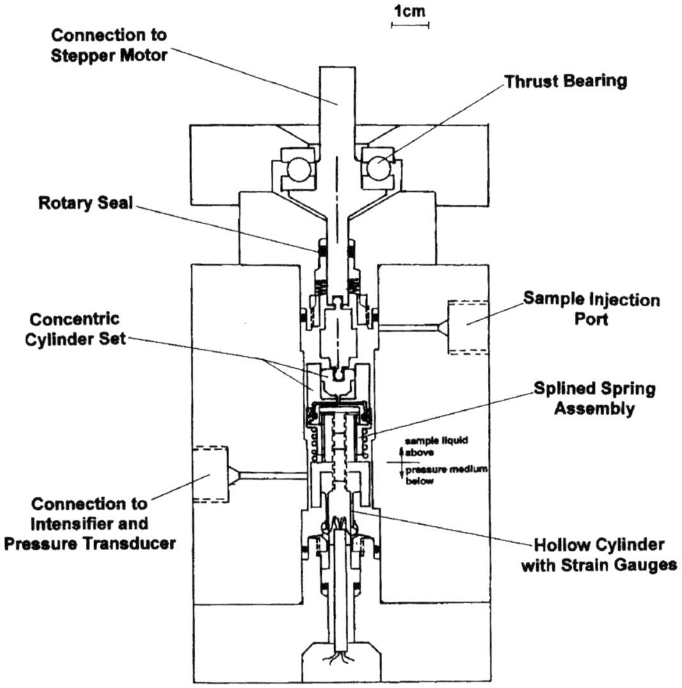

Fig. 7.8. A pressurized Couette viscometer for pressure to 700 MPa. A perfluorinated hydrocarbon occupies the lower volume to the position indicated.

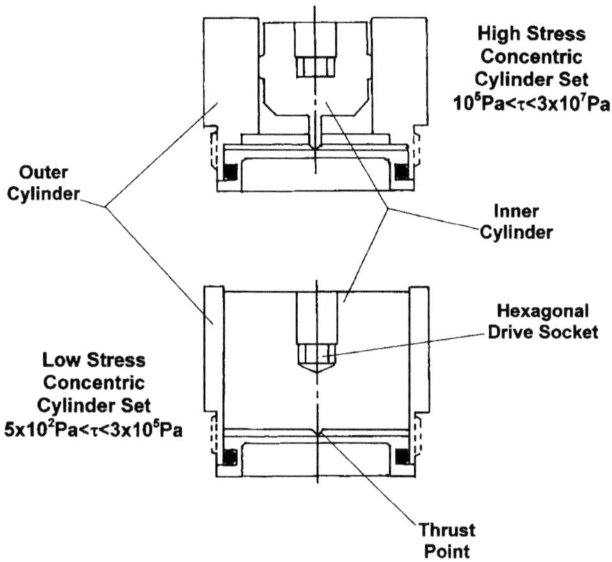

Fig. 7.9. The design of concentric cylinder pairs for the high shear stress needed for base stocks, top and for the low shear stress needed for the polymer blends, bottom.

the figure. The lower portion of the vessel is first filled to the mark with an immiscible (with respect to the sample) pressurizing medium of greater density and lower viscosity than the sample. The sample is loaded into the instrument by evacuating the space through the sample injection port. Then the sample is injected. Pressure is generated and measured with the medium though a connection to an intensifier and transducer as shown in Fig. 7.8. The pressurizing medium surrounds the torque transducer in the lower part of the vessel. This transducer responds to the torque to be measured by twisting and the twisting motion is dampened by the viscosity of the pressure medium. In early pressurized Couette viscometers, the thermal softening limitation on a measurement was, in greatest part, due to the response time of the torque measurement which was proportional to the viscosity of the liquid in which the transducer was immersed [22]. This response time has been substantially reduced by making the transducer very stiff and by reducing the viscosity of the pressurizing medium. The pressure dependencies of viscosities of candidate materials for the medium are shown in Fig. 7.10 for temperature

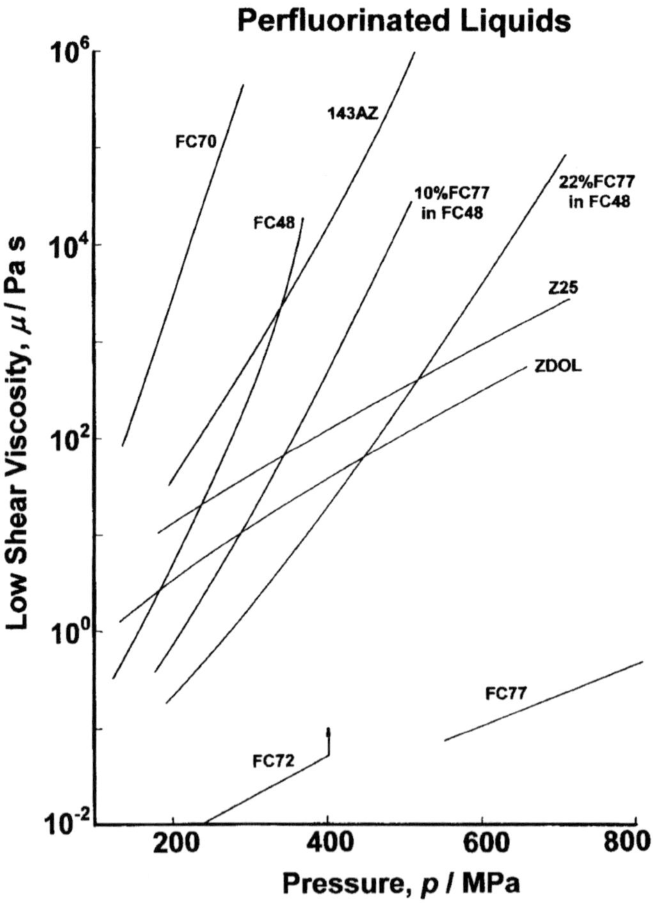

Fig. 7.10. The ambient-temperature pressure dependence of the viscosity of candidate fluorinated hydrocarbons for use as a pressurizing medium.

near ambient. These are all fluorinated oils that are immiscible in organic liquids and possess mass density about twice that of organic liquids. The liquid employed as the pressurizing medium in the present viscometer is FC77, believed to be perfluorinated octane. The FC72, perfluorohexane, solidifies at low pressure shown by the arrow in Fig. 7.10 and attacks rubber seals. The Z25 and the 143AZ are linear and branched perfluorinated polyalkylethers and the ZDOL is a hydroxyl terminated material similar to Z25. FC48 is a commercial liquid for the cooling of electronic equipment and FC70 is perfluorotripentylamine.

Aside from the obvious difficulty of the containment of high-pressure, providing for rotation within the vessel and extracting information regarding the torque acting on the cylinders, all addressed in Chapter 3, the control of thermal softening due to viscous dissipation is the most challenging design requirement. There are two aspects of the thermal problem that are somewhat independent of one another. Since the same volume of liquid is continuously receiving the mechanical work, the only mode of heat transfer out of the film is by conduction through the liquid to the solid cylinder walls. Because the heat transfer from the cylinders to the surrounding liquid outside of the shearing gap is relatively weak, the solid material of the cylinder construction must store the heat that is conducted from the film. The first aspect, of conduction through the film, is improved by providing the smallest practical shearing gap, h. The second aspect, of storing heat within the cylinder material, is optimized by the achievement of a steady shearing measurement in the shortest possible time using the greatest possible thermal conductivity, k_s, for the cylinder material. It happens that the volume heat capacity, ρC_p, does not vary sufficiently between materials to be of much importance relative to the conductivity. Materials for the construction of concentric cylinder pairs are listed in Table 7.1 with relevant properties. The stainless steel is listed for comparison only as this material has been found to be completely unsatisfactory as a cylinder material due to low thermal conductivity. The time required to obtain a steady measurement of the torque depends upon both the acceleration time of the drive motor and the response time of the torque transducer.

Shearing gaps as small as $h = 0.5\,\mu\text{m}$ [23] have been successfully employed with cylinders of about 25 mm working diameter. As a rule, decreasing the gap improves the viscous power, $\tau\dot{\gamma}$, that can be achieved without thermal softening, but repeatability is sacrificed by small gaps. The high-stress cylinder sets illustrated at the top of Fig. 7.9 require that the gap be from 1 to 4 μm and the low stress sets require from 3 to 10 μm. Larger gaps will severely restrict the useful range of viscous power. In practice, the gap

Table 7.1. Materials for concentric cylinders and parallel discs.

Alloy	k_s/W/m/K	ρc_p/MJ/m^3/K	$a_l/10^{-6}\,°\text{C}^{-1}$	Hardness/R_B
316 Stainless Steel	16	4.0	16	80
CA 240 Brass	140	3.3	19	60
Copelmet®, W72Cu28	212	3.0	10	92
Glidcop® AL60, Dispersion strengthened Cu	322	3.4	17	77

can be reduced by electrolytic plating of copper onto the inner cylinder working surface, followed by lapping the set together with diamond paste. This process is repeated many times until a uniform gap of the desired dimension is produced.

Such small gaps preclude the use of bearings to maintain concentricity of the cylinder pair. The cylinders must instead be self-centering. One cylinder is fixed to a torque measuring device while the other is rotated by a flexible coupling. The high-shear, thin film viscometer for ambient pressure was highly developed by Porter and Johnson (see [23] for a list of references) during the period 1959 to 1965 and there are at least two commercial instruments [24,25] today which use self-centering cylinders. The cylinder pair is essentially an unloaded journal bearing which in the absence of cavitation is unstable for a generalized Newtonian liquid. The force that depends only on the property of shear viscosity and results from eccentric relative rotation, acts perpendicular to a line through the cylinder centers. If there is no cavitation then, for a generalized Newtonian liquid, there is no component of force along a line through the centers [26,27]. Any initial eccentricity would cause eccentric whirl which is not self-centering. For a non-Newtonian liquid, however, possessing even a small first normal stress difference, N_1, there is a nonzero component of force along a line through the centers which will act to restore concentricity [26,27]. Thin-film Couette viscometers must then rely on non-Newtonian response for proper operation.

Cylinder sets for use with the pressurized Couette viscometer of Fig. 7.8 are shown in Fig. 7.9. The relationship between shear stress and torque, T, is

$$\tau = \frac{2T}{\pi D^2 L} \tag{7.10}$$

for thin films. The cylinder set at the bottom of Fig. 7.9 utilizes both a large length, L, and a large diameter, D, to resolve small values of τ. The set at the top utilizes a small length yet the diameter has not been made proportionately small. This is because the generation of larger stress will require larger apparent shear rate,

$$\dot{\gamma}_{app} = \frac{D\Omega}{2h} \tag{7.11}$$

and if the diameter is very small the surface velocity will likewise be small preventing the required shear rate. The solution has been to make the length to diameter ratio small for high-stress cylinder sets. This, in turn, causes a stability problem as the rotating inner cylinder can easily tilt inside the outer cylinder. Therefore, a pilot shaft is employed as shown in the set at the top of Fig. 7.9. The pilot shaft passes through a thin plate at some axial distance from the working surfaces to provide restraint against tilting.

The rotation of the inner cylinder of the cylinder set shown in Fig. 7.8 is provided by a stepper motor which is external to the pressure vessel. The torque capability of the motor must be sized for the frictional torque of the rotary seal and for acceleration; the torque required for the shearing of the sample liquid is negligible in comparison. The motor controller is programmed to provide a single revolution and to accelerate to the required velocity in at most 50 ms for the highest rotational velocity, which is usually about five revolutions per second. Continuous steady slow rotation may be required to center the cylinders at the beginning of a measurement set; however, once a measurement series has begun, there is nothing to be gained from shearing the liquid for more than one rotation and the time required to cool the cylinders following a measurement will be increased by additional rotations. If the torque transducer is not very stiff, the time required to achieve a steady shear rate may depend more on the transducer response time than the motor acceleration. Hutton and Phillips [28] described the problem of transducer "wind up" in pressurized Couette viscometers in 1973.

The pressurized Couette viscometer shown in Fig. 7.8 was used to generate the viscosities of a standard reference liquid [5] shown in Figs. 7.3 and 7.11. Calibrations such as these give confidence in the capability of accurately measuring constitutive behavior at elevated pressures.

The measurement of shear-thinning in low-viscosity liquids provides a special challenge. The viscous power may also be written as $\tau\dot{\gamma} = \tau^2/\eta$ and to observe shear-dependent viscosity, a critical stress must be exceeded, $\tau > G$. Then if the viscosity, η, is small, the large viscous power required to measure shear-dependent viscosity is likely to result in thermal softening. The obvious experimental strategy involves increasing the viscosity. Then for liquids of inherently low viscosity, shear thinning must be measured at higher pressures or lower temperatures. Low temperatures are of less interest to the EHL application. The usual way to modify an instrument for greater pressure is to make the

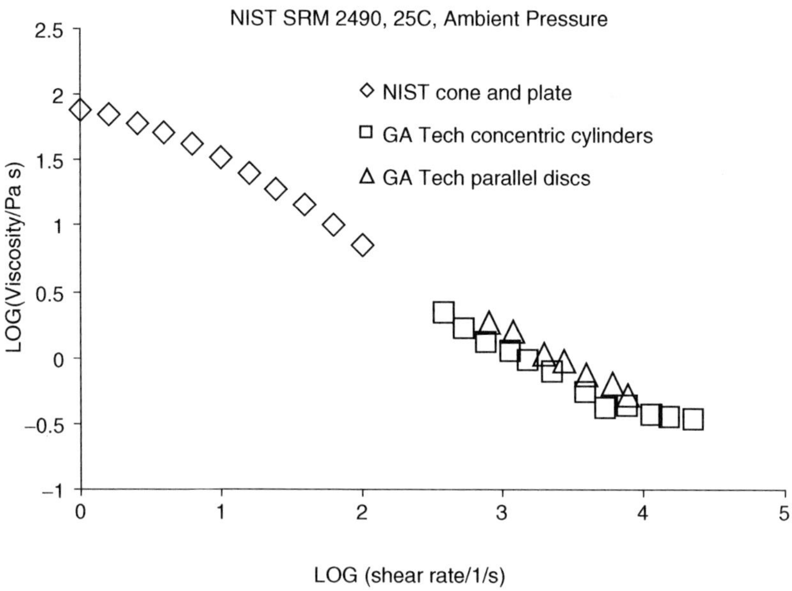

Fig. 7.11. The ambient-pressure, shear rate dependence of the viscosity of the NIST Standard Reference Material SRM2490, comparing the certified viscosities [5] with measurements by high-pressure instruments, the concentric cylinders viscometer of Fig. 7.8 and the parallel disc rheogoniometer of Fig. 7.13.

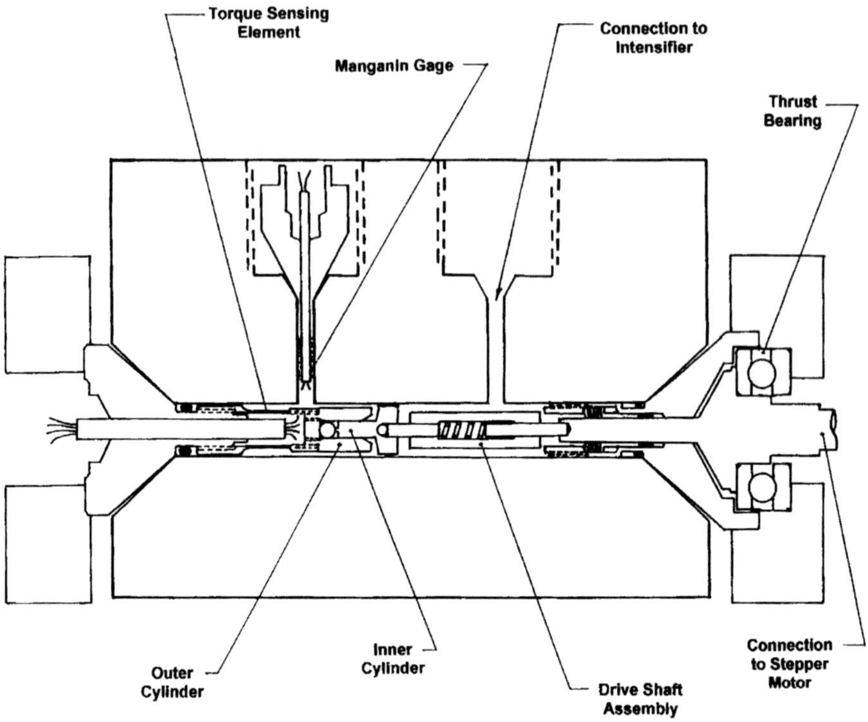

Fig. 7.12. A simplified version of the Couette viscometer of Fig. 7.8 for use to higher pressure, 1.0 GPa. Reproduced from Bair S., "The High Pressure Rheology of Some Simple Model Hydrocarbons", *Proc. Instn. Mech. Engrs.,* Vol. 216, Part J, 2002, pp. 139–149, with permission granted by the Council of the Institution of Mechanical Engineers.

structure that supports the pressure more massive and to simplify the internal features so that they may fit in a smaller bore. A pressurized Couette viscometer for pressure to 1.0 GPa is shown in Fig. 7.12. Now the sample liquid occupies the entire internal volume including the intensifier, which is directly attached. Pressure measurement is performed internally. Measurements have been obtained on very simple liquids such as heptamethylnonane [29].

7.3.2 Rotating parallel plate rheometer

The thin-film Couette viscometer, discussed above, provides measurements of the shear dependence of viscosity for the generalized Newtonian description of shear response. The more complete description of non-Newtonian behavior in simple shear will require the measurement of additional properties, the normal stress differences, N_1 and N_2, which were introduced in Chapter 2. The cone and plate rheometer provides a measurement of shear stress and first normal stress difference in a geometry that results in a uniform shear rate throughout the shearing gap. The vertex of a cone is just brought into contact with the center of a coaxial disc or, more practically, a truncated cone is positioned with the projected vertex at the disc surface. For this torsional flow instrument, one element is stationary, usually the disc, while the other element is rotated. The torque yields the shear stress and the normal thrust, which tends to separate the elements, yields the first normal stress difference. Macosko [7] gives a very complete description of the cone and plate instrument. The need to very precisely control the relative axial position of the elements makes this instrument very difficult to pressurize. The different compressibilities of the different metal alloys used in the construction causes the cone and plate separation to vary with pressure.

The parallel plate (or parallel disc) geometry shown in Fig. 7.13 does not require that a specific axial position be maintained; only that it be known so as to establish the shearing gap, h. The lower rotating disc is supported by thrust bearings and is driven at the angular velocity, Ω, by a stepper motor external to the pressure vessel. The upper disc is fastened to a two-component load cell able to measure torque, T, and thrust, Θ, as shown in Fig. 7.13. For this geometry, the shear rate is not uniform, but varies in proportion to the radial position. Measurements are reported in terms of the shear rate at the rim. From Ref. [7],

$$\dot{\gamma}_r = \frac{D\Omega}{2h} \tag{7.12}$$

and the equivalent shear stress at the rim is,

$$\tau_r = \frac{4T}{\pi D^3}\left[3 + \frac{d \ln T}{d \ln \Omega}\right] \tag{7.13}$$

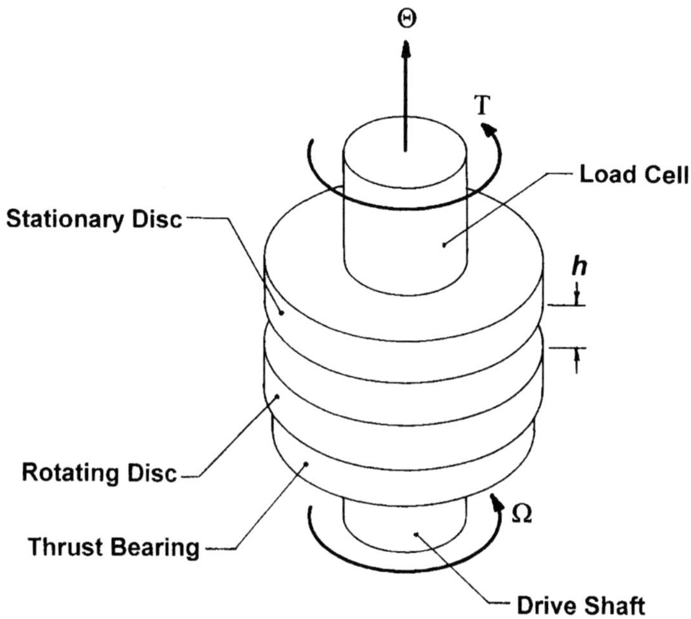

Fig. 7.13. The parallel disc rheometer.

where the second term in brackets is a correction for the nonuniform distribution of stress. The measurement of thrust in the parallel disc configuration does not directly yield the first normal stress difference as with the cone and plate [7]. Instead, the difference of the first and second normal stress differences is obtained. The equivalent value for the shear rate at the rim is

$$[N_1 - N_2]_r = \frac{4\Theta}{\pi D^2}\left[2 + \frac{d \ln \Theta}{d \ln \Omega}\right] \tag{7.14}$$

Again, the second term in brackets on the right-hand-side is a correction for the nonuniform distribution of stress.

The parallel plate rheogoniometer can be made to fit within a pressure vessel with means for driving rotation and for extracting an electrical signal for the measurement of both thrust and torque using the same techniques as were employed with the pressurized Couette viscometer described above. A miniaturized rheogoniometer [30] , which has been made to fit in a pressure vessel normally employed for a Couette viscometer, is shown in Fig. 7.14. The configuration is inverted from that of Fig. 7.13; here, the rotating disc is at the top. A similar instrument produced the viscosity data shown in Fig. 7.11 to give the same information on the shear dependence of viscosity as the concentric cylinder viscometer. The advantage of the parallel disc instrument is that additional information is recovered regarding the normal stress effect.

There are disadvantages of using the parallel disc configuration for just the measurement of viscosity under pressure. First, the gap cannot be made as small as with the concentric cylinder configuration, making this method more susceptible to thermal softening by viscous heating. Second, the different metal alloys used in the construction of a rheometer like the one in Fig. 7.14 have different compressibilities. This results in different linear compressions for different components of the structure that must maintain the gap. As a result, the gap may change considerably with changes in pressure. Third, the thrust developed by the normal stress effect can increase the gap due to the elastic deformation of the structure.

7.3.3 Viscous heating in thin-film Couette viscometers

A general exposition of the problem of thermal softening of the liquid in a viscosity measurement was given in Section 7.2.3 where it was shown that, without a great deal of care, the experimenter may mistakenly attribute an observed reduction in viscosity, which resulted from the shear induced temperature increase, to constitutive behavior. In his classic paper on the subject, Gruntfest [31], made the bold statement that because thermal processes have often been excluded from experimenters' modeling of apparent non-Newtonian flows "... complicated viscosity laws and arbitrary yield conditions must often be invented to compensate for inadequacies of their model".

The following discussion of thermal softening, resulting from viscous dissipation, assumes Newtonian response. The effective temperature–viscosity coefficient in the fully developed shear-thinning regime becomes according to Winter [32] for power-law

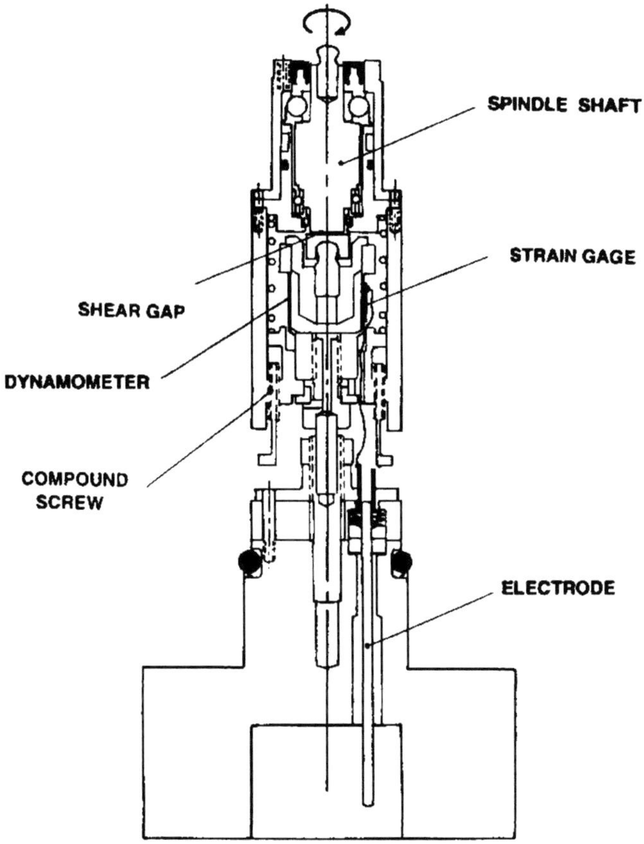

Fig. 7.14. An embodiment of the parallel disc rheometer to fit within a pressure vessel. This arrangement is inverted from that shown in Fig. 7.13. Reproduced from Figure 3 of Bair, S., "Normal Stress Difference in Liquid Lubricants Sheared under High Pressure", *Rheologica Acta,* Vol. 35, 1996, pp. 13–23, with kind permission of Springer Science and Business Media.

exponent, n,

$$\left[\frac{\partial \ln \eta}{\partial T}\right]_{\dot{\gamma},p} = n\left[\frac{\partial \ln \mu}{\partial T}\right]_{p} = n\beta \tag{7.15}$$

when rate is controlled. This is not true for stress control. For a rate-controlled process, therefore, the effect of temperature increase on the apparent viscosity is less for a shear-thinning liquid ($n < 1$). The instrument of interest, the thin film rotational Couette viscometer, is generally operated by imposing a rotation rate so that it will be conservative to apply the Newtonian assumption to a thermal analysis of shear-thinning measurements.

Gruntfest [31] analyzed the problem of plane Couette shear between thermally conducting planar solid slabs. The shear stress, τ, is uniform if inertia is neglected and the energy equation for the liquid is

$$\tau\dot{\gamma} = \frac{\tau^2}{\mu} = \rho c_p \frac{\partial T}{\partial t} - k\frac{\partial^2 T}{\partial z^2} \tag{7.16}$$

Without loss of generality, the initial temperature can be taken as zero and if exponential temperature dependence of viscosity is assumed, an analytically useful expression for the viscosity results.

$$\mu = \mu_i \exp(-\beta T) \tag{7.17}$$

The Nahme–Griffith number for plane Couette flow of thickness, h, is

$$Na = \frac{\beta\tau^2 h^2}{k\mu_i} \tag{7.18}$$

and the heat transfer Fourier number is

$$Fo = \frac{k}{\rho c_p h^2}t \tag{7.19}$$

Defining $\widehat{T} = \beta T$ and $\hat{z} = z/h$, the dimensionless energy equation becomes

$$Na\left(\frac{\tau}{\tau_i}\right)^2 \exp\left(\widehat{T}\right) = \frac{\partial \widehat{T}}{\partial Fo} - \frac{\partial^2 \widehat{T}}{\partial \hat{z}^2} \tag{7.20}$$

This equation has some interesting properties for the stress control case where $\tau \equiv \tau_i$, although this case is not entirely relevant for the present viscometer. Gruntfest [31] found that for perfectly insulated boundaries, the temperature becomes unbounded at the time for $Fo = 1/Na$ and that if heat is transferred across the boundaries, the existence of a steady solution depends upon both Na and the thermal boundary conditions. For steady flow,

$$Na \exp(\widehat{T}) = -\frac{\partial^2 \widehat{T}}{\partial \hat{z}^2} \tag{7.21}$$

Copple et al. [33], in a study of electrical dissipation in dielectric materials, found a masterful analytical solution to the above equation. Poehlein [34] recast that solution in terms of the Couette flow problem for isothermal boundaries and, because it is steady, it applies to rate control as well. For isothermal boundaries,

$$\widehat{T} = \widehat{T}_C - 2\ln\left\{\cosh\left[\sqrt{\frac{Na}{2}}\left(\frac{1}{2} - \hat{z}\right)\exp\left(\frac{\widehat{T}_C}{2}\right)\right]\right\} \tag{7.22}$$

where $\widehat{T}_C$, the maximum temperature at the midplane of the film, is given implicitly by

$$\cosh\left[\sqrt{\frac{Na}{8}}\exp\left(\frac{\widehat{T}_C}{2}\right)\right] = \exp\left(\frac{\widehat{T}_C}{2}\right) \tag{7.23}$$

When $Na < 3.5138$, two real solutions exist for the temperature distribution and the velocity ditribution given any stress. This results in two values for the boundary velocities, or two apparent shear rates (see equation (7.8)), provided that $Na < 3.5138$. The double valued apparent shear rate has been found experimentally by Sukanek and Laurence [35]. The results presented by Gruntfest [31] for low values of Na seem to show that steady state is reached at time given approximately by

$$|Fo|_{\text{steady}} \approx \frac{1}{2Na} \tag{7.24}$$

No real solution for stress control exists for $Na > 3.5138$ and the temperature runs away in what Gruntfest [31] calls the "thermal catastrophe". The shear localizes at the midplane and the velocity distribution in the liquid has the appearance of two solid slabs that are sliding at the midplane. This localization effect in stress controlled shear has been observed experimentally [36] in a liquid under elevated pressure.

The Nahme–Griffith number, Na, has great significance for the measurement of rheological properties of lubricants. If the isothermal boundaries assumption can be justified, then for measurements of subtle shear-thinning, the measurement may be regarded as being free of thermal softening due to viscous heating if $Na < 0.3$. For other effects that produce strong departure from linearity, such as thixotropy or slip, then $Na < 1$ may be sufficient. There is another dimensionless group, the Brinkman number, $Br = \left(\tau^2 h^2\right)/(T_i k \mu_i)$, with a similar definition to Na, where the temperature–viscosity coefficient has been replaced by the reciprocal of the temperature. The Brinkman number is also sometimes defined in the same way as the Nahme–Griffith number [35].

Gruntfest [31] also considered the rate-controlled case. One boundary is fixed and the other moves with velocity, u. The relationship between the temperature and the uniform shear stress, τ, is obtained from

$$u = \frac{\tau h}{\mu_i}\int_0^1 \exp(\widehat{T})d\hat{z} \tag{7.25}$$

For isothermal boundaries he found that

$$\exp\left(\widehat{T}_C\right) - 1 = \frac{Na}{8} \tag{7.26}$$

a relation that is often used to estimate temperature in lubricated sliding contacts. For the nonisothermal boundary case, Gruntfest assumed that the solid slabs were each of one-third of the thickness of the liquid film and that they were thermostated (maintained at constant temperature) at the surfaces opposite the film. While this may be an accurate description of the thermal boundary conditions for a conventional ambient-pressure

Couette viscometer, for the present case of a high-pressure, high-stress viscometer, the film is much thinner than the thickness of the cylinders and it would be very difficult to thermostat the cylinders with the conventional circulating heat transfer fluid. Instead, the viscosity measurement is performed as rapidly as possible making use of the heat capacity of the cylinder material to control the increase of temperature of the boundaries.

The author [37] has addressed the thermal description of the thin-film pressurized Couette viscometer by numerical simulation. The geometry is shown in Fig. 7.15 for a cylinder set similar to the top of Fig. 7.9. The evolution of the temperature distribution in the cylinders is important for this instrument. Only conduction in the radial direction is considered (assumes infinitely long cylinders). The energy equation for the cylinders is

$$\frac{\partial^2 T}{\partial r^2} + \frac{1}{r}\frac{\partial T}{\partial r} = \left|\frac{\rho c_p}{k}\right|_{\text{solid}} \frac{\partial T}{\partial t} \tag{7.27}$$

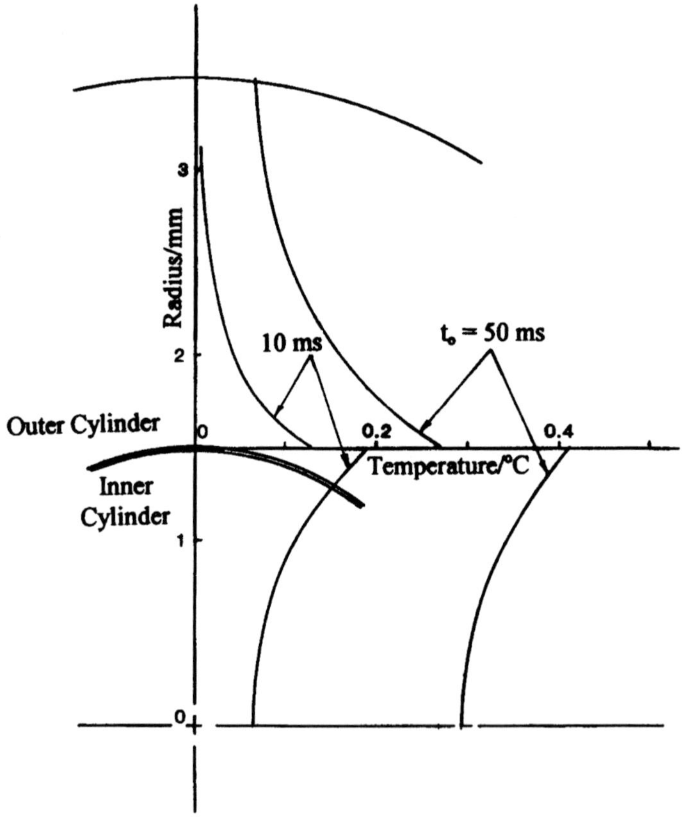

Fig. 7.15. The temperature distribution at $t = t_0$ resulting from a linear with time increase of the rotational velocity of the Couette viscometer where the apparent shear rate of $10^4\ \mathrm{s}^{-1}$ is reached at $t = t_0$. Reprinted from [37] with permission of Elsevier.

Since the film is very thin, curvature can be ignored in the equation for the film. The temperature distribution in the film evolves much more quickly than the temperature in the cylinder material, therefore the storage term in the energy equation (7.16) is dropped, resulting for the liquid film in,

$$\frac{\tau^2}{\mu} + k\frac{\partial^2 T}{\partial r^2} = 0 \tag{7.28}$$

Within the pressure vessel the outer surface of the outer cylinder is in contact with low thermal conductivity liquid. Therefore the temperature boundary condition is set to be adiabatic ($\partial T/\partial r = 0$) there. Symmetry at the axis dictates that the same adiabatic boundary condition be used here. The temperature and heat flux are made continuous at the two liquid/metal interfaces in Fig. 7.15. The temperature is initially set uniformly to $T = 0$ and equations (7.27) and (7.28) are solved in their respective domains to obtain the temperature distribution and the shear stress as a function of time. The shearing velocity is assumed to increase linearly with time to generate an apparent shear rate of $\dot{\gamma}_{app} = 10^4\ \mathrm{s}^{-1}$ after an elapsed time of t_0. For the results of Fig. 7.15, $\mu_i = 10^3$ Pa s, $\beta = 0.1°\mathrm{C}^{-1}$, $h = 1\ \mu\mathrm{m}$ and $k = 0.15$ W/m°C for the liquid. For the solid, $k = 225$ W/m°C and $\rho c_p = 3 \times 10^6$ J/m^{3}°C.

The temperature distributions shown in Fig. 7.15 were obtained for $t = t_0$. The liquid temperature reached a maximum near the surface of the inner cylinder. The reduction in stress from the isothermal case was less than 2% for $t_0 = 10$ ms and less than 4% for 50 ms. The rotational velocity in this example is about one revolution per second which can easily be achieved by a stepper motor in 10 ms. The neglect of conduction in the axial direction results in an overestimation of the temperature rise and the resulting thermal softening.

The temperature distribution was incorporated into a thermal-elastic analysis [37] of each cylinder to find the effect of thermal expansion on the shearing gap, h. No significant change in the gap was found [37].

As a practical matter, a numerical simulation of each measurement is unnecessary. If *Na* is small, the effect of the temperature variation across the film on the measured viscosity will be small. If the shear stress is observed to remain steady for a complete revolution of the cylinder, then the temperature increase of the cylinder material would not be expected to affect the measurement.

References

[1] Tanner, R.I., *Engineering Rheology*, second edition, Oxford University Press, Oxford, 2000, p. 229.

[2] Cui, S.T., Cummings, P.T., Cochran, H.D., Moore, J.D. and Gupta, S.A., "Non-equilibrium Molecular Dynamics Simulations of the Rheology of Linear and Branced Alkanes", *Int. J. Thermophys.*, Vol. 19, No. 2, 1998, pp. 449–459.

[3] Schultheisz, C.R., Flynn, K.M. and Leigh, S.D., *Certification of the Rheological Behavior of SRM2491, Polydimethylsiloxane*, NIST Special Publication 260–147, NIST, 2001.

[4] Bair, S. and Qureshi, F., "Ordinary Shear Thinning Behavior and Its Effect Upon EHL Film Thickness", *Tribological Research and Design for Engineering Systems*, Dowson, D., et al. (Ed.), Elsevier, 2002, pp. 693–699.

[5] Schultheisz, C.R. and Leigh, S.D., *Certification of the Rheological Behavior of SRM2490, Polyisobutylene Dissolved in 2,6,10,14-Tetramethylpentadecane*, NIST Special Publication 260–143, NIST, 2001.

[6] Greenwood, J.A. and Kauzlarich, J.J., "Elastohydrodynamic Film Thickness for Shear Thinning Lubricants", *Proc. Instn. Mech. Engrs.*, Vol. 212, Part J, 1998, pp. 179–191.

[7] Macosko, C.W., *Rheology: Principles, Measurements and Applications*, VCH, New York, 1994, pp. 32, 205–217, 240–244, 253.

[8] Bair, S., "The Actual Eyring Models for Thixotropy and Shear-Thinning: Experimental Validation and Application to EHD", *ASME J. Tribol.*, Vol. 126, No. 4, 2004, pp. 728–732.

[9] Bair, S. and Winer, W.O., Discussion of "The Rheology of Lubricants at High Shear Rates" by Feng and Ramnesh, *ASME J. Tribol.*, Vol. 115, No. 4, 1993, pp. 648–649.

[10] Novak, J.D. and Winer, W.O., "Some Measurements of High Pressure Lubricant Rheology", *ASME J. Lubr. Techn.*, Vol. 90, Ser. F, No. 3, 1968, pp. 580–591.

[11] Norton, A.E., Knott, M.J. and Muenger, J.R., "Flow Properties of Lubricants under High Pressure", *ASME Trans.*, Vol. 63, No. 7, 1941, pp. 631–643.

[12] Whorlow, R.W., *Rheological Techniques*, Ellis Horwood Ltd, West Sussex, 1980, p. 36.

[13] Hahn, S.J., Eyring, H., Higuchi, I. and Ree, T., "Flow Properties of Lubricating Oils Under Pressure", *NLGI Spokesman*, Vol. 21, No. 3, 1958, pp. 123–128.

[14] Bell, J.C., Kannel, J.W. and Allen, C.M., "The Rheological Behavior of the Lubricant in the Contact Zone of a Rolling Contact System", *ASME J. Basic Eng.*, Vol. 86, 1964, pp. 423–435.

[15] Sukanek, P.C. and Laurence, R.L., "An Experimental Investigation of Viscous Heating in Some Simple Shear Flows", *AIChE Journal*, Vol. 20, No. 3, 1974, pp. 474–484.

[16] Winer, W.O. and Bair, S., "Influence of Ambient Pressure on the Apparent Shear-Thinning of Liquid Lubricants – An Overlooked Phenomenon", I. Mech. E. Conference Proceeding Paper C190–87, 1987, pp. 395–398.

[17] Kottke, P., *Rheological Implications of Tension in Liquids*, Ph.D. Thesis, Georgia Institute of Technology, 2004, p. 16.

[18] Kottke, P.A., Bair, S. and Winer, W.O., "Cavitation in Creeping Shear Flows", *AIChE Journal*, Vol. 51, No. 8, 2005, pp. 2150–2170.

[19] Semjonow, V., "Über ein Rotationsviskosimeter zur Messung der Druckabhängigkeit der Viskosität hochpolymerer Schmelzen", *Rheologica Acta*, Vol. 2, No. 2, 1962, pp. 138–143.

[20] Cogswell, F.N., "The Influence of Pressure on the Viscosity of Polymer Melts", *Plastics and Polymers*, Vol. 41, 1975, pp. 39–45.

[21] Binding, D.M., Couch, M.A. and Walters, K., "The Rheology of Multigrade Oils at Elevated Pressures", *J. Non-Newt. Fluid Mech.*, Vol. 87, 1999, pp. 155–164.

[22] Bair, S. and Winer, W.O., "The High Shear Stress Rheology of Liquid Lubricants at Pressures of 2 to 200 MPa", *Trans. ASME Journal of Tribol.*, Vol. 112, No. 2, 1990, pp. 246–253.

[23] Manrique, L.A. and Porter, R.S., "An Improved Couette High Shear Viscometer", *Rheologica Acta*, Vol. 14, 1975, pp. 926–930.

[24] ASTM standard, D 4741-00, "Standard Test Method for Measuring Viscosity at High Temperature and High Shear Rate by Tapered-Plug Viscometer", 2000.

[25] ASTM standard, D 4683-96, "Standard Test Method for Measuring Viscosity at High Temperature and High Shear Rate by Tapered Bearing Simulator", 1996.

[26] Abdel-Wahab, M., Giesekus, H. and Zidan, M., "A New Eccentric-Cylinder Rheometer", *Rheologica Acta*, Vol. 29, 1990, pp. 16–22.

[27] Davies, M.J. and Walters, K., "The Behavior of Non-Newtonian Lubricants in Journal Bearings – A Theoretical Study", in *The Rheology of Lubricants*, Davenport, T.C., (Ed.), Appl. Sci. Pub., London, 1973, pp. 65–80.

[28] Hutton, J.F. and Phillips, M.C., "High Pressure Viscosity of a Polyphenyl Ether Measured with a New Couette Viscometer", *Nature Phys. Sci.*, Vol. 245, 1973, pp. 15–16.

[29] Bair, S., "The High Pressure Rheology of Some Simple Model Hydrocarbons", *Proc. Instn. Mech. Engrs.*, Vol. 216, Part J, 2002, pp. 139–149.

[30] Bair, S., "Normal Stress Difference in Liquid Lubricants Sheared under High Pressure", *Rheologica Acta*, Vol. 35, 1996, pp. 13–23.

[31] Gruntfest, I.J., "Apparent Departures from Newtonian Behavior in Liquids Caused by Viscous Heating", *Trans. Soc. Rheology*, Vol. 9, No. 1, 1965, pp. 425–441.

[32] Winter, H.H., "Viscous Dissipation in Shear Flows of Molten Polymers", *Advances in Heat Transfer*, Vol. 13, 1977, pp. 205–267.

[33] Copple, C., Hartree, D.R., Porter, A. and Tyson, H., "The Evaluation of Transient Temperature Distributions in a Dielectric in an Alternating Field", *Institution of Electrical Engineers*, Vol. 85, 1939, pp. 56–63.

[34] Poehlein, G., "Temperature Effects in Couette Viscometers", *Trans. Soc. Rheology*, Vol. 12, Issue 2, 1968, pp. 351–353.

[35] Sukanek, P.C. and Laurence, R.L., "An Experimental Investigation of Viscous Heating in Some Simple Shear Flows", *AIChE Journal*, Vol. 20, No. 3, 1974, pp. 474–484.

[36] Bair, S., Qureshi, F. and Khonsari, M., "Adiabatic Shear Localization in a Liquid Lubricant under Pressure", *Trans. ASME J. Tribol.*, Vol. 116, 1994, pp. 705–709.

[37] Bair, S., Khonsari, M. and Winer, W.O., "High-Pressure Rheology of Lubricants and Limitations of the Reynolds Equation", *Tribol. Int.*, Vol. 31, No. 10, 1999, pp. 573–586.

CHAPTER 8

The Shear Dependence of Viscosity at Elevated Pressure

8.1 Introduction

In Section 7.1 we saw examples of organic liquids of low molecular weight, $M < 400$ kg/kmol, that shear thin in the same manner as a liquid polymer with $M = 28{,}000$. The shear stress required to obtain the nonlinear response in Fig. 7.1, however, is greater for the low-molecular weight material by the same ratio that the molecular weight is smaller. The classical Newtonian film thickness formulas such as equation (1.18) were validated with low-viscosity mineral oils with similarly low-molecular weight. For the automotive applications, in particular, the use of neat low-viscosity mineral oil to establish the viscosity of a lubricant is becoming uncommon. Mineral oils are generally blended with polymer thickener to reduce the temperature dependence of the viscosity and frequently now the base stock is of higher molecular weight than before. Both of these practices may result in shear-thinning within the inlet zone of the lubricated contact where the film thickness for the contact is established.

Examples of measurements of film thicknesses for which the Newtonian assumption overestimates the thicknesses were shown in Fig. 1.6. The shear-thinning response of the same liquids can be seen in Fig. 8.1 where the data were obtained with a pressurized Couette viscometer described in the previous chapter. The interval of shear stress of 0.1–1.0 MPa is important to the generation of pressure within the inlet zone and the pressure profile in this region establishes the film thickness for the Hertz region as well. The low molecular weight ($M = 423$ kg/kmol) branched alkane, squalane, can be seen in Fig. 8.1 to be reasonably constant in viscosity to $\tau = 0.30$ MPa and in other work [1], the Newtonian limit has been found to be reached at about 6 MPa. The Newtonian assumption embodied in the Hamrock and Dowson film thickness equation (1.18) provides an accurate prediction of the film thickness in Fig. 1.6 for the low-molecular weight squalane. This film thickness calculation is accurate for the Newtonian case,

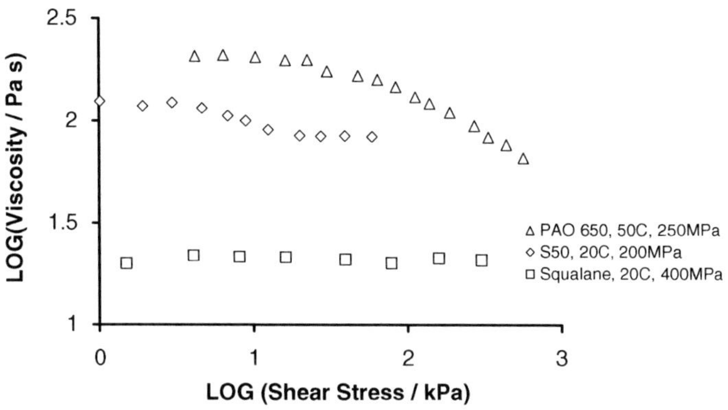

Fig. 8.1. Flow curves at elevated pressure for the liquids used in the film thickness measurements shown in Fig. 1.6. The interval of shear stress from 10^2 to 10^3 kPa is apparently very important to the film-forming ability of a liquid.

even when the pressure dependence of the viscosity is very different from that assumed by Hamrock and Dowson [2] for equation (1.18), because the particular pressure–viscosity coefficient, α^* defined by equation (1.3), is a reasonably accurate means to quantify the piezoviscous response of a Newtonian liquid for the purpose of film formation in the inlet zone.

The PAO650 oil, a high-viscosity polyalphaolefin, generates a substantially thinner film than the Newtonian prediction and the discrepancy increases with increasing velocity in Fig. 1.6. The viscosity in Fig. 8.1 for this liquid can be seen to decrease with shear stress in the important interval of stress above $\tau = 0.1$ MPa. Another type of shear-thinning is seen in the traction fluid, S50, in Fig. 8.1. The viscosity decreases from its low-shear value, μ, in the range of $0.003 < \tau < 0.02$ MPa to a second Newtonian value, μ_2. Measurements at still higher shear stress [3] have shown that the second Newtonian plateau of S50 extends to shear stress greater than 10 MPa. The film thickness of S50 in Fig. 1.6 is less than the Newtonian prediction but the velocity dependence is the same as the Newtonian prediction, suggesting that the film is governed by the Newtonian response of the second plateau. Then the thickness of an EHL film depends upon some average of the viscosity in the interval of stress important to film-forming and the velocity dependence of the thickness of an EHL film is governed by the shear dependence of the viscosity in the important stress interval.

Compared with film thickness, the effect of shear-thinning on traction is somewhat more complicated to describe, while at the same time more simple to calculate and in some circumstances more subtle in its effect. At low contact pressures, for highly shear-thinning liquids, the departure from Newtonian traction behavior can be dominated by the shear dependence of viscosity as is illustrated by Fig. 8.2 for the PAO650. In this figure the experimental data were generated at INSA, Lyon [4] in a point contact between a steel ball and glass plate at Hertz pressure of $p_H = 0.528$ GPa. The shear-thinning traction

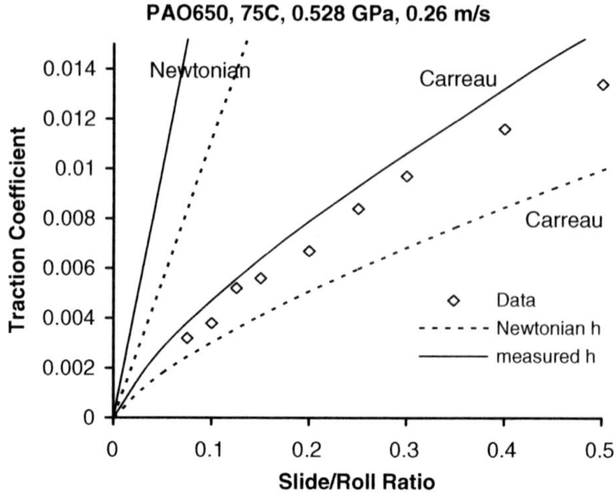

Fig. 8.2. The concentrated contact traction response of the PAO650 in Fig. 8.1. Calculations for both the Newtonian assumption and the Carreau equation (8.1) fitted to the data of Fig. 8.1 were based on uniform shear rate within the contact using film thicknesses from the Newtonian calculation of equation (1.18), broken curves, and from measurements, solid curves.

calculation utilizes the single Newtonian version of the Carreau [5] equation,

$$\eta = \mu \left[1 + \left(\frac{\mu \dot{\gamma}}{G} \right)^2 \right]^{\frac{(n-1)}{2}} \tag{8.1}$$

fitted to the data of Fig. 8.1 to give $G = 3.1 \times 10^4$ Pa and $n = 0.74$. The shear rate was assumed to be uniform throughout the Hertz region and related to the slide-to-roll ratio, Σ, by $\dot{\gamma} = \bar{u}\Sigma/h_c$. The dependence of the low-shear viscosity on pressure, $\mu(p)$, was described by the isothermal Tait–Doolittle free volume model, equations (4.6) and (6.44), with parameters given in Ref. [4]. In Fig. 8.2, the value of film thickness used in the calculation was obtained by both interferometric measurement (solid curves) and the Newtonian formula (1.18) of Hamrock and Dowson (broken curves). The traction coefficient was found by dividing the average shear stress, $\bar{\tau}$, by the average pressure, $\bar{p} = 2p_H/3$. The average shear stress was found by integration of the local shear stress over the circular contact area assuming the hemispherical Hertz pressure distribution.

Clearly, the Newtonian assumption is not helpful for any easily measurable value of traction coefficient in Fig. 8.2. The slide-to-roll ratio must be reduced to $\Sigma < 3 \times 10^{-4}$, where the traction coefficient based on a Hertz zone calculation is about $\bar{\tau}/\bar{p} = 5 \times 10^{-5}$, to achieve a traction coefficient within 10% of the Newtonian calculation. For these conditions it is likely that churning of the liquid outside of the contact region would result in a traction force comparable to that generated within the Hertz zone. As of this writing, the author is not aware of any traction measurements that have been sufficiently sensitive to extract an accurate value of the averaged low-shear viscosity from contact-based measurements, even for very low-molecular-weight liquids. It can also be seen in Fig. 8.2, that when the shear-thinning substantially affects the film thickness, as it does in this case, the Newtonian based film thickness calculation will result in reduced accuracy of the traction prediction.

This example is typical only for low-pressure contacts. As the contact pressure is increased, the effect that shear-thinning has on traction is reduced. At the highest contact pressures a linear regime develops, for small Σ, where the traction varies linearly with Σ as a result of the elastic response of the solid rollers. The traction coefficient at large Σ for high pressures is dominated by a liquid failure that has come to be known as *limiting shear stress* where the film cannot support a shear stress greater than some characteristic fraction of the average pressure. Mechanical shear localization and the resulting limit to the shear stress will be the subject of Chapter 10. For intermediate pressures, the sinh-law traction behavior of equation (1.19) presents itself as a result of the interplay of the pressure–viscosity response and the limiting stress.

The parameters of equation (8.1) are listed in Table 8.1 for measurements made in the author's laboratory on some low-molecular-weight liquids using a pressurized Couette viscometer. The remainder of this chapter is devoted to providing a suitable description of the shear-thinning which has been observed in liquid lubricants so that analyses of lubricated contact behavior may incorporate realistic shear response of the liquid.

8.2 Normal Stress Differences at Elevated Pressures

Non-Newtonian behavior in lubrication is almost exclusively described by the generalized Newtonian model, equation (2.7). In a simple shear experiment, $\partial u/\partial z = \dot{\gamma}$, and the other

Table 8.1. Carreau parameters for low-molecular weight single-component liquids.

Designation	Type	$\mu_0(40°\text{C})/\text{Pa s}$	G/Pa	n
SHF 8001	PAO 800	6.6	4.5×10^3	0.666
PAO-650	PAO 650	6.2	3.1×10^4	0.74
SHF 1001	PAO 100	1.1	1.5×10^6	0.625
SHF 403	PAO 40	0.38	6.0×10^6	0.40
PG460	Polyglycol	0.44	4.0×10^5	0.66
143AD	Branched PFPE	0.92	5×10^5	0.666
Z25	Linear PFPE	0.23	7×10^4	0.82
BS	Bright Stock	0.29	3.5×10^6	0.65
LVI260	Mineral Oil	0.26	5.6×10^6	0.34
ISO100 POE	Polyol Ester	0.78	4.0×10^6	0.56
Heptamethylnonane	High Purity Hydrocarbon	0.0024	5.3×10^6	0.44
Squalane	High Purity Hydrocarbon	0.016	6.6×10^6	0.46
Dibenzylethylbenzene	High Purity Hydrocarbon	0.014	6.5×10^6	0.41
Dicyclohexyl Methylpentane	High Purity Hydrocarbon	0.017	1.3×10^7	0.41

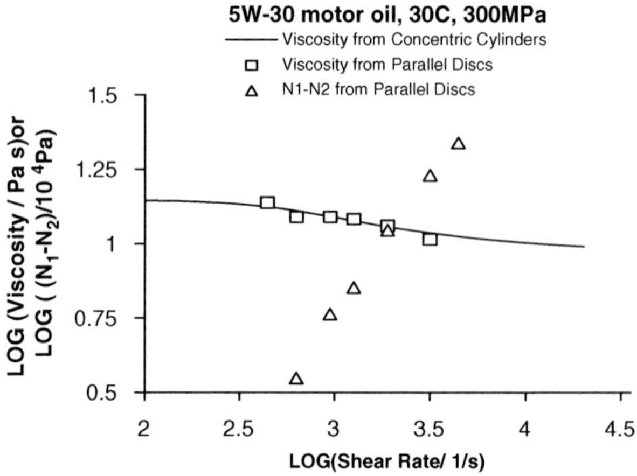

Fig. 8.3. The viscosity, η, and the normal stress difference, $N_1 - N_2$, for a multigrade motor oil at elevated pressure generated in the parallel disc rheometer of Fig. 7.13. The curve is the Carreau equation fitted to the data generated in a pressurized Couette viscometer from [6] for the same oil.

velocity gradients are zero. For a generalized Newtonian liquid, $\sigma_{xz} = \sigma_{zx} = \tau = \eta(\dot{\gamma})\dot{\gamma}$ or $\tau = \eta(\tau)\dot{\gamma}$. The other shear stresses are zero and the first and second normal stress differences are $N_1 = \sigma_{xx} - \sigma_{zz} = 0$ and $N_2 = \sigma_{zz} - \sigma_{yy} = 0$. See Fig. 2.1 for a definition of the stresses. This is, however, an incomplete description of the stress field.

In general, when the shear stress becomes sufficiently large that the viscosity becomes shear dependent, differences in the normal stresses become measurable. Their magnitude becomes comparable to the magnitude of the shear stress. This can be illustrated by the measurements shown in Fig. 8.3 for a pressure of 300 MPa. The subject of these normal stress measurements is a 5W-30 multigrade motor oil which has been investigated [6] for shear-thinning using a pressurized Couette viscometer and for which the parameters of the double-Newtonian Carreau equation,

$$\eta = \mu_2 + (\mu - \mu_2)\left[1 + \left(\frac{\mu\dot{\gamma}}{G}\right)^2\right]^{\frac{(n-1)}{2}} \tag{8.2}$$

have been published [6]. This equation is plotted as the curve in Fig. 8.3 while the data points in the figure were generated with the parallel disc rheometer of Fig. 7.13.

The very existence of differences between the normal stresses in the principle shear rate directions during simple shear may seem foreign to lubrication engineers, and it is appropriate to discuss the implications of these stress differences here. Consider the case of a bearing surface that is infinitely wide in the y-direction. The one-dimensional

Reynolds equation for a Newtonian liquid is,

$$\frac{d}{dx}\left(\frac{\rho h^3}{\mu}\frac{dp}{dx}\right) = 12\bar{u}\frac{d(\rho h)}{dx} \tag{1.11}$$

Integrating and applying the condition that h_o is the film thickness at any position for which the pressure gradient vanishes, $dp/dx = 0$, gives for the incompressible case,

$$\frac{h^3}{\mu}\frac{dp}{dx} = 12\bar{u}(h - h_0) \tag{8.3}$$

Then for locations along the flow for which $h > h_0$ the pressure is increasing in the flow direction and for $h < h_0$ the pressure is decreasing. When the same pressure boundary conditions are applied at both ends of the bearing surface, $x = 0$ and L, the position for $h = h_0$ is within the bearing surface and (8.3) can be integrated to provide a pressure distribution. If the film thickness is always converging in the flow direction, $dh/dx < 0$, then the average pressure, $\bar{p}$, is greater than the pressure at the boundaries and the pressure distribution can be integrated along the flow direction to find the bearing load that is supported by the pressure profile,

$$W = \int_0^L (-\sigma_{zz})\,dx = \int_0^L p dx \tag{8.4}$$

The load support from the converging gap of a Newtonian hydrodynamic bearing comes about from the viscous shear stress that generates a pressure gradient which gradually boosts the pressure as the pressure gradient acts along the bearing surface, the *wedge effect*.

Tanner [7] analyzed the specific case of a plane slider bearing lubricated by the theoretical second-order liquid. For a second-order liquid, the viscosity is independent of shear, the second normal stress difference is $N_2 = 0$ and the first normal stress difference is $N_1 = \Psi_1\dot{\gamma}^2$. The load equation must now be written in a different way [7],

$$W = \int_0^L (-\sigma_{zz})\,dx = \int_0^L p dx + \frac{1}{4}\int_0^L N_1 dx \tag{8.5}$$

where p is the pressure generated for the Newtonian case by the wedge effect. The first normal stress difference, N_1, acts to directly add to the load in a hydrodynamic bearing. Then the relative additional load support provided by the first normal stress difference is of the order of $N_1/\bar{p}$. The largest measured value of $N_1 = 13$ MPa appears to have been obtained at elevated pressure in a polymer thickened mineral oil [8]. Therefore, for a hydrodynamic bearing operating with average pressure, $\bar{p}$, no greater than about 100 MPa, lubricated by polymer blended oil, it is possible to derive a small but useful portion of the load from normal stress effects. On the other hand, the load supported by the liquid film in an EHL contact can be assumed to be entirely due to the pressure distribution owing to the much larger value of $\bar{p}$ for these contacts.

The most interesting aspect of the load supported by a non-Newtonian lubricated slider bearing may be that a nonzero load can be generated in a parallel film. For a parallel film, the wedge effect vanishes, $\bar{p} = 0$, yet equation (8.5) predicts load support if $N_1 > 0$.

An interesting implication of the first normal stress difference is that the principal shear stress axes are not coincident with the principal shear rate axes. This will be important later in the discussion of the liquid failure which results from slip within the film at a critical shear stress.

8.3 The Origin of Non-Newtonian Behavior in Low-Molecular-Weight Liquids at Elevated Pressures

8.3.1 The viscoelastic hypothesis

In Chapter 2, the corotational Maxwell model was introduced to illustrate an example of a theoretical model which was capable of at least qualitatively describing shear-thinning and a normal stress effect. Viscoelastic liquids can both store energy elastically and dissipate energy viscously. These models can be used to explain the analogy between continuous and oscillatory shear. Viscoelastic models were widely explored as candidates for the description of the constitutive behavior of liquids in EHL contacts during the 1970s and early 1980s. Elastic compliance of the liquid lubricant film was invoked to explain the linear small slide/roll ratio traction behavior at high contact pressures and the large strain behavior of the viscoelastic model was used to explain nonlinear traction response. (Other explanations for these phenomena have since been found.) The monograph by Harrison [9] provides an excellent review of the experimental rheology techniques of the period that involve the propagation of a plane shear wave into the liquid. Measurements were reported at pressures to 1.4 GPa. These acoustic methods were performed at ultrasound frequencies of up to $\omega = 10^9\ \mathrm{s}^{-1}$ and provide a means to investigate the small strain oscillatory shear response which was modeled by the single mode Maxwell model in Section 2.4.1.

For a time varying shear strain of $\gamma = \gamma_A \sin \omega t$, the Maxwell model (2.13) gives a time-dependent shear stress of

$$\tau = \frac{\dfrac{\mu^2 \omega \gamma_A}{G}}{1 + \left(\dfrac{\mu\omega}{G}\right)^2} \left[\frac{G}{\mu} \cos \omega t + \omega \sin \omega t\right] \tag{2.25}$$

The component of stress in phase with the strain is

$$\tau'' = \frac{\dfrac{\mu^2 \omega^2}{G} \gamma_A \sin \omega t}{1 + \left(\dfrac{\mu\omega}{G}\right)^2} \tag{8.6}$$

The component of the dynamic shear modulus in phase with the shear strain is

$$G''(\omega) = \frac{\tau''}{\gamma} = \frac{\dfrac{\mu^2\omega^2}{G}}{1 + \left(\dfrac{\mu\omega}{G}\right)^2} \tag{8.7}$$

and the limiting high-frequency value of G'' is then

$$G_\infty = \lim_{\omega\to\infty} G''(\omega) = G \tag{8.8}$$

Therefore, for a single mode Maxwell liquid, a small-strain oscillatory shear experiment can be used to measure the shear modulus property if sufficiently high frequency is available to observe a limiting high frequency value, G_∞, of the component of the dynamic shear modulus in phase with the shear strain, G''. Generally G'' approaches the limiting value more slowly [9, p. 145] than expected from the single mode Maxwell model, equation (8.7).

Hereafter in this book, G_∞ will denote the elastic shear modulus property and G will be reserved for the (critical) shear stress at the limit of Newtonian response or the material modulus associated with λ, the characteristic time employed in a shear-thinning model, by $G = \mu/\lambda$. The reason for making this distinction will be apparent from the following example where experimental measurements of G_∞ and G have been made for the same material, di(2-ethylhexyl)phthalate (DOP), over a similar range of state variables.

Measurements of G_∞ for DOP were obtained by Hutton and Phillips [10] for $-30 \leq T \leq 30°\text{C}$ and to pressures to 0.815 GPa using high-frequency transverse sound. They fitted their results to

$$G_\infty = \frac{1.29 \times 10^{13}\ \text{Pa}}{(T - 92.8\ \text{K})^2} + \frac{1.74 \times 10^5}{(T + 16.7\ \text{K})^2}p \tag{8.9}$$

which is in good agreement with the results obtained by Barlow et al. [11] who obtained at 30°C, $G_\infty = 0.5\ \text{GPa} + 2.45p$ for pressures to 1.4 GPa. Where comparison can be made, the ultrasonic measurements of Hutton [12] compare very well with direct mechanical measurements from the author's laboratory of the shear modulus for the model mineral oil, LVI260, under high pressure near to the glass transition. Then we can accept equation (8.9) as an accurate representation of the shear modulus of a well-defined liquid, DOP.

The shear-dependent viscosity of DOP has also been measured by the author [1] for temperatures and pressures similar to those for the shear modulus measurements presented above. These data are shown as the points in Fig. 8.4. When these results were originally presented [1], the author fitted the Carreau equation (8.1) to obtain $n = 0.41$ and $G = 6$ MPa. Slightly different values of n and G, to be introduced in the next section, were used to draw the Carreau curves in Fig. 8.4 for a purpose to be explained in the next section. Experimentally, then, the limit of Newtonian behavior is approximately at $\tau \approx G = 6$ MPa. For the temperatures and pressures of Fig. 8.4, equation (8.9)

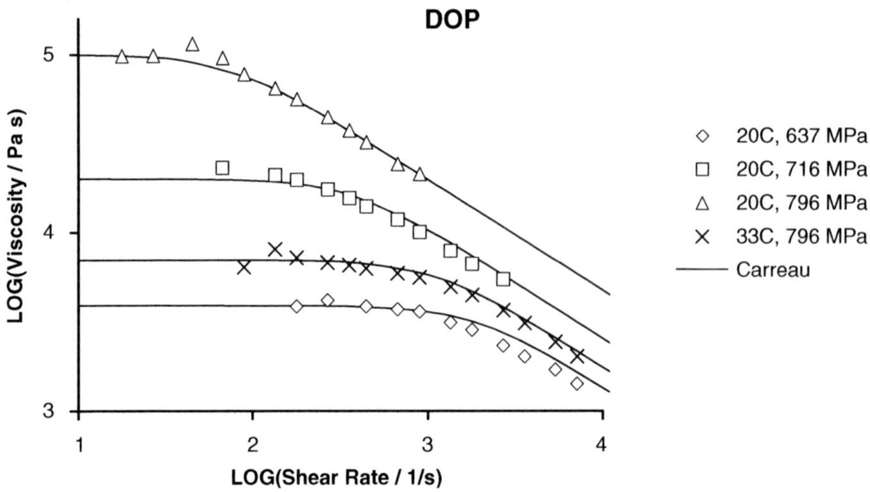

Fig. 8.4. Flow curves for di(2-ethylhexyl)phthalate for pressures and temperatures similar to those for which the elastic shear modulus, G_∞, has been measured. The curves are the Carreau equation with the usual characteristic time, $\lambda = \mu/G$, set equal to the Einstein–Debye rotational relaxation time, λ_{EB}.

gives $1480 < G_\infty < 1770$ MPa. Clearly the magnitudes of G and G_∞ are unrelated for DOP and the viscoelastic model using directly measured values of G_∞ cannot result in a quantitative description of shear-thinning. For the large-strain viscoelastic model, equation (2.21), the limit of Newtonian behavior is roughly at $\tau \approx G_\infty$, at least two orders-of-magnitude too large.

8.3.2 The kinetic theory of molecules

The Einstein–Debye relation [13] for the rotational relaxation time of a molecule is

$$\lambda_{EB} = \frac{\mu V_m}{k_B T} \tag{8.10}$$

where V_m is a molecular volume and k_B is the Boltzmann constant. Debye [14] developed a theory, from Einstein's explanation [15] of Brownian movement, for the liquid molecules that are electrically polar, having an electrical moment. In liquid lubricants, polarity usually arises from the chemical addition of oxygen to a hydrocarbon structure. Thermal (Brownian) motions constantly reorient molecules that may have had, at some instant, some specific orientation. The reorientation is retarded by the viscosity. Debye [14] defined the rotational relaxation time "as the time required for the moments of the molecules to revert to practically a random distribution after the removal of the impressed field".

Debye, in his development of the relaxation time, clearly meant that the field should be electrical; however, there is nothing in the development of equation (8.10) that should restrict its application to only electrical effects. Suppose instead that a viscous shear stress provides the "impressed field". Also suppose that the impressed field has associated with it a periodic character arising from the interaction of individual molecules and that the frequency of these interactions may reasonably be approximated by the shear rate, $\dot{\gamma}$. It should be expected that the rotational alignment of a molecule by the impressed field would serve to ease the flow, to reduce the viscosity. Then, if there is insufficient time for the orientation of a molecule arising from an interaction with another molecule to be removed by thermal motion during the interaction, the aligning process will accumulate and the viscosity will be reduced. As a result, a transition to shear-thinning would be observed when the product $\lambda_{EB}\dot{\gamma}$ is of order one.

This hypothesis has been tested in Fig. 8.4 where the curves are the Carreau equation,

$$\eta = \mu \left[1 + (\lambda_{EB}\dot{\gamma})^2\right]^{\frac{(n-1)}{2}} \tag{8.11}$$

with the characteristic time $\lambda = \lambda_{EB}$. The power-law exponent has been adjusted to $n = 0.38$ to provide the good fit to the data in Fig. 8.4. It was assumed that the molecular volume, V_m, in equation (8.10) is given by the total liquid volume per molecule,

$$V_m = \frac{M}{\rho}\frac{k_B}{R_g} \tag{8.12}$$

where the ratio, R_g/k_B, is Avogadro's number and the formula molecular weight of DOP is $M = 390.56$ kg/kmol. An equation of state for density, ρ, was obtained by fitting the temperature modified Tait framework, equations (4.6), (4.12) and (4.14), to the density data for DOP reported in the ASME Pressure–Viscosity Report [16] with $K_0' = 10.647$, $a_V = 8.33 \times 10^{-4}\ \mathrm{K}^{-1}$, $K_{00} = 11.005$ GPa, $\beta_K = 6.327 \times 10^{-3}\ \mathrm{K}^{-1}$, $T_R = 25°\mathrm{C}$ and $\rho_R = 1/V_R = 981\ \mathrm{kg/m^3}$. In the case of DOP, Fig. 8.4 shows that the Einstein–Debye rotational relaxation time is an excellent predictor of the onset of shear-thinning when used with the Carreau model and

$$\lambda = \lambda_{EB} = \frac{\mu M}{\rho R_g T} \tag{8.13}$$

8.3.3 Nonequilibrium molecular dynamic simulations

Contributed by Clare McCabe, Vanderbilt University

The same information regarding viscosity and normal stress differences as is found through experimental measurements can be obtained from molecular simulations. Molecular simulation methods for calculating viscosity can be broadly classified into *equilibrium molecular dynamics* (EMD) and *nonequilibrium molecular dynamics* (NEMD) methods. In an EMD simulation, Newton's equations (or a variant thereof) describing the motion of the atoms in the model system are solved as a function of time.

The viscosity η is then given as an integral of the stress–stress auto-correlation functions determined during the simulation, viz.,

$$\eta = \frac{V}{k_B T} \lim_{\lambda \to \infty} \int_0^{\lambda} dt \langle \sigma_{xy}(t) \sigma_{xy}(0) \rangle \tag{8.14}$$

where V is the volume of the system, k_B is Boltzmann's constant, T is temperature and t is time. The quantity $\sigma_{xy}(t)$ is the value of the xy component of the traceless symmetric stress tensor at time t, and so $\sigma_{xy}(t)\,\sigma_{xy}(0)$ is the stress–stress autocorrelation function and $\langle \sigma_{xy}(t)\,\sigma_{xy}(0) \rangle$ is its ensemble average measured during the course of the simulation. The time over which the stress–stress autocorrelation function needs to be computed in order to obtain high-accuracy results from equation (8.14) is at least several hundred relaxation times [17]. EMD methods based on the application of equation (8.14) (and similar approaches [18]) have been applied routinely and successfully to low-molecular-weight fluids [19–21].

While EMD has been used extensively for calculating transport properties, in recent years the NEMD method has become increasingly popular since the key algorithm, SLLOD [22], is a direct implementation of the experimental method for measuring viscosity, and because it can also probe the non-Newtonian regime (Fig. 8.5). The NEMD SLLOD algorithm for viscosity involves applying a planar Couette flow field at strain rate $\dot{\gamma} = \partial u / \partial y$, which characterizes the constant change in streaming velocity u_x in the x direction with vertical position y. The viscosity at strain rate $\dot{\gamma}$ is then computed from

$$\eta = -\frac{\langle \sigma_{xy}^{0,s} \rangle}{\dot{\gamma}} \tag{8.15}$$

where $\sigma_{xy}^{0,s}$ is the traceless symmetric stress tensor computed during the course of the simulation. The result is a strain-rate-dependent shear-thinning viscosity. The critical

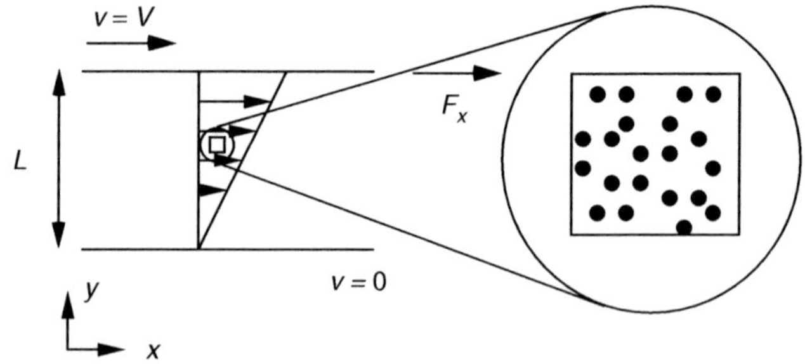

Fig. 8.5. Schematic of an NEMD simulation. The fluid is sheared between two surfaces – the upper surface moves at velocity V relative to a fixed lower surface. At steady state this maintains a velocity profile across the gap of width, L, which is characterized by the strain rate, $\dot{\gamma}$. The force/unit area required to maintain the steady state velocity of the moving plate has the units of pressure and is the xy component of the stress tensor. The viscosity is simply the constant of proportionality between σ_{xy} and $\dot{\gamma}$.

strain rate transition above which we see shear-thinning and below which we see a Newtonian plateau (η independent of $\dot{\gamma}$) typically occurs at $\dot{\gamma} \approx \lambda^{-1}$, where λ is the longest relaxation time at equilibrium (i.e. in the absence of shear). This is usually the rotational relaxation time. In systems with long relaxation times (i.e. $O(0.1$ ns$)$) for which EMD appears to be impractical, NEMD is the only practical route to compute the Newtonian viscosity. However, when applied to systems of high molecular weight at extreme states ($\lambda \geq 10$ ns), the computational burden associated with NEMD increases dramatically [23,24]. While the viscosity converges rapidly at high strain rate, the length of simulation needed to obtain reliable averages for $\langle \sigma_{xy} \rangle$ increases nonlinearly as the strain rate is reduced; in particular, for $\dot{\gamma} < \lambda^{-1}$, $\langle \sigma_{xy} \rangle$ must be averaged through at least several, and preferably many, rotational relaxation times. Snapshots from a molecular dynamics simulation of a C_{100} melt are given in Fig. 8.6 and clearly demonstrate the alignment of the molecules at high shear rates, the molecular origin of shear thinning.

The NEMD method has been employed successfully to study the viscosity and transport properties of a wide range of fluids [24]. In particular, Kioupis and Maginn [25] have examined the rheological properties of three isomers of C_{18} to high pressures by EMD and NEMD and McCabe et al. have studied the high-pressure rheology of 9-octylyheptadecane and squalane (2,6,10,15,19,23-hexamethyltetracosane) [23,26,27]. This work provided the first experimental verification of NEMD simulation in the non-Newtonian regime through a comparison of NEMD simulation results and experiment using the standard rheological analysis technique of time–temperature–pressure superposition.

8.4 Time–Temperature–Pressure Superposition

Flow curves of true shear-thinning behavior, when plotted as log of viscosity versus log of shear rate or shear stress or when plotted as log of shear stress versus log of shear rate have a useful property. Any curve can be shifted vertically and horizontally without rotation to superimpose over another curve, often within experimental accuracy. The only exception appears to be the portion of the curve influenced by a second Newtonian plateau. The ratio of the first Newtonian viscosity to the second Newtonian viscosity, μ/μ_2, often is found to vary with temperature and pressure [28]. This graphical shifting of flow curves to superimpose has been called time–temperature–pressure superposition, or alternatively the method of reduced variables, and is dramatically illustrated by the reduction of the disparate data for squalane in Fig. 8.7(a) to the master curve of Fig. 8.7(b) using reduced variables. These data were obtained from the NEMD technique of the previous section and experimental measurement [27]. See Section 2.5 as well.

Consider a plot of $\log \eta$ versus $\log \dot{\gamma}$ as in Fig. 8.4. If the various curves are to superimpose, the terminal regimes must superimpose and the terminal behavior is determined only by the low-shear viscosity. The reduced variable for viscosity is therefore, $\hat{\eta} = \eta/\mu$. The shear rate at the transition to shear-thinning is, for all states, $1/\lambda$ or G/μ and therefore the choice for the reduced shear rate is $\hat{\dot{\gamma}} = \dot{\gamma}\lambda = \dot{\gamma}\mu/G$. The reduced shear stress must then be $\hat{\tau} = \hat{\eta}\hat{\dot{\gamma}} = \tau/G = \tau\lambda/\mu$. These are the definitions, assuming that $\lambda = \lambda_{EB}$, of the reduced stress and rate that were successfully applied to the data of Fig. 8.7(b).

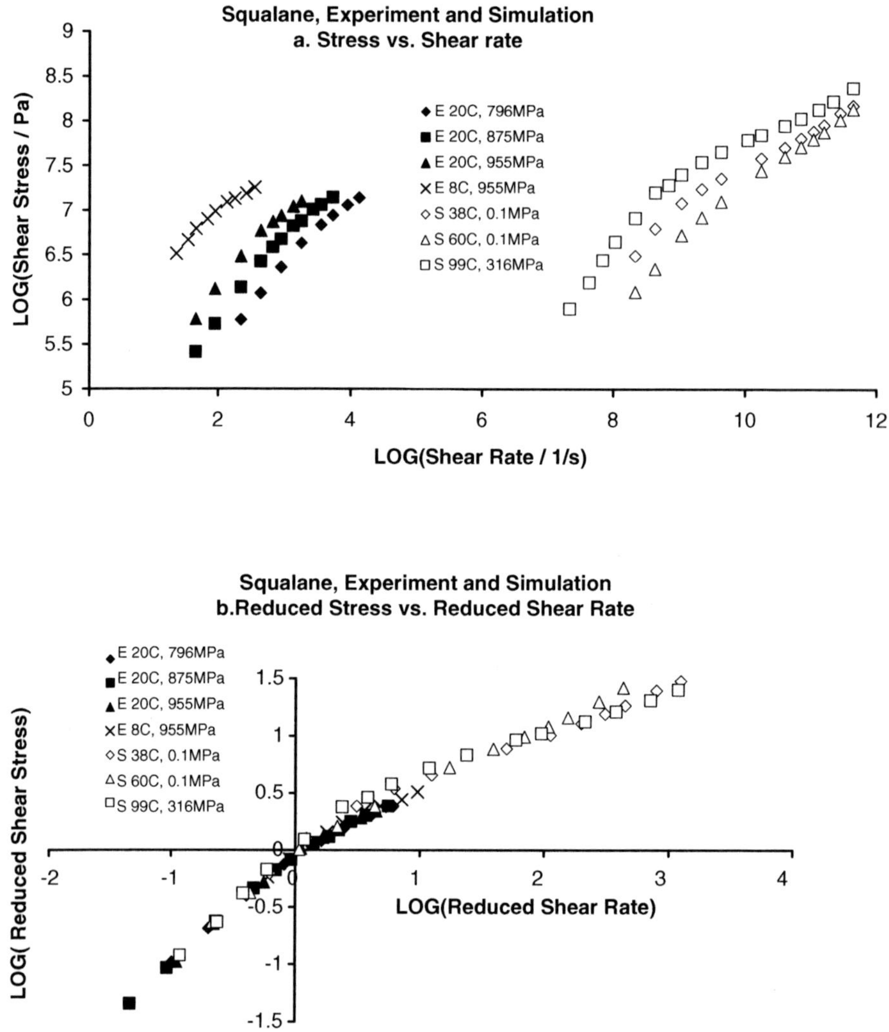

Fig. 8.7. (a) Flow curves plotted as shear stress, τ, versus shear rate, $\dot{\gamma}$, for squalane obtained from experiment (E) with a pressurized Couette viscometer (solid points) and from simulation (S) using nonequilibrium molecular dynamics (open points). (b) The same data replotted as reduced shear stress, $\tau\lambda_{EB}/\mu$, versus reduced shear rate, $\dot{\gamma}\lambda_{EB}$, to obtain a master flow curve.

The principle that has been laid out above is extremely important for EHL calculations. Flow curves (even a single flow curve) that have been obtained at temperatures and pressures of experimental convenience may be used to describe the shear dependence of viscosity at the often more severe conditions of the lubricated contact. It is only necessary to find $\mu(T, p)$ and $G(T, p)$ to provide the shifting rules, and the various models for $\mu(T, p)$ have been presented in detail in Chapter 6. (When a second Newtonian is

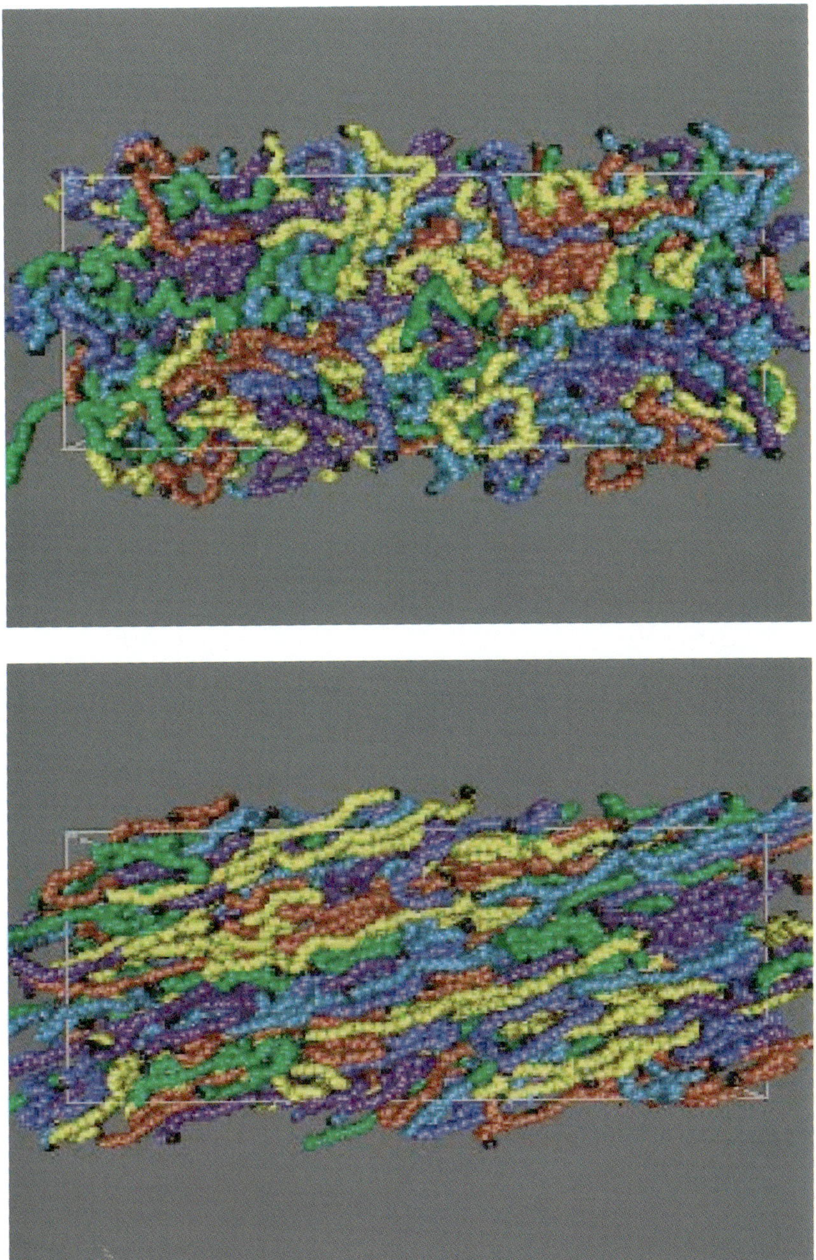

Fig. 8.6. Snapshot of a configuration from a molecular dynamics simulation of a $C_{100}H_{202}$ melt at equilibrium (top) and undergoing shear at a high shear rate (bottom). The molecules are colored randomly for ease of identification. J.D. Moore, Private Communication.

present, an expression for $\mu_2(T, p)$ will also be required.) For the examples of Figs. 8.4 and 8.7(b), the Einstein–Debye relation was used to set $G(T, p) = \rho R_g T/M$.

More often, equation (8.13) is not taken as an exact relationship for shifting but utilized for the temperature and pressure dependence only to give another shifting rule,

$$G = G_R \frac{\rho T}{\rho_R T_R} \tag{8.16}$$

the standard or Ferry rule [29]. Here, the subscript R indicates a value obtained for a reference state, T_R and p_R. A very simple rule that is often accurate is

$$G = G_R \tag{8.17}$$

the Vinogradov–Malkin rule [29]. Measurements made over a wide range of temperature and pressure such as the capillary measurements of Fig. 7.7 often show that G decreases with temperature, which is in contradiction to the increase predicted by the Einstein–Debye relation (8.10) if molecular volume increases with temperature at a rate less than in proportion to temperature. In these cases

$$G = G_R \left(\frac{\mu}{\mu_R}\right)^m \tag{8.18}$$

with $0 \leq m \leq 0.25$ may be helpful [30,31].

The pressure–viscosity and the temperature–viscosity coefficients for the low-shear viscosity were defined in equations (5.1) and (5.2) respectively.

$$\alpha(T, p) = \frac{1}{\mu}\frac{\partial \mu}{\partial p} \tag{5.1}$$

$$\beta(T, p) = -\frac{1}{\mu}\frac{\partial \mu}{\partial T} \tag{5.2}$$

It should be possible to define similar coefficients for the generalized viscosity, η. For the terminal regime the definitions must be identical to those above.

$$\left[\frac{1}{\eta}\frac{\partial \eta}{\partial p}\right]_{T,\dot{\gamma}\to 0} = \alpha \quad \text{and} \quad -\left[\frac{1}{\eta}\frac{\partial \eta}{\partial T}\right]_{p,\dot{\gamma}\to 0} = \beta \tag{8.19}$$

As was the case for thermal softening in Section 7.3.3, it will be necessary to specify the nature of the process to determine the temperature and pressure sensitivity. The single-Newtonian generalized Newtonian models of Table 2.1 reduce, in the limit of high shear rate or stress, to

$$\eta \to \mu^n \left(\frac{G}{\dot{\gamma}}\right)^{1-n} = \mu \left(\frac{G}{\tau}\right)^{\frac{1}{n}-1} \tag{8.20}$$

If the shifting rule of equation (8.18) is invoked, in the limit of high shear rate or stress,

$$\eta \to \mu^{(n+m-mn)} \left(\frac{G_R}{\mu_R^m \dot{\gamma}} \right)^{1-n} = \mu^{\left(1+\frac{m}{n}-m\right)} \left(\frac{G}{\mu_R^m \tau} \right)^{\frac{1}{n}-1} \tag{8.21}$$

and the pressure–viscosity coefficient for η is in the limit of high shear

$$\alpha_{\eta,\tau} = \left[\frac{1}{\eta} \frac{\partial \eta}{\partial p} \right]_{T,\tau} \to \left(1 + \frac{m}{n} - m \right) \alpha \tag{8.22}$$

for stress control and

$$\alpha_{\eta,\dot{\gamma}} = \left[\frac{1}{\eta} \frac{\partial \eta}{\partial p} \right]_{T,\dot{\gamma}} \to (n + m - mn)\,\alpha \tag{8.23}$$

for rate control. Similar relationships can be derived for the temperature–viscosity coefficients by replacing α by β. For the case of Vinogradov–Malkin shifting, $m = 0$ and $\alpha_{\eta,\tau} = \alpha$ and $\alpha_{\eta,\dot{\gamma}} = n\alpha$. This is in agreement with the result, $\beta_{\eta,\dot{\gamma}} = n\beta$, obtained experimentally by Winter [32].

8.5 The Competition between Thermal Softening and Shear-Thinning

Two sections of the previous chapter dealt with the thermal softening of the liquid that results from viscous heating in viscometers. The effect is clearly of major concern for the measurement of shear-dependent viscosity of low-molecular weight liquids as it is usually the first limitation that is encountered as a measurement is extended toward and into the shear-thinning regime. Thermal softening is often so much more severe in the reduction of viscosity with shear than is the constitutive behavior that it is not even possible to correct with accuracy for the viscous heating effect. In Couette shear, thermal softening will impose a limit to the steady stress, corresponding to $Na = 3.5138$, the limit to real solutions of equation (7.23), which may be less than the stress at which shear-thinning begins. This competition between thermal softening and shear-thinning has implications beyond just the measurement of viscosity. In steady Couette shear, if the Nahme–Griffith number, *Na*, becomes larger than unity before the Weissenberg number, *Wi*, approaches one, then regardless of the circumstances, it will be unlikely that shear-thinning will play a significant role. This is true of liquid films in lubricated contacts as well as in viscometers.

To provide guidance in deciding whether or not shear-thinning will be operative in any Couette film, the author [33,34] has introduced a new dimensionless group,

$$B = \frac{Na}{Wi^2} = h^2 G^2 \frac{\beta}{\mu k} \tag{8.24}$$

Dividing *Na* by the square of *Wi* cancels the shear rate (or shear stress) so that B is independent of the kinematics. For $B > 1$, thermal softening will overwhelm the shear-dependence of viscosity and a Newtonian description will be sufficient; however if $B < 1$, it still cannot be said that thermal softening will not be important at any magnitude of shear. If typical values, for EHL pressures, of the temperature viscosity coefficient, β, and the liquid thermal conductivity, k, are assumed and if $G = 3$ MPa is typical for a mineral oil, then thermal softening will preclude shear-thinning when the film is thicker than $h = 0.3\,\mu$m, where $\mu = 1$ Pa s and when the film is thicker than $h = 30\,\mu$m where $\mu = 10^4$ Pa s.

8.6 Multi-Component Systems

The single Newtonian Carreau model (8.1) has proven to be an accurate form for monodisperse liquids [33]. Practical lubricants, however, are seldom comprised of a single monodisperse liquid. The significance of the value of the shear stress that sets the limit of Newtonian behavior, G, for an individual liquid material often tends to be retained when that liquid becomes a part of a two-component system. This is dramatically illustrated in Fig. 8.8 where dibenzylethylbenzene (DBEB) has been blended with 10% of a nearly monodisperse polystyrene ($M \approx 700$ kg/kmol) of polydipersity of 1.06. (A monodisperse liquid has a polydipersity of one.) The flow curve for the neat DBEB shows a transition to shear-thinning at a shear stress of 6.2 MPa and this transition is also apparent in the flow curve for the mixture. The mixture shows an additional transition

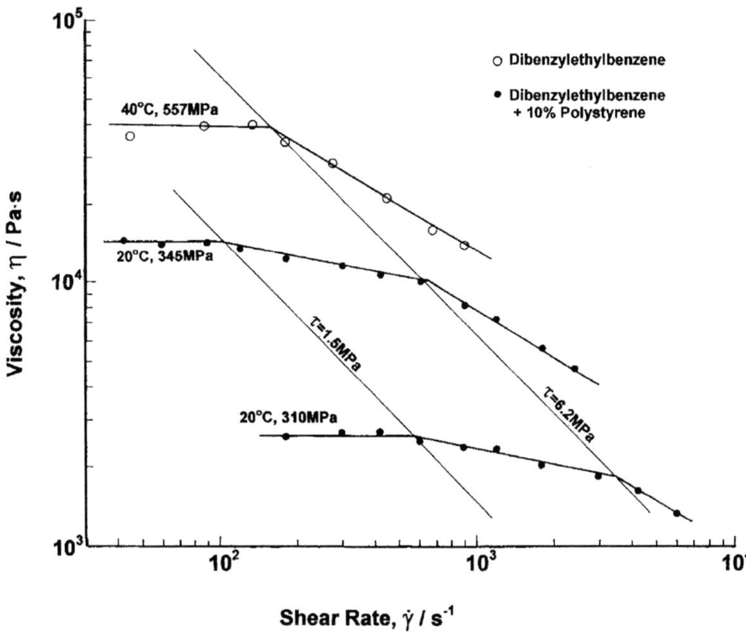

Fig. 8.8. Flow curves for DBEB and for DBEB with 10% by weight of a nearly monodisperse polystyrene. The critical stress (6.2 MPa) for transition to shear-thinning of DBEB is retained in the mixture. Reproduced from Bair, S., "The High-Pressure Rheology of Mixtures", *ASME J. Tribol.*, Vol. 126, No. 4, 2004, pp. 697–702, with permission of the American Society of Mechanical Engineers.

at $\tau = 1.5$ MPa which must be attributed to the presence of the polystyrene additive [35]. The curves composed of connected straight lines in Fig. 8.8 represent the Spriggs truncated power-law model [36] introduced in Table 2.1.

Most lubricant base stocks cannot be described as monodisperse. Their molecular weights are broadly distributed. Figure 8.8 illustrates how mixing of liquids of different molecular weights broadens the transition from Newtonian to power-law response. A viscosity function that accommodates broadening of the transition is the Carreau–Yasuda [37] model,

$$\eta = \mu \left[1 + \left(\frac{\mu \dot{\gamma}}{G} \right)^{a} \right]^{\frac{(n-1)}{a}} \tag{8.25}$$

See Table 2.1. For $a = 2$, this is the Carreau model and as a decreases, the transition broadens.

The shear response of blends of polydisperse liquids may become complicated as illustrated by Fig. 8.9. The PAO100 is a 100 cS grade polyalphaolefin. It differs from a previously studied [38] PAO100 (see Table 8.1) in having a high molecular weight "tail" in the distribution which apparently causes the first departure from constant viscosity at $\tau \approx 10^5$ Pa in the second from top flow curve in Fig. 8.9. The PAO650 was introduced in Fig. 8.1. When the PAO650 is blended with the PAO100, the transition at $\tau \approx 3 \times 10^4$ Pa, which is clearly present in the flow curve for the PAO100, appears in the curve for the blend. The author has suggested a viscosity function for mixtures that

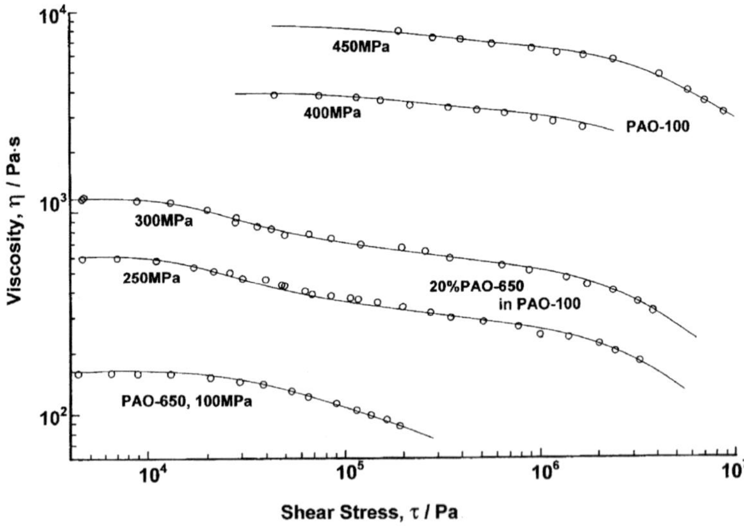

Fig. 8.9. Flow curves at 20°C for three liquids; a 100 cS grade PAO, a 650 cS grade PAO and a blend of the two. The curves represent equation (8.26). Reproduced from Bair, S., Vergne, P. and Querry, M., "A Unified Shear-Thining Treatment of both Film Thickness and Traction in EHD," *Tribol. Lett.*, Vol. 18, No. 2, 2005, pp. 145–152, with kind permission of Springer Science and Business Media.

Table 8.2. Parameters of equation (8.26) for the liquids shown in Fig. 8.9.

Designation	N	i	f_i	n_i	G_i/Pa
PAO-100	2	1	0.5	0.8	1×10^5
		2	0.5	0.5	4×10^6
PAO-650	1	1	1	0.74	3.1×10^4
20% PAO-650 in PAO-100	3	1	0.4	0.28	3.1×10^4
		2	0.3	0.75	1×10^5
		3	0.3	0.4	4×10^6

is written

$$\eta = \mu \sum_{i=1}^{N} f_i \left[1 + \left(\frac{\tau}{G_i}\right)^2\right]^{(1-1/n_i)/2} \tag{8.26}$$

where f_i is analogous to the areal fraction in the Ree–Eyring model [39] and also, as in the Ree–Eyring model, $\sum_{i=1}^{N} f_i = 1$. Unlike Ree–Eyring there is no restriction to $N > 1$. It is remarkable that the values for G_i that are to be used in equation (8.26) for the blend can be obtained from the flow curves for the individual components as shown in Table 8.2. For $N = 2$, $f_1 = 1 - \mu_2/\mu$, $f_2 = \mu_2/\mu$ and $n_2 = 1$, equation (8.26) becomes the more usual model [28] that is used for a polymer oil solution,

$$\eta = \mu_2 + (\mu - \mu_2)\left[1 + \left(\frac{\tau}{G}\right)^2\right]^{(1-1/n)/2} \tag{8.27}$$

8.7 The Power-Law Exponent and the Second Newtonian Viscosity

We have seen in Section 8.3.2 that the value of the material modulus, G, for single-component, low-molecular-weight liquids can be estimated from the ratio of the low-shear viscosity to the Einstein–Debye rotational relaxation time and in Section 8.6 that the description of shear-thinning of a multi-component system can be estimated from the moduli, G_i, of the individual components. For an EHL analysis, it will be necessary to have a numerical value for the power-law exponent, n, as well and if the lubricant is a polymer solution, a value for the second Newtonian viscosity, μ_2, will also be required. The means to estimate these parameters have not yet been derived; however a few trends have been established.

Some preliminary associations have been made between the power-law exponent and branching of the molecule. Wood-Adams [40] made steady and dynamic shear viscosity measurements on samples of polyethylene, all having molecular weights of about 9×10^4 kg/kmol. She found that as the number of long chain branches increased, the value of n decreased.

The NEMD simulations of Jabbarzadeh et al. [41] appear to contradict this trend; however, the flow curves displayed in the reference do not appear to have fully developed into the power-law regime. The values of n in Table 8.1 seem to confirm that branching decreases the power-law exponent. For the branched perfluorinated polyether, 143AD, $n = 0.67$, whereas for the linear PFPE, $n = 0.82$.

In addition, the author [42] has plotted the measured values of n against the measured values of G for lubricant base oils. For whatever reason, low values of n only appeared for large values of G. This may be more closely related to the experience of manufacture of synthetic lubricants where a base oil with small values of both n and G would have poor film-forming ability, silicone oils for example. Alternatively, the more highly branched structures may have lower molecular weight at a given viscosity so that G will be larger for the highly branched structures with inherently low n.

In order to meet multi-grade requirements for lubricants, polymeric viscosity index improvers are added to base oils. The objective is to minimize the viscosity–temperature variation for base oils. The conventional model assumes that the molecular configuration of the oil soluble VII varies with temperature. At low temperature, the dissolved polymer molecule curls up in a roughly spherical form, with minimum contribution to base oil viscosity. As temperature is raised, the sphere volume increases owing to the increased solubility and results in a corresponding increase in viscosity of the solution.

The multi-grade lubricants are polymer solutions which often display a second Newtonian plateau as in Figs. 7.3, 7.7, 8.1 and 8.3, with second Newtonian viscosity, μ_2. Often, μ_2 can be approximated by the viscosity, μ_b, of the solvent [43]. For lubricants the solvent is the base oil and this is not always a useful approximation. Novak and Winer [44] investigated the shear-dependent viscosity of polymer blended mineral oils at 4% polymer concentration in a high-pressure capillary viscometer. For a polyalkylmethacrylate polymer $\mu/\mu_2 = 1.67$ and $\mu/\mu_b = 2.0$ and here the assumption, $\mu_2 \approx \mu_b$, is accurate. For a polyalkylstyrene, $\mu/\mu_2 = 1.63$ whereas $\mu/\mu_b = 4.9$ and the assumption is not useful here.

The solvent in SRM 2490 of Fig. 7.3 is pristane with viscosity, $\mu_b = 0.0029$ Pa s at 50°C and ambient pressure, which is very different from the second Newtonian viscosity that appears in the figure to be of the order of $\mu_2 = 1$ Pa s. Van Krevelen [45] has associated the viscosity of the second Newtonian with the viscosity of the unentangled polymer when entanglement contributes to the solution viscosity. This may explain the large ratio μ_2/μ_b for the SRM 2490 where the molecular weight of the polyisobutylene, 10^6 kg/kmol, is likely above the entanglement limit. For lubricants the polymer is not likely to be entangled and the appearance of the second Newtonian is more likely a result of a saturation of the molecular alignment process. If the solvent or base oil begins to shear-thin at a level of shear stress for which the polymer continues to contribute to the viscosity then, of course, the second Newtonian cannot appear. This has been reported [46] by the author's laboratory for a polybutene/mineral oil solution and has also been observed in formulated gear oils [46]. In the absence of data showing a second Newtonian plateau, the safest approach may be to set μ_2 equal to the viscosity of the base oil.

References

[1] Bair, S., "The High Pressure Rheology of Some Simple Model Hydrocarbons", *Proc. Instn. Mech. Engrs.*, Vol. 216, Part J, 2002, pp. 139–149.

[2] Hamrock, B.J. and Dowson, D., *Ball Bearing Lubrication – The Elastohydrodynamics of Elliptical Contacts*, Wiley-Interscience, New York, 1981.

[3] Bair, S., "Recent Developments in High-Pressure Rheology of Lubricants", *Lubricants and Lubrication*, Dowson, D., Taylor, C.M., Childs, T.H.C. and Dalmaz, G., (Eds.), Elsevier Science, 1995, pp. 169–187.

[4] Bair, S., Vergne, P. and Querry, M., "A Unified Shear-Thinning Treatment of both Film Thickness and Traction in EHD", *Tribol. Lett.*, Vol. 18, No. 2, 2005, pp. 145–152.

[5] Carreau, P.J., "Rheological Equations from Molecular Network Theories", *Trans. Soc. Rheology*, Vol. 16, No. 1, 1972, pp. 99–127.

[6] Bair, S., "A More Complete Description of The Shear Rheology of High-Temperature, High-Shear Journal Bearing Lubrication", *STLE Tribol. Trans.*, Vol. 39, No. 1, 2006, pp. 39–45.

[7] Tanner, R.I., *Engineering Rheology*, second edition, Oxford University Press, Oxford, 2000, pp. 18–19, 55, 292–293, 325.

[8] Bair, S., "Normal Stress Difference in Liquid Lubricants Sheared under High Pressure", *Rheologica Acta*, Vol. 35, 1996, pp. 13–23.

[9] Harrison, G., *The Dynamic Properties of Supercooled Liquids*, Academic Press, London, 1976.

[10] Hutton, J.F. and Phillips, M.C., "Shear Modulus of Liquids at Elastohydrodynamic Pressures", *Nature Phys. Sci.*, Vol. 238, 1972, pp. 141–142.

[11] Barlow, A.J., Harrison, G., Irving, J.B., Kim, M.G., Lamb, J. and Pursley, W.C., "The Effect of Pressure on the Viscoelastic Properties of Liquids", *Proc. Roy. Soc. London*, Series A, Vol. 327, 1972, pp. 403–412.

[12] Hutton, J.F., "Reassessment of Rheological Properties of LVI 260", *ASME J. Lubr. Techn.*, Vol. 106, 1984, pp. 536–537.

[13] Paluch, M., Sekula, M., Pawlus, S., Rzoska, S.J., Ziolo, J. and Roland, C.M., "Test of the Einstein-Debye Relation in Supercooled Dibutylphthalate at Pressures up to 1.4 GPa", *Phys. Rev. Lett.*, Vol. 90, No. 20, 2003, pp. 175702-1–175702-4.

[14] Debye, P., *Polar Molecules*, Dover, New York, 1929, pp. 77–85.

[15] Einstein, A., "On the Movement of Small Particles Suspended in a Stationary Liquid Demanded by the Molecular-Kinetic Theory of Heat", *Annalen der Physik*, Vol. 17, 1905, pp. 549–556.

[16] Kleinschmidt, R.V., Bradbury, D. and Mark, M., *Viscosity and Density of Over Forty Lubricating Fluids of Known Composition at Pressures to 150,000 psi and Temperatures to 425 F*, ASME, New York, 1953.

[17] Mondello, M. and Grest, G.S. "Viscosity Calculations of n-Alkanes by Equilibrium Molecular Dynamics", *J. Chem. Phys.*, Vol. 106, 1997, pp. 9327–9336.

[18] Daivis, P.J. and Evans, D.J., "Comparison of Constant-Pressure and Constant Volume Nonequilibrium Simulations of Sheared Model Decane", *J. Chem. Phys.*, Vol. 100, 1994, pp. 541–547.

[19] Cui, S.T., Cummings, P.T. and Cochran, H.D. "The Calculation of Viscosity of Liquid n-Decane and n-Hexadecane by the Green-Kubo Method", *Mol. Phys.*, Vol. 93, 1998, pp. 117–121.

[20] Haile, J.M., *Molecular Dynamics Simulation*, Wiley, New York, 1992.

[21] Mondello, M. and Grest, G.S., "Molecular-Dynamics of Linear and Branched Alkanes", *J. Chem. Phys.*, Vol. 103, 1995, pp. 7156–7165.

[22] Evans, D.J. and Morriss, G.P., "Nonlinear-Response Theory for Steady Planar Couette-Flow", *Phys. Rev. A*, Vol. 30, 1984, pp. 1528–1530.

[23] McCabe, C., Cui, S.T., Cummings, P.T., Gordon, P.A. and Saeger, R.B., "Examining the Rheology of 9-Octylheptadecane to Giga-Pascal Pressures", *J. Chem. Phys.*, Vol. 114, 2001, pp. 1887–1891.

[24] Cummings, P.T. and Evans, D.J., "Nonequilibrium Molecular Dynamics Approaches to Transport Properties and Non-Newtonian Fluid Rheology", *Ind. Eng. Chem. Res.*, Vol. 31, 1992, pp. 1237–1252.

[25] Kioupis, L.I. and Maginn, E.J. "Molecular Simulation of Poly-alpha-olefin Synthetic Lubricants: Impact of Molecular Architecture on Performance Properties", *J. Phys. Chem. B*, Vol. 103, 1999, pp. 10781–10790.

[26] McCabe, C., Manke, C.W. and Cummings, P.T., "Predicting the Newtonian Viscosity of Complex Fluids from High Strain Rate Molecular Simulations", *J. Chem. Phys.*, Vol. 116, 2002, pp. 3339–3342.

[27] Bair, S., McCabe, C. and Cummings, P.T., "Comparison of Non-Equilibrium Molecular Dynamics with Experimental Measurements in the Nonlinear Shear-Thinning Regime", *Phys. Rev. Lett.*, Vol. 88, No. 5, 2002, pp. 058302-1–058302-4.

[28] Bair, S. and Khonsari, M., "An EHD Inlet Zone Analysis Incorporating the Second Newtonian", *ASME J. Tribol.*, Vol. 118, No. 2, 1996, pp. 341–343.

[29] Schurz, J., Vollrath-Rodiger, M. and Krassig, H., "Rheologische Untersuchungen an Polacrylnitril-Spinnlosungen: Erweiterung HV 6 fur hohere Temperaturen", *Rheologica Acta*, Vol. 20, 1981, pp. 569–578.

[30] Greenwood, J.A. and Kauzlarich, J.J., "Elastohydrodynamic Film Thickness for Shear Thinning Lubricants", *Proc. Instn. Mech. Engrs.*, Vol. 212, Part J, 1998, pp. 179–191.

[31] Bair, S., "Reference Liquids for Quantitative Elastohydrodynamics: Selection and Rheological Characterization", *Tribol. Lett.*, Vol. 22, No. 2, 2006, pp. 197–206.

[32] Winter, H.H., "Viscous Dissipation in Shear Flows of Molten Polymers", *Advances in Heat Transfer*, Vol. 13, 1977, pp. 205–267.

[33] Bair, S., "The Shear Rheology of Thin Compressed Liquid Films", *Proc. Instn. Mech. Engrs.*, Vol. 216, Part J, 2002, pp. 1–17.

[34] Bair, S. and Qureshi, F., "Ordinary Shear Thinning Behavior and Its Effect Upon EHL Film Thickness", *Proceedings of the 29th Leeds/Lyon Symposium on Tribology*, Elsevier Tribology Series No. 41, Elsevier, 2002, pp. 693–699.

[35] Bair, S., "The High-Pressure Rheology of Mixtures", *ASME J. Tribol.*, Vol. 126, No. 4, 2004, pp. 697–702.

[36] Florez, W.F., Power, H. and Janna, F.C., "Multi-Domain Dual Reciprocity for the Solution of Inelastic Non-Newtonian Flow Problems at Low Reynolds Number", *Comput. Mech.*, Vol. 27, 2001, pp. 396–411.

[37] Yasuda, K., Armstrong, R.C. and Cohen, R.E., "Shear Flow Properties of Concentrated Solutions of Linear and Star Branched Polystyrenes", *Rheologica Acta*, Vol. 20, No. 2, 1981, pp. 163–178.

[38] Bair, S., Khonsari, M. and Winer, W.O., "High-Pressure Rheology of Lubricants and Limitations of the Reynolds Equation", *Tribol. Int.*, Vol. 31, No. 10, 1998, pp. 573–586.

[39] Ree, F.H., Ree, T. and Eyring, H., "Relaxation Theory of Transport Problems in Condensed Systems", *Ind. Eng. Chem.*, Vol. 50, 1958, pp. 1036–1038.

[40] Wood-Adams, P.M., "The Effect of Long Chang Branches on the Shear Flow Behavior of Polyethylene", *J. Rheology*, Vol. 45, No. 1, 2001, pp. 203–210.

[41] Jabbarzadeh, A., Atkinson, J.D. and Tanner, R.I., "The Effect of Branching on Slip and Rheological Properties of Lubricants in Molecular Dynamics Simulation of Couette Shear Flow", *Tribol. Int.*, Vol. 35, 2002, pp. 35–46.

[42] Bair, S., "Shear-Thinning Correction for Rolling/Sliding Elastohydrodynamic Film Thickness", *Proc. I. Mech. E.*, Vol. 219, Part J, 2005, pp. 69–73.

[43] Tanner, R.I., *Engineering Rheology*, second edition, Oxford University Press, Oxford, 2000, p. 19.

[44] Novak, J.D. and Winer, W.O., "Some Measurements of High Pressure Lubricant Rheology", *ASME J. Lubr. Techn.*, Vol. 90, 1968, pp. 580–590.

[45] Van Krevelen, D.W., *Properties of Polymers: Their Correlation with Chemical Structure: Their Numerical Estimation and Prediction*, third edition, Elsevier, Amsterdam, 1990, p. 477.

[46] Bair, S. and Qureshi, F., "The High-Pressure Rheology of Polymer–Oil Solutions", *Tribol. Int.*, Vol. 36, 2003, pp. 637–645.

CHAPTER 9

The Glass Transition and Related Transitions in Liquids Under Pressure

9.1 Measurements of Glass Transition at Elevated Pressure

In the previous chapter, we saw that a transition in the rheological response to shear occurs when the molecule has insufficient time to attain its equilibrium orientation for an applied rate of shear. In this case the generalized viscosity, η, begins to decrease with increased rate of shear. There are other processes for which transitions occur as a result of the molecule being unable to respond to the change in its environment because of a large value of low shear viscosity, μ. The most important transition for EHL, aside from the transition to shear-thinning, is the glass transition.

The shear viscosity of simple organic liquids at the melting temperature is of the order of 10^{-2} Pa s [1, p. 4]. Melting of a sample, which has solidified in the conventional sense, may be observed during a viscosity measurement in a falling body viscometer by decreasing the pressure until the body begins to fall [2]. Some very approximate observed melting state points and viscosities are listed in Table 9.1. Consider the suggestion that the nucleation and growth of crystals always occur in the time scale of a viscosity measurement. This suggestion leads to the incorrect conclusion that the viscosities of the liquids studied in Chapter 5, all of which exceed 1 Pa s, represent measurements performed with crystalline solid materials. The melting point is well-defined in an experiment since liquids do not readily exist in the solid state at temperatures above the melting point; however, the liquids with complicated structure that are used as lubricants do supercool below the melting temperature.

Trimethylolpropane trilaurate can exist as a liquid at 20°C for years in spite of the melting temperature for ambient pressure being 29°C. The viscosity may be measured to about 0.45 GPa at 50°C and to about 1.2 GPa at 100°C. At these pressures the sample begins to slowly crystallize and a viscosity measurement can be resumed only after the liquid is heated to a temperature greater than the melting temperature. The viscosity of trimethylolpropane trilaurate at the ambient pressure melting temperature

Table 9.1. Melting pressures measured in a falling body viscometer.

Liquid	Temperature/°C	Approximate Melting Pressure/GPa	Approximate Melting Viscosity/Pa s
n–Octane	20	0.6	0.020
n–Dodecane	40	0.20	0.007
n–Dodecane	100	0.6	0.013
Dicyclohexyl	40	0.1	0.010
Dicyclohexyl	100	0.3	0.011
Cyclododecane	40	0.2	0.06
Cyclododecane	100	0.4	0.07
Acetone	30	0.8	0.003
Perfluorohexane	20	0.3	0.020

is about 0.07 Pa s; however, this material will remain liquid for sufficient time that viscosities of up to 10 Pa s may be measured. The liquids for which viscosity measurements were reported in Chapter 5 supercool easily. They exist as liquids for long times at temperatures below the melting temperature. The behavior of liquids at temperatures below the melting temperature (or above the melting pressure) can be understood by considering the structural changes required to place the molecules in a configuration favorable to crystallization. These structural changes are slowed by the viscosity. As the temperature is decreased or the pressure is increased, the tendency for crystallization is strengthened but, at the same time, the viscosity which slows the structural rearrangement is increased.

If a supercooled liquid is cooled or compressed further the structural relaxation time will be further increased. The volume change in response to a change in temperature or pressure, which may be calculated from an equation of state such as (4.6) or (4.8), can be decomposed into an instantaneous solid-like component and a liquid-like relaxing component. If during an isothermal compression step the experimental observation time of the measurement of volume is shorter than the structural relaxation time, the measured change in volume will be that of the instantaneous solid-like component only. There will be insufficient time for the relaxation of volume to occur and the equilibrium volume will not be achieved. The thermodynamic pressure, p, that is used in a liquid equation of state (Chapter 4) will no longer be equal to the mechanical pressure, p_m, in equation (2.2). Further compressions are now accompanied by the solid-like compressibility of a glass. This is the widely accepted explanation for the glass transition [3].

The glassy material is not at equilibrium and the volume or density depends upon the history. A glass formed by a glass transition that was traversed under pressure will have a greater density than a glass formed at ambient pressure. A glass formed by a rapid traverse of the glass transition will have a lower density than a glass that was cooled or compressed slowly. Plazek and Bero [3] found that increasing the cooling rate of an epoxy resin by three orders-of-magnitude raised the glass transition temperature at atmospheric pressure by 5°C and decreased the density of the resulting glass by 0.14%. Therefore, all properties that depend upon density, including perhaps viscosity, will be history dependent. That is to say, these properties will depend upon the pressure at the glass transition, the *fictive pressure* [4,5], and upon the rate at which the glass transition was approached.

Some glass transition conditions are listed in Table 9.2 for four well-defined organic liquids and two synthetic lubricant base oils and a mineral oil. Both isothermal compression and isobaric cooling histories were utilized. The compression rate was roughly 10 MPa/min and the cooling rate was roughly 1 K/min. A rapid change in compressibility or expansivity, as measured in a dilatometer of the types shown in Fig. 4.1, was used to detect the glass transition. Examples of compression curves and cooling curves are shown in Figs. 4.2 and 4.3 respectively.

Table 9.2. Glass transition temperatures or pressures by isobaric cooling or isothermal compression, respectively.

Liquid	Method	Temperature/°C	Pressure/GPa
11,13-Dioctyl 11-methyltricosane	compression	−18	0.71
	compression	2	0.92
Dibenzylethylbenzene, DBEB	compression	26	0.58
	compression	45	0.79
	compression	65	0.88
	compression	85	1.01
	cooling	−27	0.21
	cooling	−19	0.28
	cooling	−7	0.35
	cooling	−2	0.42
1,1-(1,1,3-Trimethyl-1, 3-propanediyl)bis-cyclohexane, BCH	compression	14	0.60
	compression	39	0.83
	compression	55	0.95
	cooling	−26	0.28
	cooling	−15	0.35
	cooling	−9	0.41
Bis(phenoxyphenoxy)benzene, 5P4E	compression	63	0.41
	compression	87	0.58
	compression	99	0.67
	compression	123	0.78
	cooling	56	0.36
	cooling	64	0.44
	cooling	93	0.61
PAO-40	compression	0	0.80
	compression	9	0.92
	compression	16	1.00
	compression	25	1.29

Continued

Table 9.2. —Cont'd

Liquid	Method	Temperature/°C	Pressure/GPa
Pentaerythritol ester	compression	4	1.27
Mineral oil, N1	compression	34	0.58
	compression	45	0.70
	compression	52	0.79
	compression	60	0.90
	cooling	–18	0.35
	cooling	–5	0.41

In Section 6.5, the viscosity divergence temperature, T_∞, was fitted to a logarithmic function of pressure in analogy to the glass transition temperature, T_g. The form applicable to $T_g(p)$ is

$$T_g = T_{g0} - \frac{T'^2_{g0}}{T''_{g0}} \ln\left(1 - \frac{T''_{g0}}{T'_{g0}} p\right) \tag{9.1}$$

where $T'_{g0} = \left|dT_g/dp\right|_{p=0}$ and $T''_{g0} = \left|d^2T_g/dp^2\right|_{p=0}$. For bis(phenoxyphenoxy)benzene, 5P4E, $T_{g0} = -24.9°\text{C}$, $T'_{g0} = 275\,\text{K/GPa}$, and $T''_{g0} = -420\,\text{K/GPa}^2$ using the data of Table 9.2.

The glass transition at atmospheric pressure is most often measured by calorimetry. This technique has been extended to elevated pressures by Richmond et al. [7] who obtained, for 5P4E, a glass transition pressure of 0.21 GPa at 25°C. Their result is in excellent agreement with equation (9.1) which yields 0.209 GPa. The glass transition in lubricants under pressure has been studied by Brillouin scattering of light, a method whereby the frequency shift in scattered light from elastic waves is used to measure the sound velocity [8]. A change in the pressure or temperature coefficient of the sound velocity is used to detect the glass transition.

9.2 Measurements of Dielectric Transition at Elevated Pressure

There are other techniques for the measurement of the transition from a relaxing to a nonrelaxing response aside from dilatometry, calorimetry and light scattering that have the advantage of easily varying and quantifying the rate of change of the environment. One technique that was proposed for this purpose is the measurement of the dielectric relaxation time, λ_D [1, p. 186]. In Section 8.3.2 we discovered that the Einstein–Debye relation (8.10), which was derived for the rotational relaxation time of a polar molecule in an electric field, was useful for a prediction of the shear rate at the transition to shear-thinning. Dielectric response of polar molecules may be studied at elevated pressures by applying a periodic electric field to a polar liquid sample used as the dielectric medium in a capacitor. For convenience and sensitivity, the capacitor is formed by placing the sample in the narrow space between metal concentric cylinders. The dynamic impedance of the capacitor, which is excited at a specified frequency, is measured yielding the permittivity, a measure of the electrically stored energy relative to a vacuum, and the loss tangent, a measure of electrical dissipation.

Lubricant base oils are sometimes made polar by the presence of oxygen in the chemical structure. Even when the nominal composition is that of a nonpolar hydrocarbon, a low concentration of polar molecules in a nonpolar solvent may provide a transition in the permittivity that is controlled by the solvent viscosity [1, p. 179]. Therefore, a relatively simple measurement of dielectric response may be used to study the temperature and pressure dependence of viscosity for mineral oils and other nonpolar liquids [9].

Examples of dielectric measurements on the polar liquid, 5P4E, are shown in Figs. 9.1 and 9.2 for the permittivity and loss tangent, respectively. The pressure is 207 MPa and the

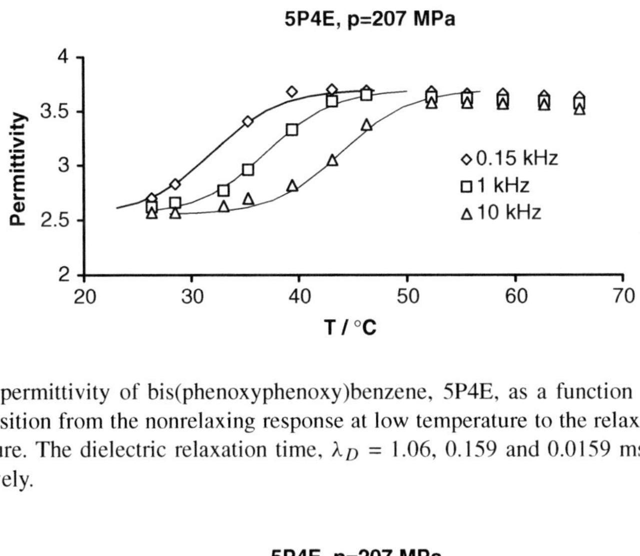

Fig. 9.1. The permittivity of bis(phenoxyphenoxy)benzene, 5P4E, as a function of temperature showing a transition from the nonrelaxing response at low temperature to the relaxing response at high temperature. The dielectric relaxation time, λ_D = 1.06, 0.159 and 0.0159 ms at 32, 37 and 44°C respectively.

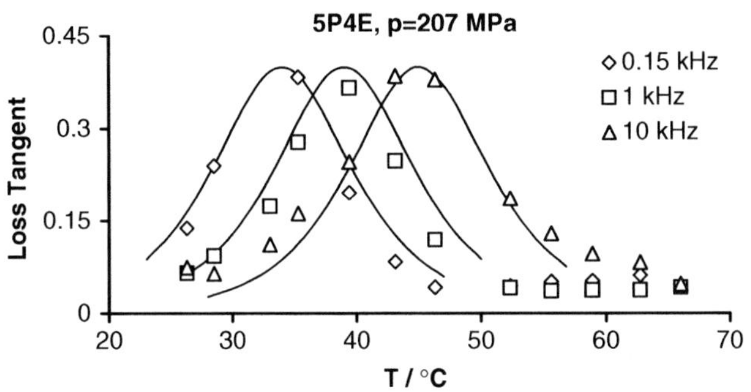

Fig. 9.2. The loss tangent of bis(phenoxyphenoxy)benzene, 5P4E, as a function of temperature showing the peak in electrical dissipation where the dielectric relaxation time coincides with the period of the excitation.

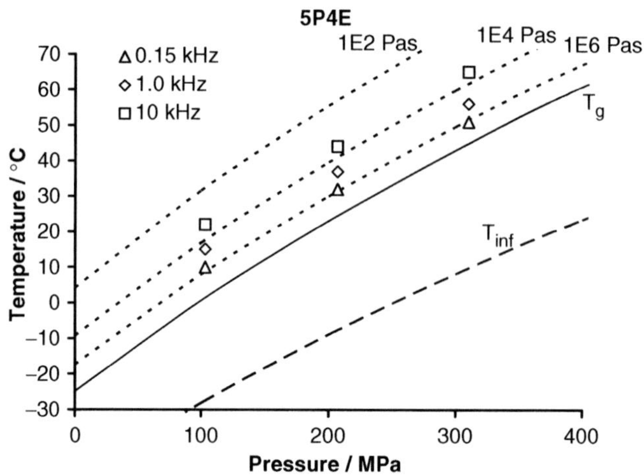

Fig. 9.3. The dielectric transition temperature (data points), the glass transition temperature, the divergence temperature and the isoviscous temperatures of bis(phenoxyphenoxy)benzene, 5P4E, as a function of pressure.

temperature is varied to detect the dielectric transition which is placed at the temperature at the midpoint of the change in permittivity and the temperature at the peak in the loss. At high temperatures the molecule may freely rotate to align with the electric field and the dielectric "constant", the permittivity, is large on the right-hand side of Fig. 9.1. At low temperatures the viscosity inhibits molecular rotation and the permittivity is low, more like that of a nonpolar molecule. At these extremes the loss is minimal. When the frequency of the field is tuned to the dielectric relaxation time, $\omega = 2\pi f = 1/\lambda_D$, the loss is maximized as shown in Fig. 9.2. Similar results were obtained at pressures of 103 and 310 MPa. All of the measured dielectric transition temperatures are plotted in Fig. 9.3 at the indicated frequencies.

9.3 The Transitions as Isoviscous States

The low-shear viscosity, μ, of 5P4E was measured at temperatures of 40, 60, 80 and 100°C for pressures from atmospheric to the pressure for a viscosity of about 10^4 Pa s. The variation of viscosity with both temperature and pressure was fitted to the VTF equation,

$$\mu = \mu_\infty \exp\left[\frac{D_F T_\infty}{T - T_\infty}\right] \tag{6.3}$$

assuming the pressure dependence of the divergence temperature that was introduced in Section 6.5 as

$$T_\infty = T_0 - \frac{T_0'^2}{T_0''} \ln\left(1 - \frac{T_0''}{T_0'} p\right) \tag{6.42}$$

Here $T_0' = |dT_\infty/dp|_{p=0}$ and $T_0'' = \left|d^2T_\infty/dp^2\right|_{p=0}$ and the resulting parameters are $\mu_\infty = 7.704 \times 10^{-5}\,\text{Pa s}$, $D_F = 3.438$, $T_0 = -50.23°\text{C}$, $T_0' = 238\,\text{K/GPa}$ and $T_0'' = -378\,\text{K/GPa}^2$. The temperatures resulting in viscosities of 10^2, 10^4 and 10^6 Pa s are plotted as the three isoviscous curves in Fig. 9.3.

It is immediately obvious from the figure that the dielectric transition for a given frequency occurs at approximately a constant value of viscosity. One can go even farther by dividing the viscosity by the dielectric relaxation time, resulting in $\mu/\lambda_D \approx 4 \times 10^8$ Pa within a factor of two over the entire range of measurement.

The divergence temperature, $T_\infty(p)$, sometimes called the idealized glass transition temperature, is also plotted in Fig. 9.3. This is the extrapolated temperature for unbounded viscosity. The glass transition temperature, $T_g(p)$, from equation (9.1) is plotted in the figure as well. The difference, $T_g(p) - T_\infty(p)$, varies from 30 to 40 K as pressure increases. Solving equations (6.3) and (6.42) for the temperature that yields a viscosity of 10^8 Pa s approximates the glass transition temperature from equation (9.1) within about one degree over the pressure interval of 100 to 400 MPa. Therefore, to a good approximation, the glass transition of 5P4E is an isoviscous state with a characteristic viscosity, $\mu_g = 10^8$ Pa s.

The concept of the glass transition as an isoviscous state may be used to construct a temperature–pressure–viscosity correlation. Yasutomi et al. [6] modified a free-volume temperature–viscosity relation, the WLF equation (6.4), using μ_g as a reference viscosity at the reference temperature chosen as T_g and given by equation (9.1). It was necessary to assume a relation for the pressure-dependence of the thermal expansion of the free volume. This framework can reproduce the inflection in the log viscosity versus pressure curve and the increase in temperature–viscosity coefficient with pressure; however, the pressure–viscosity coefficient at the glass transition, that is, the pressure fragility, m_p, is badly underestimated [10]. A useful product of the Yasutomi model has been that the value of μ_g has been estimated from extrapolation of the viscosity to the transition state for many lubricants. The value of μ_g appears to be dependent upon the material and falls in the range of $10^7 \le \mu_g \le 10^{12}$ Pa s.

Remaining with 5P4E as the example, a flow curve for this liquid has been fitted [11] to the Carreau equation (8.1),

$$\eta = \mu \left[1 + \left(\frac{\mu\dot{\gamma}}{G}\right)^2\right]^{\frac{(n-1)}{2}} \tag{8.1}$$

for the temperature of 40°C and pressure of 260 MPa, yielding $n = 0.68$ and $G = 4\,\text{MPa}$ [11]. The transition from linear Newtonian shear response to shear-thinning occurs for the Weissenberg number, $\mu\dot{\gamma}/G$, equal to unity. If the modulus, G, is relatively independent of temperature and pressure, as it seems to be, then the transition from Newtonian shear response to shear-thinning will occur for a given shear rate, $\dot{\gamma}$, at an isoviscous state with viscosity equal to $G/\dot{\gamma}$. Therefore a plot of temperature against pressure, for the transition to shear-thinning, will appear for constant $\dot{\gamma}$ as curves parallel to the isoviscous curves in Fig. 9.3. For example, the shear-thinning transition for $\dot{\gamma} = 400\,\mathrm{s}^{-1}$ is the same as the curve marked 1×10^4 Pa s in Fig. 9.3.

9.4 The Pressure Variation of Viscosity across the Transition

The data of Table 9.2 clearly show that, for the usual operating temperatures of EHL contacts, there are at least some high-pressure contacts operating with pressures that exceed the glass transition pressure of ordinary liquid lubricants, especially considering that an increase in the rate of compression will reduce the glass transition pressure. Then a thorough examination of the rheology of concentrated contact lubrication should include a framework for specifying the variation of viscosity with temperature and pressure across the glass transition and to the highest pressures of the contact.

The measurement of the low-shear viscosity across the glass transition presents a very challenging experimental problem. If the glass transition viscosity is even as low as $\mu_g = 10^7$ Pa s, and the Newtonian limit is $G = 10^6$ Pa, then the shear rate must be less than $0.1\ \mathrm{s}^{-1}$ for a valid measurement at the glass transition. For ordinary liquid lubricants, in order to thoroughly define the pressure–viscosity response at pressures above the glass transition, shear rates as low as $10^{-7}\ \mathrm{s}^{-1}$ would be required. Most workers in this area would have a conceptual difficulty with the notion of a low-shear viscosity of a glass. Viscosity measurements have been reported, however, for conditions that cross the glass transition for isobaric cooling [12] and isothermal compression [13,14] and it is not clear that these measurements were at sufficiently low stress to represent the limiting low-shear behavior. In any event, these measurements indicate a reduced dependence on temperature [12] and pressure [13,14] in going from liquid to glass. Such observations would be consistent with the free volume theory which associates changes in viscosity with changes in free volume since the compressibility and expansivity are substantially reduced by the glass transition.

Returning now to the problem of specifying the low-shear viscosity for $p > p_g$, the present author has found that for simple traction calculations [14,15] and film thickness calculations [16], the results are insensitive to the assumption that either:

1. the low-shear viscosity of the glass is continuous with the low-shear viscosity of the liquid and increases exponentially with pressure with the same coefficient, α_g, or

2. the low-shear viscosity of the glass is a constant equal to the low-shear viscosity of the liquid at the transition, μ_g.

The value that is assumed for μ_g is not critical as long as it is greater than about 10^7 Pa s.

The reason that the central film thickness is insensitive to the glass viscosity assumption is simply that the film thickness appears to be established in the inlet zone when the viscosity has increased by a factor of about 20. Then further changes in viscosity have little effect on the distribution of the liquid within the contact because the Poiseuille component of flow has disappeared. Traction calculations become insensitive to the assumption of the glass viscosity, at low slide-to-roll ratio, because the traction, under sufficiently high pressure to require consideration of the glass viscosity, is governed by the elasticity of the rollers. For higher slide-to-roll ratio, with the viscosity at the level of the glass transition, the shear stress will reach large values relative to the pressure for even small shear rate. Then the traction is governed by a form of liquid failure known as limiting shear stress that is the topic of the following chapter.

References

[1] Harrison, G., *The Dynamic Properties of Supercooled Liquids*, Academic Press, London, 1976, p. 4, 186.

[2] Akki, R., Bair, S. and Abhiraman, A.S., "Low Shear Viscosity and Crystallization in Dilute Solutions of Polymers at High Pressures: Falling Body Viscometry of High Molecular Weight Polyethylene Solutions", *Polym. Eng. Sci.*, Vol. 35, No. 22, 1995, pp. 1781–1784.

[3] Plazek, D.J. and Bero, C.A., "Precise Glass Temperatures", *J. Phys.: Condens. Matter*, Vol. 15, 2003, pp. S798–S802.

[4] Tool, A.Q., "Relation between Inelastic Deformability and Thermal Expansion of Glass in Its Annealing Range", *J. Amer. Ceramic Soc.*, Vol. 29, No. 9, 1946, pp. 240–253.

[5] Hirai, N. and Eyring, H., "Bulk Viscosity of Polymeric Systems", *J. Polym. Sci.*, Vol. 37, 1959, pp. 51–70.

[6] Yasutomi, S., Bair, S. and Winer, W.O., "An Application of a Free Volume Model to Lubricant Rheology 1-Dependence of Viscosity on Temperature and Pressure", *ASME J. Tribol.*, Vol. 106, 1984, pp. 291–303.

[7] Richmond, J., Nilsson, O. and Sandberg, O., "Thermal Properties of Some Lubricants under High Pressure", *J. Appl. Phys.*, Vol. 56, No.7, 1984, pp. 2065–2067.

[8] Alsaad, M., Bair, S., Sanborn, D. and Winer, W.O., "Glass Transitions in Lubricants: Its Relation to Elastohydrodynamic Lubrication (EHD)", *ASME J. Lubr. Techn.*, Vol. 100, 1978, pp. 418–422.

[9] Bair, S. and Winer, W.O., "Some Observations on the Relationship between Lubricant Mechanical and Dielectric Transitions under Pressure", *ASME J. Lubr. Techn.*, Vol. 102, No. 2, 1980, pp. 229–235.

[10] Bair, S. and Kottke, P., "Pressure-Viscosity Relationships for Elastohydrodynamics", *STLE Tribol. Trans.*, Vol. 46, 2003, pp. 289–295.

[11] Bair, S. "Complete Isothermal Solution for the Viscous Regime of Concentrated Contact Lubrication", *Int. J. Appl. Mech. Eng.*, Vol. 7, No. 3, 2002, pp. 719–727.

[12] Kobayashi, H., Takahashi, H. and Hiki, Y., "*Viscosity of Glassy Polymers Near and Below the glass Transition Temperature*", Proc. XIII International Congress Rheology, Cambridge, UK, 2000, pp. 4-91–4-92.

[13] Brazhkin, V.V. and Lyapin, A.G., "Universal Viscosity Growth in Metallic Melts at Megabar Pressures: The Vitreous State of the Earth's Inner Core", *Physics Uspekhi*, Vol. 43, No. 5, 2000, pp. 493–508.

[14] Bair, S. and Winer, W.O., "The High Pressure High Shear Stress Rheology of Liquid Lubricants", *ASME J. Tribol.*, Vol. 114, 1992, pp. 1–13.

[15] Bair, S. and Kotzalas, M., "A Simple Calculation of EHL Traction at Low Slide/Roll Ratio Considering Roller Elasticity", *Submitted to Tribol. Int.*, 2006.

[16] Bair, S., Liu, Y. and Wang, Q.J., "The Pressure-Viscosity Coefficient for Newtonian EHL Film Thickness with General Piezoviscous Response", *ASME J. Tribol.*, Vol. 123, No. 3, 2006, pp. 624–631.

CHAPTER 10

Shear Localization, Slip and the Limiting Stress

10.1 Introduction

In 1960 Smith [1] wrote of the concentrated contact lubrication problem: "An important consideration is that there is a limit to the shear stress that a fluid can transmit, even when under pressure great enough to prevent tensile failure. At this "limiting shear stress" the material is assumed to deform plastically, so that an increase in rate of shear produces no increase in shearing stress". This concept, that a liquid possesses a plastic yield strength (not to be confused with the yield behavior that prevents flow at low stress), which results in rate-independent shearing when the shear stress reaches the limiting value, was somewhat obvious from the appearance of measured traction curves. These traction curves, plots of traction force, shear stress or traction coefficient, often show a curious plateau where the traction coefficient is independent of the sliding velocity at a value, Λ, of the traction coefficient of about from 0.03 to 0.12 which appears to be material dependent and which is only slightly temperature dependent and independent of the average contact pressure. The effect is unlike the thermal maximum in a Couette flow curve with apparent shear rate in which there is no plateau, only a local maximum. The plateau can extend for more than a decade in shear rate [2,3]. Johnson, in 1986 [3], appears to be first to associate this form of liquid failure with the operation of a shear band.

Examples of thick-film traction curves which clearly show limiting stress response are shown in Fig. 10.1. The liquid lubricant is a well-studied polybutene, L100, having an average molecular weight of about 510 kg/kmol [4]. The instrument is the twin disc machine of Fig. 1.5 employing skewed rollers to measure transverse traction. The rolling velocity is $\bar{u} = 2.5$ m/s and the inlet temperature is 50°C. Figures 10.1(a–c) correspond to Hertz contact pressures of $p_H = 1.93, 2.52$ and 3.31 GPa, respectively. The traction coefficients for the plateaus are $\Lambda = \bar{\tau}/\bar{p} = 0.103, 0.107$ and 0.107, respectively and the theoretical Newtonian central film thicknesses are $h_c = 848$, 804 and 761 nm. The load increases by a factor of five and the pressure increases by 71% over the three measurements, whereas the traction coefficient for the plateau remains within a range of only 4%. For EHL, these are relatively thick films, yet the friction in Fig. 10.1 for $|\Sigma| > 0.007$

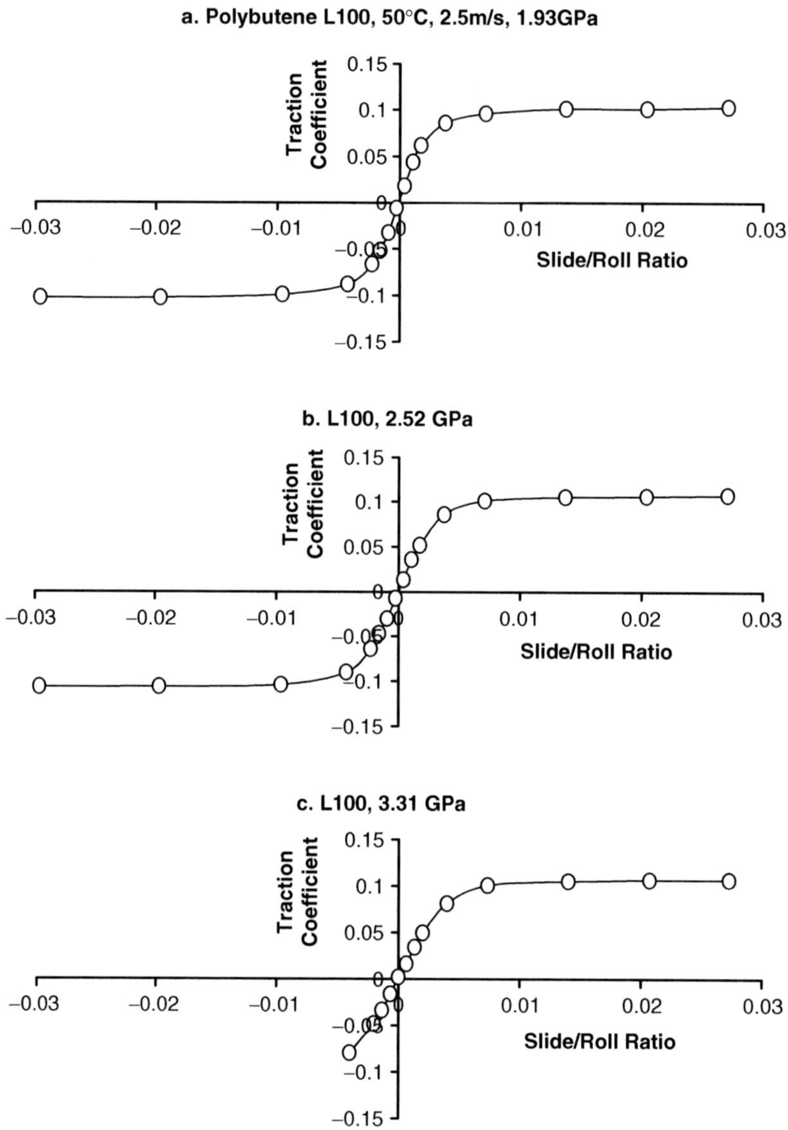

Fig. 10.1. Traction curves for a polybutene, L100, at 50°C and Hertz pressure of 1.93, 2.52 and 3.31 GPa. The rolling velocity is 2.5 m/s.

behaves in the manner of an unlubricated contact with a traction coefficient nearly independent of load and sliding velocity. The liquid must, therefore, possess a kind of internal friction coefficient that is related to Λ.

This chapter addresses the most important aspect, for EHL traction, of the liquid response to a high shear stress at high-pressure. Unfortunately, it is also the least studied and most poorly understood. The very existence of a limiting stress, $\tau_L = \Lambda p$, may still be controversial within EHL and is relatively unknown outside the field. The combination of a very high hydrostatic pressure with a very high shear stress in a liquid-filled geometry that allows for the efficient removal of the dissipated heat is probably unique to this field. Within the last fifteen years, experimental observations have shown that the limit to shear stress is not constitutive. It results from a localization of the deformation into bands where the shear rate is much greater than in the bulk of the liquid and these bands may be regarded essentially as slip planes just as Johnson predicted [3]. The greatest unresolved problem in modeling of EHL may be the development of a framework for incorporating the inhomogeneous shear that results in a stress limit.

10.2 Measurements of Rate Independent Shear Stress

The traction curves of Fig. 10.1 alone do not validate the existence of a limiting stress. A traction curve may have the appearance of a flow curve, but it must be emphasized that the quantities that are presented in a traction curve are the averaged values over the contact area. There is precedence for the misinterpretation of a traction curve as constitutive behavior [5]. The local pressure within the contact covers the interval of ambient pressure to approximately the Hertz contact pressure, resulting in orders-of-magnitude variation in the low-shear viscosity. Therefore it may be possible to postulate a rate-dependent, temperature-dependent constitutive law that will result in the appearance of a rate-independent traction plateau as seen in Fig. 10.1. This has not been done, but the possibility cannot be discounted on the basis of traction data alone.

Flow curves generated in a viscometer where the temperature, pressure and shear stress are homogeneous are the sort of evidence required to validate the limiting stress concept. Figure 10.2 provides just such evidence for a branched perfluoropolyalkylether, 143 AD. A pressurized Couette viscometer was employed [6] and the shear-thinning parameters are listed in Table 8.1. The solid straight curves in Fig. 10.2 represent power-law shear-thinning. The conditions for the Nahme–Griffith number, $Na = 0.3$, are shown by a thin straight curve and all data were generated at low values of Na. Then the effect of temperature variation across the film on viscosity may be neglected. The stress histories that provided the data in the figure also showed no time dependence; therefore the effect of heating of the cylinders on the viscosity may be ignored. The flow curves at $p = 172$ MPa indicate a curious departure from the power-law behavior at a shear stress of about 7 MPa, which is not present in the $p = 207$ MPa data until the stress reaches 10 MPa. In particular, the last three data points for 7°C have a nearly constant shear stress of $\tau \approx 12.8$ MPa while the shear rate doubles. This is the manifestation of the limiting shear stress in a flow curve and an increase in pressure will move the transition to a greater stress. Then for the PFPE here, $\Lambda = \tau_L/p = 0.75$. A similar flow curve is shown for the mineral oil LVI 260 in Fig. 10.3 where $\Lambda = \tau_L/p = 0.53$.

It can be appreciated that it is experimentally very difficult to generate a shear stress of several megapascals in a liquid while maintaining a nearly constant temperature. This is only possible with thin films of approximately 1 μm and very large viscosity. Early measurements of limiting shear stress outside of a contact were performed on the liquid above the glass transition pressure [7]. These measurements were relatively simple

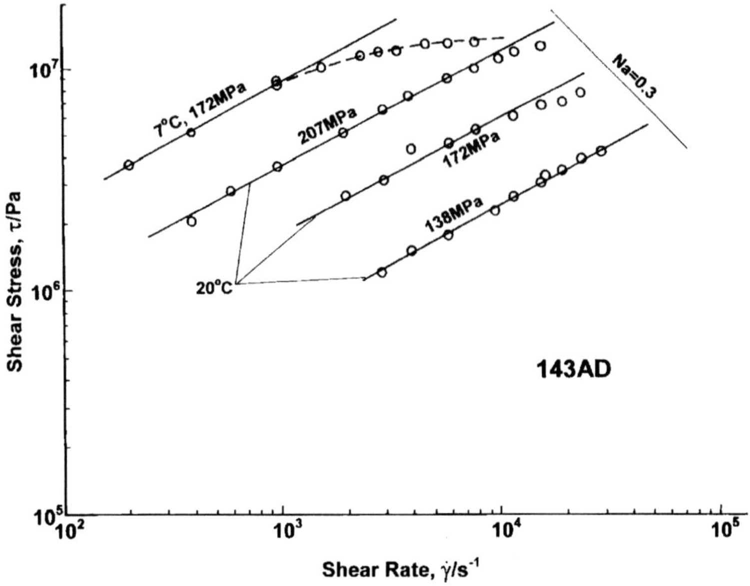

Fig. 10.2. Flow curves for a branched PFPE showing a transition to rate-independent response at 172 MPa and 7°C. Reproduced from Ref. [6] with permission of Elsevier.

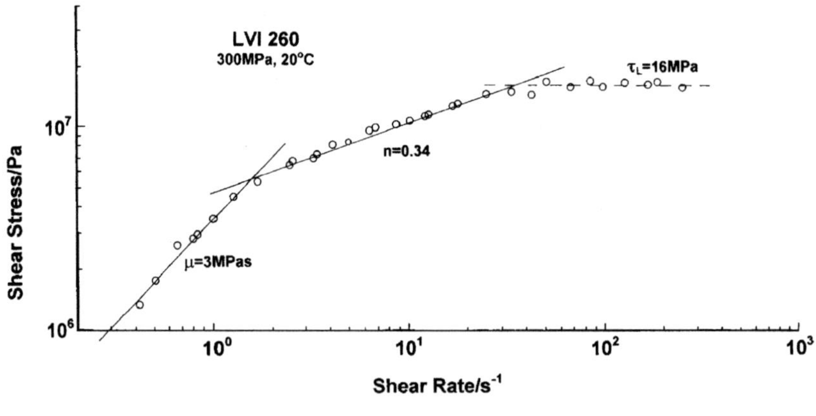

Fig. 10.3. The log of shear stress versus the log of apparent shear rate for the model mineral oil, LVI 260 showing a transition to rate-independent response. Reproduced from Ref. [6] with permission of Elsevier.

and easy to accomplish using an axially translating concentric cylinder device shearing a thick annular slab of glassy liquid. The shear rates were low, less than $1\,s^{-1}$. The limiting stress was clearly shown to rise linearly with pressure with a value of Λ which was characteristic of the material [7]. This technique also provided the measurements of the elastic shear modulus, G_∞, that compared well with acoustic measurements [8].

Later, the thin-film rotating concentric cylinder instrument was developed [5] to extend the measurements to lower pressures, below the glass transition. Capillary instruments cannot generate the necessary shear stress. When the pressure is very low, which was the case for early pressurized Couette viscometers, the effect of the onset of the limiting stress may obscure the ordinary shear-thinning which is weak by comparison. At these low pressures the following empirical expression for the apparent shear rate may apply for $\tau \geq 0$ [5].

$$\dot{\gamma}_{app} = \frac{\tau_L}{\mu} \ln\left(\frac{1}{1 - \frac{\tau}{\tau_L}}\right) \tag{10.1}$$

The development of a higher pressure viscometer [9] raised the value of τ_L in a typical experiment and revealed the ordinary shear-thinning that is present at lower shear stress. The shear-thinning was first evident as a broadening of the transition to rate independence as pressure was increased [9]. The broadening effect, as should be expected, was more pronounced for higher molecular weight liquids.

One of the puzzling features of the limiting stress is that the proportionality between τ_L and pressure remains unchanged as the glass transition is traversed. This suggests that the limiting stress is not so much a rheological property related to the density as it is a property of an interface.

10.3 Flow Visualization of Shear Bands

An alternative explanation to the internal slip hypothesis for rate independent shear of liquids is the possible failure of the no-slip boundary condition. Fluid mechanics generally assumes that the velocity of the liquid continuum is continuous with the velocity of a contacting solid. One might posit that a difference in velocity of the liquid and the solid at the boundary exactly accommodates any increase of the boundary velocity once a limiting shear stress has been attained. Flow visualization provides a definitive test.

Bair and Qureshi [10] developed a flow visualization cell for high pressure (280 MPa) that generated a shear stress up to about 40 MPa in planar Couette shear. The instrument is pictured in Fig. 10.4 with one fourth of the device cut away. A moving shaft was driven axially by a hydraulic cylinder. A flat machined on the side of the moving shaft and the end of a stationary pin, shown in Fig. 10.4, defined a volume of liquid being sheared. The shearing gap, the gap between the moving shaft and the end of the pin, could be varied from 40 to 200 mm. The shaft was sealed to the housing using the technique of Fig. 3.2(d). Two windows of the type shown in Fig. 3.5 were provided with axes perpendicular to the axes of the shaft and pin. The shearing gap was illuminated through the lower window in Fig. 10.4 and viewed through the upper window by a microscope with a long working distance objective. The apparent shear rate was less than 10 s^{-1}.

Tracer particles of 2 μm diameter were added to the liquid sample. No wall slip, within the 1 μm resolution of measurement of the particle position, was detected during shear.

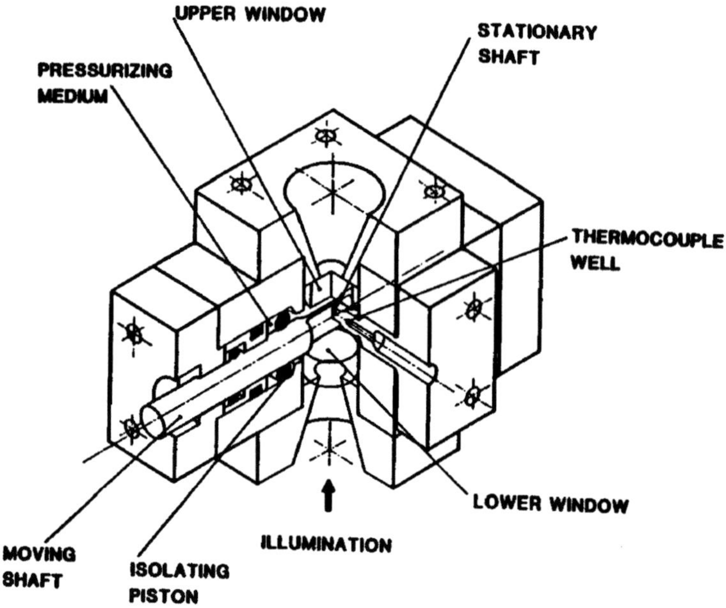

Fig. 10.4. Flow visualization cell. Reproduced from Bair, S., Qureshi, F. and Winer, W.O., "Observations of Shear Localization in Liquid Lubricants under Pressure", *Trans. ASME J. Tribol.*, Vol. 115, 1993, pp. 507–514, by permission of the American Society of Mechanical Engineers.

Clearly detected were individual slip events within the liquid of up to 10 μm displacement [10]. These slip events were not entirely in the direction of flow. Two different slip directions were eventually identified [11] by the appearance of shear bands within the liquid. The slip occurs in a short time, generally less than the smallest resolvable time of 30 ms.

Micrographs of shear bands in the polyphenyl ether, 5P4E, and a cycloaliphatic hydrocarbon traction fluid, MCS 1218, are shown in Fig. 10.5 from Ref. [12]. The lower dark solid in the micrographs is the stationary pin and the upper dark solid is the moving shaft which is moving from right to left as shown by the arrow in the schematic above. Dark bands of two types, inclined to the surfaces at angles θ_1 and θ_2, are visible in the micrographs. Band angles are listed in Table 10.1. Bands form when the shear velocity exceeds a critical rate [10]. The rate of production of bands increases with shear velocity above this critical value. The image of a band persists for a time which is proportional to the viscosity and the image persistence time for the polyphenyl ether is approximately $\mu/10^8$ Pa [13] or approximately the shear relaxation time. From the appearance of the band and its persistence, the image is more likely the result of a mechanically induced dilation than a thermal expansion which would have a much shorter persistence [13]. Ohno et al. [14] have also observed shear bands in a liquids sheared at elevated pressure using a similar optical cell to Fig. 10.4.

Shear bands in liquids have been shown to occur in squeeze films at pressures above 1 GPa [15]. Two types of bands are also found in squeeze films inclined to the surfaces

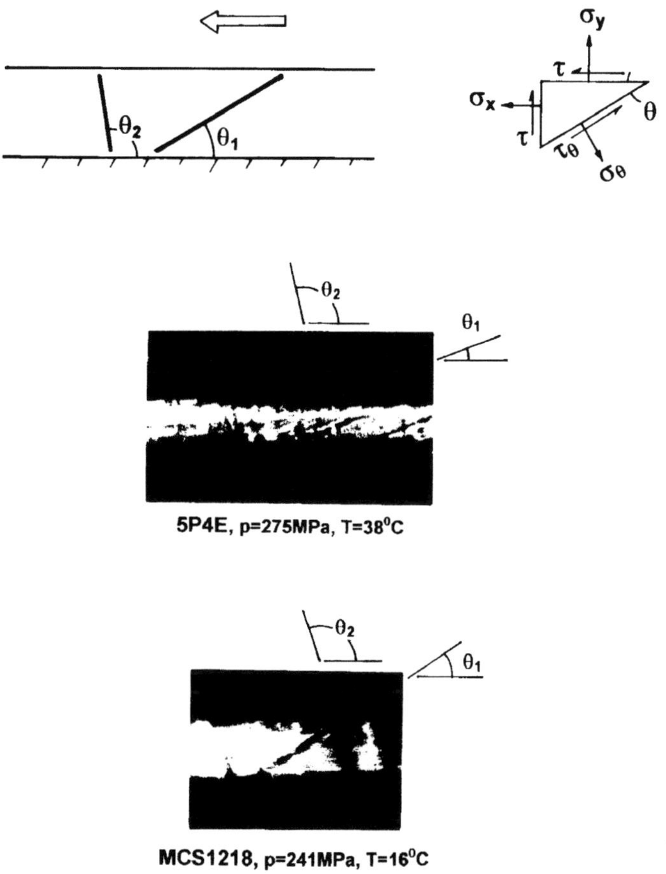

Fig. 10.5. Micrographs of shear bands shown below and inclination angles and stresses defined at top. Reproduced from Ref. [12] with permission of Elsevier.

Table 10.1. Shear band properties for the micrographs of Fig. 10.5.

Liquid	$\theta_1/°$	$\theta_2/°$	Λ	ς	φ	N_1/τ
5P4E	18	102	0.091	0.0525	0.106	1.16
MCS 1218	32	107	0.17	0.195	0.268	2.29

at slightly greater than 45° in opposing directions. They were found to be similar in appearance to shear bands photographed from the plane strain compression of solid polymethylmethacrylate [16] at atmospheric pressure.

The observation of shear bands operating at high shear stress suggests that the rate-independent response in EHL traction and high-pressure viscometer measurements is a result of intermittent slip along bands inclined to the flow direction in a manner similar to the yield behavior of some solid polymers. A similar pressure variation of stress at yield is found for solid polymers. For polymethylmethacrylate at room temperature, the shear stress at yield is 50 MPa $+ 0.2p$ [16]. The association of the limiting stress with slip along planes inclined to the flow complicates the application of the limiting stress model to EHL. Shear bands have been visualized in Couette flow where the shear stress is initially homogeneous. Do shear bands operate in the inhomogeneous shear of Poiseuille flow? What would be the orientation? Models for shear banding are considered next.

10.4 Mohr–Coulomb Failure Criterion

Bowden [16] has expressed the Mohr–Coulomb failure criterion as "failure will occur when the shear stress on any plane in the material reaches a critical value which varies linearly with the stress normal to that plane". Using the definition of stresses at the top of Fig. 10.5, the critical shear stress is [16]

$$|\tau_\theta| = \tau_{\theta 0} - \phi \sigma_\theta \tag{10.2}$$

where ϕ is the *coefficient of internal friction* and σ_θ is the stress normal to the cutting plane (compressive is negative).

For the geometry in Fig. 10.5, if the flow is in the x-direction, $N_1 = \sigma_x - \sigma_y$ and we define $\varsigma = N_1/2p$. If the pressure is equal to the mechanical pressure, then $p = -(\sigma_x + \sigma_y)/2$. The compressive normal stress on an arbitrary inclined plane is

$$-\sigma_\theta = p + \varsigma p \cos 2\theta - \tau \sin 2\theta \tag{10.3}$$

and the shear stress on an arbitrary inclined plane is

$$\tau_\theta = \tau \cos 2\theta + \varsigma p \sin 2\theta \tag{10.4}$$

If limiting stress results from slip at planes defined by the Mohr–Coulomb criterion, then for slip, $\tau = \tau_L = \Lambda p$, where Λ is the limiting stress pressure coefficient (or the limiting value of the traction coefficient in a full-film traction curve displaying a plateau). If we can neglect the $\tau_{\theta 0}$ in equation (10.2), then

$$\frac{|\tau_\theta|}{-\sigma_\theta} = \phi \tag{10.5}$$

Since slip must first occur on the plane defined by equation (10.5), this equation applies to the plane for which $|\tau_\theta|/(-\sigma_\theta)$ is maximized. There can be no greater values of this ratio and therefore,

$$\frac{d}{d\theta}\left(\frac{\tau_\theta}{-\sigma_\theta}\right) = 0 \tag{10.6}$$

Dividing (10.4) by (10.3) and differentiating, (10.6) becomes

$$\Lambda^2 + \varsigma^2 - \Lambda \sin 2\theta + \varsigma \cos 2\theta = 0 \tag{10.7}$$

There are two solutions for θ, the two shear band angles, θ_1 and θ_2. The internal friction coefficient can then be found from either θ_1 or θ_2 as

$$\phi = \frac{\Lambda \cos 2\theta + \varsigma \sin 2\theta}{1 + \varsigma \sin 2\theta - \Lambda \cos 2\theta} \tag{10.8}$$

Results obtained from the measured shear band angles are listed in Table 10.1. Shear band angles seem to be independent of the operating conditions [10], therefore the values of Λ, ς, φ, and $N_1/\tau = 2\varsigma/\Lambda$ in Table 10.1 should be material constants. This is sensible for Λ since the plateau traction coefficients in Figs. 10.1(a–c) are reasonably independent of Hertz pressure. Nonequilibrium molecular dynamics simulations [12] have provided results that show that N_1/τ may be reasonably constant at a value of about 1 to 2 at very high τ in low-molecular-weight liquids.

Flow curves obtained in a pressurized Couette viscometer for the two liquids of Fig. 10.5 are shown in Fig. 10.6. Each of these curves display Newtonian, power-law shear-thinning and rate-independence. For the shear-thinning regime, $n = 0.68$ and 0.41 and $G = 4$ and 10 MPa for the 5P4E and the MCS1218 respectively. The limiting shear stresses predicted from $\tau_L = \Lambda p$ using the tabulated values of Λ in Table 10.1 are plotted as the broken horizontal lines in Fig. 10.6 and they agree reasonably with the observed rate-independent plateau. The MCS1218 displays an overshoot before the limiting stress is reached. This behavior has not been observed in other materials.

There is one other type of prediction for the limiting stress in Fig. 10.6 shown as a dot–dash line which represents the condition for $\tau = 1/\alpha$. This prediction is based on a change in character of the equations of motion which will be discussed next.

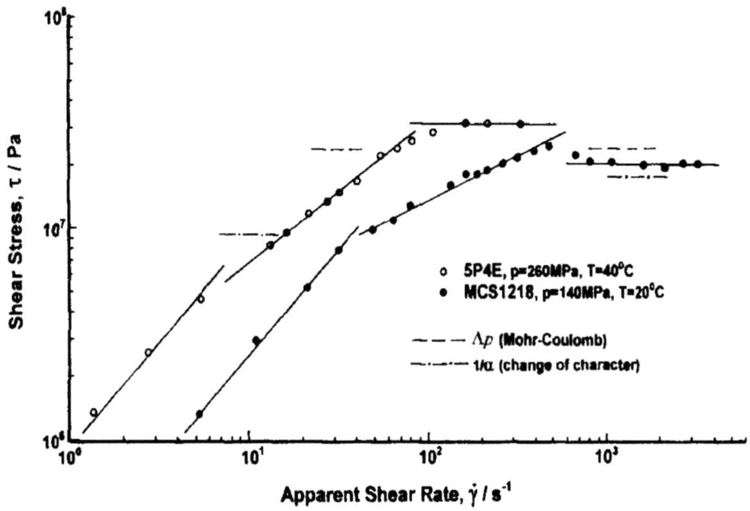

Fig. 10.6. Flow curves for the liquids pictured in Fig. 10.5. Reproduced from Ref. [12] with permission of Elsevier.

10.5 Change of Character of the Piezoviscous Navier–Stokes Equations

Shear banding is often modeled in solids using the method of characteristics. Shear banding is considered to be the result of a change of character of the governing field equations from elliptic to hyperbolic [17]. The piezoviscous effect can produce a loss of ellipticity of the Navier–Stokes equations.

Consider the two-dimensional creeping flow of an incompressible Newtonian liquid. The Navier–Stokes equations reduce to

$$\frac{\partial p}{\partial x} = 2\frac{\partial}{\partial x}\left(\mu\frac{\partial u}{\partial x}\right) + \frac{\partial}{\partial y}\left[\mu\left(\frac{\partial v}{\partial x} + \frac{\partial u}{\partial y}\right)\right] \tag{10.9a}$$

$$\frac{\partial p}{\partial y} = 2\frac{\partial}{\partial y}\left(\mu\frac{\partial v}{\partial y}\right) + \frac{\partial}{\partial x}\left[\mu\left(\frac{\partial v}{\partial x} + \frac{\partial u}{\partial y}\right)\right] \tag{10.9b}$$

The continuity equation is

$$\frac{\partial u}{\partial x} + \frac{\partial v}{\partial y} = 0 \tag{10.10}$$

and the viscosity gradients in equation (10.9) are $\partial\mu/\partial x = \alpha\mu\partial p/\partial x$ and $\partial\mu/\partial y = \alpha\mu\partial p/\partial y$, where α is defined in equation (5.1). The Navier–Stokes equations for an incompressible piezoviscous Newtonian liquid are for creeping flow,

$$\left(2\alpha\mu\frac{\partial u}{\partial x} - 1\right)\frac{\partial p}{\partial x} + \alpha\mu\left(\frac{\partial u}{\partial y} + \frac{\partial v}{\partial x}\right)\frac{\partial p}{\partial y} + \mu\left(\frac{\partial^2 u}{\partial x^2} + \frac{\partial^2 u}{\partial y^2}\right) = 0 \tag{10.11a}$$

$$\left(2\alpha\mu\frac{\partial v}{\partial y} - 1\right)\frac{\partial p}{\partial y} + \alpha\mu\left(\frac{\partial u}{\partial y} + \frac{\partial v}{\partial x}\right)\frac{\partial p}{\partial x} + \mu\left(\frac{\partial^2 v}{\partial x^2} + \frac{\partial^2 v}{\partial y^2}\right) = 0 \tag{10.11b}$$

Differentiating equation (10.10) gives two more governing equations,

$$\frac{\partial^2 u}{\partial x^2} + \frac{\partial^2 v}{\partial y \partial x} = 0 \tag{10.12a}$$

$$\frac{\partial^2 v}{\partial y^2} + \frac{\partial^2 u}{\partial y \partial x} = 0 \tag{10.12b}$$

The system of equations (10.11a) to (10.12b) has the characteristic equation

$$\left[1 + \left(\frac{dy}{dx}\right)^2\right]\left[C_2\left(\frac{dy}{dx}\right)^2 - 2C_1\frac{dy}{dx} + C_3\right] = 0 \tag{10.13}$$

where $C_1 = \alpha\left[(\partial u/\partial y) + (\partial v/\partial x)\right]$, $C_2 = (2\alpha\partial u/\partial x) - (1/\mu)$ and $C_3 = (2\alpha\partial v/\partial y) - (1/\mu)$.

The characteristic directions are $dy/dx = \pm i, \left(C_1 \pm \sqrt{C_1^2 - C_2C_3}\right)/C_2$ and there are always two imaginary characteristics. The equations of motion are elliptic when there are no real characteristics. Therefore

$$\mu\left[\left(\frac{\partial u}{\partial y}+\frac{\partial v}{\partial x}\right)^2-4\frac{\partial u}{\partial x}\frac{\partial v}{\partial y}\right]^{\frac{1}{2}}<\frac{1}{\alpha} \tag{10.14}$$

for ellipticity. The left-hand-side of the equation above is the principal shear stress. Then when the principal shear stress exceeds the reciprocal of the local pressure–viscosity coefficient, two real characteristics appear with characteristic directions along which discontinuities in the velocity may propagate.

The hypothesis, that shear bands result from a change in character of the governing equations, can be tested by comparing plateau stresses with the shear stresses for $\tau = 1/\alpha$ in Fig. 10.6. These are plotted as the dot–dash lines. The correlation with rate-independent behavior is not as good as for the Mohr–Coulomb model. Another difficulty for this explanation is that the quantity, $1/\alpha$, decreases with pressure as shown in Fig. 6.11, whereas τ_L must increase.

A surprising property of the incompressible liquid with exponential pressure-dependent viscosity is impossibility of unidirectional slit flow that was first discovered by Bair et al. in 1998 [18] and confirmed by Hron et al. in 2001 [19]. As the shear stress approaches $1/\alpha$, cross film pressure gradients and velocities develop. It has not yet been shown that compressibility will remove these secondary flows.

The piezoviscous Navier–Stokes equations (10.11) contain additional terms with pressure gradients in multiple directions that are not present in the constant viscosity Navier–Stokes equations from which the Reynolds equation was derived. Some workers [18,20] have reached the surprising conclusion that the Reynolds equation (1.11) is not appropriate for piezoviscous problems.

10.6 Thermal Localization, Adiabatic Shear Bands

There is another type of shear band that results from localization of the shear deformation. When a liquid is sheared between parallel plates, in the absence of a pressure gradient, under a controlled stress at high Nahme–Griffith number, a steady flow is not possible. Recall that the Nahme–Griffith number for plane Couette flow of thickness, h, is

$$Na = \frac{\beta\tau^2h^2}{k\mu_i} \tag{7.18}$$

where k is the thermal conductivity of the liquid and μ_i is the initial viscosity. Stress controlled flow for $Na > 3.5138$ results in what Gruntfest [21] termed the thermal catastrophe. Conduction cannot keep pace with the viscous dissipation and the temperature distribution rapidly evolves from near parabolic to the shape of a cusp with an unbounded temperature at the midplane. The velocity distribution rapidly evolves from linear to a step distribution resembling the sliding of two solid slabs. A thermal shear band has been observed in a flow visualization cell at high-pressure [22]. This type of shear band is often called an *adiabatic shear band*. In reality, for adiabatic shear, there is no temperature gradient, and therefore the deformation is homogeneous.

In plane Couette shear, a single band forms. In plane Poiseuille flow, two bands are predicted, symmetrically spaced about the midplane [22]. The critical value of the Nahme–Griffith number for this flow is $Na = 90.37$ [23] where the definition for plane Poiseuille flow is

$$Na = \frac{\beta h^4}{k\mu_i}\left(\frac{dp}{dx}\right)^2 \tag{10.15}$$

10.7 Interfacial Slip

For high-molecular-weight polymeric liquids, the usual (no slip) assumption, that the velocity of the liquid at vanishingly small distance from a solid boundary is equal to the velocity of the boundary, is sometimes violated for flows above a critical wall shear stress [24]. The critical stress is often approximately 0.1 MPa for experiments at atmospheric pressure. This critical stress is interestingly similar to the stress for shear cavitation discussed in Section 7.2.4.

Tanner [25] has reviewed the extensive literature on wall-slip and concluded that slip does not occur when the surface roughness scale of the wall is larger than the molecular dimensions and that pressure probably reduces the tendency for slip. For EHL therefore, where very high pressures are usual and the surfaces have the ordinary engineering roughness of at least 10 nm, wall-slip can be ignored.

References

[1] Smith, F.W., "Lubricant Behavior in Concentrated Contact-Some Rheological Problems", *ASLE Trans.*, Vol. 3, No. 1, 1960, pp. 18–25.

[2] Johnson, K.L. and Tevaarwerk, J.L., "Shear Behavior of Elastohydrodynamic Oil Films", *Proc. Roy. Soc. London A*, Vol. 356, 1977, pp. 215–236.

[3] Evans, C.R. and Johnson, K.L., "The Rheological Properties of Elastohydrodynamic Lubricants", *Proc. Instn. Mech. Engrs.*, Vol. 200, No. C5, 1986, pp. 303–312.

[4] Bair, S., "The High-Pressure, High-Shear-Stress Rheology of a Polybutene", *J. Non-Newt. Fluid Mech.*, Vol. 97, 2001, pp. 53–65.

[5] Bair, S. and Winer, W.O., "The High Shear Stress Rheology of Liquid Lubricants at Pressures of 2 to 200 MPa", *ASME J. Tribol.*, Vol. 112, No. 2, 1990, pp. 246–253.

[6] Bair, S., "The Nature of the Logarithmic Traction Gradient", *Tribol. Int.*, Vol. 35, 2002, pp. 591–597.

[7] Bair, S. and Winer, W.O., "Some Observations in High Pressure Rheology of Lubricants", *ASME J. Lubr. Techn.*, Vol. 104, No. 3, 1982, pp. 357–364.

[8] Hutton, J.F., "Reassessment of Rheological Properties of LVI 260", *ASME J. Lubr. Techn.*, Vol. 106, 1984, pp. 536–537.

[9] Bair, S. and Winer, W.O., "A New High-Pressure, High-Shear Stress Viscometer and Results for Lubricants", *STLE Tribol. Trans.*, Vol. 36, No. 4, 1993, pp.721–725.

[10] Bair, S., Qureshi, F. and Winer, W.O., "Observations of Shear Localization in Liquid Lubricants under Pressure", *Trans. ASME J. Tribol.*, Vol. 115, 1993, pp. 507–514.

[11] Bair, S., Winer, W.O. and Distin, K.W., "Experimental Investigations into Shear Localization in an Operating EHD Contact", *Thin Films in Tribology*, Dowson, D., Priest, M., Taylor, C.M., Ehret, P., Childs, T.H.C., Dalmaz, G., Lubrecht, A.A., Berthier, Y., Flamand, L. and Gerreges, J.M. (Eds.), 1993, pp. 383–388.

[12] Bair, S. and McCabe, C., "A Study of Mechanical Shear Bands in Liquids at High Pressure", *Tribol. Int.*, Vol. 37, 2004, pp. 783–789.

[13] Bair, S. and Winer, W.O., "A Rheological Basis for Concentrated Contact Friction", *Dissipative Processes in Tribology*, Dowson, D., Taylor, C.M., Childs, T.H.C., Godet, M. and Dalmaz, G. (Eds.), Elsevier Science, 1994, pp. 37–44.

[14] Ohno, N., Shiratake, A., Kumamoto, S., Kuwano, N. and Hirano, F., "Occurrence of Shear Bands at High-Pressure", *Proc. JAST Tribol. Conf.*, 1997, pp. 515–517 (in Japanese).

[15] Bair, S., "Recent Developments in High-Pressure Rheology of Lubricants", *Lubricants and Lubrication*, Dowson, D., Taylor, C.M., Childs, T.H.C. and Dalmaz, G. (Eds.), Elsevier, 1995, pp. 169–187.

[16] Bowden, P.B., "The Yield Behavior of Glassy Polymers", *The Physics of Glassy Polymers*, Haward (Ed.), Halstead Press, 1973, pp. 279–339.

[17] Lee, Y.K., "Conditions for Shear Banding and Material Instability in Finite Elastoplastic Deformation", *Int. J. Plast.*, Vol. 5, 1989, pp. 197–226.

[18] Bair, S., Khonsari, M. and Winer, W.O., "High-Pressure Rheology of Lubricants and Limitations of the Reynolds Equation", *Tribol. Int.*, Vol. 31, No. 10, 1998, pp. 573–586.

[19] Hron, J., Malek, J. and Rajagopal, K.R., "Simple Flows of Fluids with Pressure-Dependent Viscosities", *Proc. Roy. Soc. London A*, Vol. 457, 2001, pp. 1603–1622.

[20] Rajagopal, K.R. and Szeri, A.Z., "On an Inconsistency in the Derivation of the Equations of Elastohydrodynamic Lubrication", *Proc. Roy. Soc. London A*, Vol. 459, 2003, pp. 2771–2786.

[21] Gruntfest, I.J., "Apparent Departures from Newtonian Behavior in Liquids Caused by Viscous Heating", *Trans. Soc. Rheology*, Vol. 9, No. 1, 1965, pp. 425–441.

[22] Bair, S., Qureshi, F. and Khonsari, M., "Adiabatic Shear Localization in a Liquid Lubricant Under Pressure", *Trans. ASME J. Tribol.*, Vol. 116, 1994, pp. 705–709.

[23] Murthy, H.B.D., *Limitations to Non-Isothermal Flow in Lubricant Films Due to Viscous-Shear Heating*, Thesis, Delft University, 1985, p. 324.

[24] Archer, L.A., Larson, R.G. and Chen, Y.L., "Direct Measurements of Slip in Sheared Polymer Solutions", *J. Fluid Mech.*, Vol. 301, 1995, pp. 133–151.

[25] Tanner, R.I., *Engineering Rheology*, second edition, Oxford University Press, Oxford, 2000, pp. 133–139.

CHAPTER 11

The Reynolds Equation

11.1 Background

The hydrodynamic lubrication problem might be addressed by a solution of the constitutive relation combined with the stress equilibrium equations (That combination is the Navier–Stokes equations in the case of a generalized Newtonian liquid.) and the continuity equation. Such solutions are very difficult. Fortunately, the thin-film nature of the lubrication problem allows for simplifications, which along with the Newtonian assumption result in an equation that has been the foundation of hydrodynamic lubrication, the Reynolds [1] equation. Pressure-dependent viscosity was apparently not considered by Reynolds. Even so, EHL has made extensive and successful use of this relationship between the pressure gradient, the film shape, surface velocity and viscosity.

In the following, the coordinate system is placed so that the film shape does not vary in time, the surface velocities have no component in the y-direction and the z-direction is across the film. The origin is placed at the center of the film and the film being very thin in comparison to other dimensions allows that curvature may be neglected. Begin with the incompressible generalized Newtonian equation (2.7). Later on, the liquid will be made compressible by allowing the density to vary while neglecting volume relaxation. The equations of stress equilibrium, ignoring body forces, are

$$\rho\frac{Du}{Dt} = \frac{\partial\sigma_{xx}}{\partial x} + \frac{\partial\sigma_{yx}}{\partial y} + \frac{\partial\sigma_{zx}}{\partial z} \tag{11.1}$$

and so forth for the y- and z-directions. The inertia term on the LHS can be ignored as these are creeping flows. Then the expressions for the components of stress are substituted from equation (2.7) to obtain

$$\frac{\partial p}{\partial x} = \frac{\partial}{\partial x}\left(2\eta\frac{\partial u}{\partial x}\right) + \frac{\partial}{\partial y}\left[\eta\left(\frac{\partial u}{\partial y} + \frac{\partial v}{\partial x}\right)\right] + \frac{\partial}{\partial z}\left[\eta\left(\frac{\partial w}{\partial x} + \frac{\partial u}{\partial z}\right)\right] \tag{11.2}$$

and so forth for the y- and z-directions. The assumption is made [2] that owing to the thin film geometry, that the only significant velocity gradients are $\partial u/\partial z$ and $\partial v/\partial z$ and the others may be ignored. The resulting set of equations is

$$\frac{\partial p}{\partial x} = \frac{\partial \eta}{\partial z}\frac{\partial u}{\partial z} + \eta\frac{\partial^2 u}{\partial z^2} \tag{11.3a}$$

$$\frac{\partial p}{\partial y} = \frac{\partial \eta}{\partial z}\frac{\partial v}{\partial z} + \eta\frac{\partial^2 v}{\partial z^2} \tag{11.3b}$$

$$\frac{\partial p}{\partial z} = \frac{\partial \eta}{\partial x}\frac{\partial u}{\partial z} + \frac{\partial \eta}{\partial y}\frac{\partial v}{\partial z} \tag{11.3c}$$

It is customary at this point in the derivation to make the argument that $\partial p/\partial z$ must be small because of the small dimension of the film thickness [2]. Here we are only concerned with hard elastohydrodynamic films where the piezoviscous response is not only important but essential. An inconsistency has been revealed in the use of the iso-viscous Reynolds equation for EHL [3,4]. To evaluate the effect of a pressure-dependent viscosity on the assumption that the cross-film gradient may be ignored, consider a two-dimensional flow, where $v = 0$ and $\partial p/\partial y = 0$, of a liquid in the terminal regime where $\eta = \mu$. Recalling the definition of the local pressure–viscosity coefficient, α, equation (5.1), $\partial \mu/\partial x = \alpha\mu\partial p/\partial x$ and $\partial \mu/\partial z = \alpha\mu\partial p/\partial z$. Equations (11.3a) and (11.3c) become

$$\frac{\partial p}{\partial x} = \alpha\mu\frac{\partial p}{\partial z}\frac{\partial u}{\partial z} + \mu\frac{\partial^2 u}{\partial z^2} \tag{11.4a}$$

$$\frac{\partial p}{\partial z} = \alpha\mu\frac{\partial p}{\partial x}\frac{\partial u}{\partial z} \tag{11.4c}$$

These equations can be combined to yield an expression for $\partial p/\partial x$,

$$\frac{\partial p}{\partial x} = \frac{\mu\dfrac{\partial^2 u}{\partial z^2}}{1 - \left(\alpha\mu\dfrac{\partial u}{\partial z}\right)^2} \tag{11.5}$$

When the cross-film pressure gradient is neglected, as it must be to arrive at a Reynolds equation, the denominator on the RHS side of the above equation should be close to unity. This denominator may be approximated by $1 - (\alpha\tau)^2$. Therefore, the Reynolds equation is valid for a Newtonian liquid only when $\alpha\tau \ll 1$. In the case of a fragile liquid (see Section 6.3), α may be as large as 0.1 MPa^{-1} at high pressure and, for a strong liquid, α may be as small as 0.01 MPa^{-1} at high pressure. Then the Reynolds equation will be valid at high pressure in the terminal regime for the fragile liquid if $\tau \ll 10$ MPa and for the strong liquid if $\tau \ll 100$ MPa. These stresses are often attained in the Hertz zone in either case. Recall (Section 8.4) that for a generalized Newtonian liquid in the fully developed power-law regime, for a rate controlled process the effective pressure–viscosity coefficient is reduced to approximately $n\alpha$, where n is the power-law exponent.

If however, $\partial u/\partial z$ in the denominator of equation (11.5) is primarily the result of a pressure gradient, then the process is stress controlled and the operative coefficient is still simply α. Also complicating matters, for a generalized Newtonian liquid there will be additional viscosity gradients due to the shear dependence of viscosity.

The discussion above of limitations of the Reynolds equation is related to the change in character of the piezoviscous incompressible Navier–Stokes equations that was examined in Section 10.5. There is the possibility of secondary flows which do not occur when the viscosity does not depend exponentially on pressure [3,5]. The generalized Newtonian model is itself a simplification which neglects normal stress differences among other non-Newtonian phenomena. The Reynolds equation, for example, does not predict the thrust that is generated in the parallel disc rheometer of Section 7.3.2 and therefore may underestimate the load generated in a thrust washer lubricated by polymer thickened oil.

Henceforth, the possibility of a significant cross-film pressure gradient will be ignored so that derivations of appropriate forms of the Reynolds equation may be obtained. Remaining with the Newtonian assumption, the viscosity gradients in the z-direction vanish and the simplified Navier–Stokes equations are then

$$\frac{1}{\mu}\frac{\partial p}{\partial x} = \frac{\partial^2 u}{\partial z^2} \tag{11.6a}$$

$$\frac{1}{\mu}\frac{\partial p}{\partial y} = \frac{\partial^2 v}{\partial z^2} \tag{11.6b}$$

These may be integrated twice to obtain expressions for the velocities,

$$u = \frac{1}{2\mu}\frac{\partial p}{\partial x}z^2 + C_1(x, y)z + C_2(x, y) \tag{11.7a}$$

$$v = \frac{1}{2\mu}\frac{\partial p}{\partial y}z^2 + C_3(x, y)z + C_4(x, y) \tag{11.7b}$$

The surface velocities provide the boundary conditions. At $z = -h/2$, $u = u_1$, $w = w_1$ and $v_1 = 0$. At $z = h/2$, $u = u_2$, $w = w_2$ and $v_2 = 0$. Therefore,

$$u = \frac{1}{2\mu}\frac{\partial p}{\partial x}\left(z^2 - \frac{h^2}{4}\right) + \frac{u_2 - u_1}{h}z + \frac{u_2 + u_1}{2} \tag{11.8a}$$

$$v = \frac{1}{2\mu}\frac{\partial p}{\partial y}\left(z^2 - \frac{h^2}{4}\right) \tag{11.8b}$$

The continuity equation is

$$\frac{\partial \rho}{\partial t} + \frac{\partial(\rho u)}{\partial x} + \frac{\partial(\rho v)}{\partial y} + \frac{\partial(\rho w)}{\partial z} \tag{11.9}$$

Assume that the local density is independent of time and integrate across the film to get

$$\int_{-h/2}^{h/2} \frac{\partial(\rho u)}{\partial x} dz + \int_{-h/2}^{h/2} \frac{\partial(\rho v)}{\partial y} dz + \rho(w_2 - w_1) = 0 \tag{11.10}$$

Applying Leibnitz's rule yields

$$\frac{\partial}{\partial x} \int_{-h/2}^{h/2} (\rho u) dz - \rho \bar{u} \frac{\partial h}{\partial x} + \frac{\partial}{\partial y} \int_{-h/2}^{h/2} (\rho v) dz + \rho(w_2 - w_1) = 0 \tag{11.11}$$

and recognizing that $w_2 - w_1 = \bar{u}\,(\partial h/\partial x)$, where $\bar{u} = (u_1 + u_2)/2$ cancels the second and fourth terms on the LHS. The expressions for u and v, equations (11.8) can be inserted into (11.11) to arrive at a two-dimensional Reynolds equation introduced in Chapter 1,

$$\frac{\partial}{\partial x}\left(\frac{\rho h^3}{\mu} \frac{\partial p}{\partial x}\right) + \frac{\partial}{\partial y}\left(\frac{\rho h^3}{\mu} \frac{\partial p}{\partial y}\right) = 12\bar{u} \frac{\partial(\rho h)}{\partial x} \tag{1.17}$$

There are more complete forms available with fewer restrictions [6]. The reader is directed to Dowson [2] for a generalized Reynolds equation which considers the possibility that density and viscosity (but not pressure) may vary across the film.

The Reynolds equation above applies only to the terminal regime and, therefore, is very limited in application to EHL. In the following section, some one-dimensional Reynolds forms are presented for common generalized Newtonian liquid models.

11.2 Reynolds Equations for Generalized Newtonian Fluids

It has become accepted that the full description of non-Newtonian behavior is not required for an accurate film calculation [7]. Normal stress differences and extensional viscosity are not expected to play a significant role in the load support in EHL. The shear dependence of the viscosity is the most important non-Newtonian effect and the inelastic model is sufficient. A significant complication, however, is that the viscosity must depend upon some combination of the three scalar invariants of either the strain rate tensor or of the deviatoric stress tensor. See Section 2.3.1. Suitable definitions are

$$\dot{\gamma}^* = \sqrt{\left(\frac{\partial u}{\partial y} + \frac{\partial v}{\partial x}\right)^2 + \left(\frac{\partial u}{\partial z} + \frac{\partial w}{\partial x}\right)^2 + \left(\frac{\partial w}{\partial y} + \frac{\partial v}{\partial z}\right)^2 + 2\left(\frac{\partial u}{\partial x}\right)^2 + 2\left(\frac{\partial v}{\partial y}\right)^2 + 2\left(\frac{\partial w}{\partial z}\right)^2} \tag{2.8}$$

and

$$\tau^* = \sqrt{\frac{(\sigma_{xx} + p)^2}{2} + \frac{(\sigma_{yy} + p)^2}{2} + \frac{(\sigma_{zz} + p)^2}{2} + (\sigma_{xy})^2 + (\sigma_{xz})^2 + (\sigma_{yz})^2} \tag{2.9}$$

If, as was assumed in the previous section, the only significant velocity gradients are $\partial u/\partial z$ and $\partial v/\partial z$, equation (2.8) reduces to

$$\dot{\gamma}^* = \sqrt{\left(\frac{\partial u}{\partial z}\right)^2 + \left(\frac{\partial v}{\partial z}\right)^2} \tag{11.12}$$

and if the only significant deviatoric stresses are σ_{xz} and σ_{yz}, then

$$\tau^* = \sqrt{\sigma_{xz}^2 + \sigma_{yz}^2} \tag{11.13}$$

A two-dimensional Reynolds equation for a generalized Newtonian liquid must be written for a viscosity function, $\eta(\dot{\gamma}^*)$ or $\eta(\tau^*)$. The coupling of the expressions for the pressure gradients in the two directions makes the derivation of a two-dimensional Reynolds equation for a generalized Newtonian liquid quite difficult and we begin with the one-dimensional forms where $v = 0$ and $\partial p/\partial y = 0$.

11.2.1 One-dimensional Reynolds equations for generalized Newtonian fluids

The shear stress distribution across the film is known from the stress equilibrium equations (11.1) to be governed by

$$\frac{\partial \sigma_{xz}}{\partial z} = \frac{\partial \tau}{\partial z} = \frac{dp}{dx} \tag{11.14}$$

Then, if the shear stress at the midplane is $\tau_m = \tau(z = 0)$,

$$\tau = \tau_m + \frac{dp}{dx} z \tag{11.15}$$

No such simple relationship exists for the shear rate, $\partial u/\partial z = \dot{\gamma}$, as a function of z. It is for this reason that the exact forms have been written only for the viscosity functions of the form $\eta(\tau)$, that is, the forms listed on the RHS of Table 2.1. These are the Newtonian, Ostwald–deWaele, Spriggs, Ellis, Ferry, Rabinowitsch and two models by this author. Of these, only the models that can be integrated twice analytically have provided exact Reynolds equations.

If $v = 0$ and density does not vary across the film, then equation (11.11) becomes

$$\frac{d}{dx}(\rho Q) = 0 \tag{11.16}$$

where the volume flow rate, Q, is

$$Q = \int_{-h/2}^{h/2} u dz = \int_{-h/2}^{h/2} \left[u_1 + \int_{-h/2}^{z} \dot{\gamma} dz \right] dz \tag{11.17}$$

and $\dot{\gamma} = \tau/\eta(\tau)$. For pure rolling, $u_2 = u_1$, the midplane stress is $\tau_m = 0$. For sliding contacts, τ_m is nonzero and if the viscosity depends upon stress, a relation for τ_m

will be required to calculate the flow rate. Derivations of exact Reynolds equations for generalized Newtonian liquid models have been published [8–10].

1. For example, the Newtonian model, $\eta(\tau) = \mu$, yields

$$\tau_m = \mu \frac{(u_2 - u_1)}{h} \tag{11.18}$$

$$Q = \bar{u}h - \frac{h^3}{12\mu}\frac{dp}{dx} \tag{11.19}$$

and applying (11.16), the result is the Newtonian Reynolds equation from Chapter 1, which does not depend upon τ_m,

$$\frac{d}{dx}\left(\frac{\rho h^3}{\mu}\frac{dp}{dx}\right) = 12\bar{u}\frac{d(\rho h)}{dx} \tag{1.11}$$

2. The Ostwald–deWaele model is

$$\eta = \mu \left|\frac{\tau}{G}\right|^{1-\frac{1}{n}} \tag{11.20}$$

and the mid-plane stress, τ_m, may be found from [10]

$$\frac{\mu}{G^2}\frac{dp}{dx}(u_2 - u_1)\left(1 + \frac{1}{n}\right) - \left|\frac{\tau_m}{G} + \frac{h}{2G}\frac{dp}{dx}\right|^{1+\frac{1}{n}} + \left|\frac{\tau_m}{G} - \frac{h}{2G}\frac{dp}{dx}\right|^{1+\frac{1}{n}} = 0 \tag{11.21}$$

The one-dimensional Reynolds equation for the Ostwald–deWaele model for $dp/dx \neq 0$ is

$$\frac{d}{dx}\left\{\frac{\rho G^2}{\mu}\left[h\frac{\left|\frac{\tau_m}{G}+\frac{h}{2G}\frac{dp}{dx}\right|^{1+\frac{1}{n}}+\left|\frac{\tau_m}{G}-\frac{h}{2G}\frac{dp}{dx}\right|^{1+\frac{1}{n}}}{2\frac{dp}{dx}\left(1+\frac{1}{n}\right)} - G\frac{\left(\frac{\tau_m}{G}+\frac{h}{2G}\frac{dp}{dx}\right)\left|\frac{\tau_m}{G}+\frac{h}{2G}\frac{dp}{dx}\right|^{1+\frac{1}{n}}-\left(\frac{\tau_m}{G}-\frac{h}{2G}\frac{dp}{dx}\right)\left|\frac{\tau_m}{G}-\frac{h}{2G}\frac{dp}{dx}\right|^{1+\frac{1}{n}}}{\left(\frac{dp}{dx}\right)^2\left(2+\frac{1}{n}\right)\left(1+\frac{1}{n}\right)}\right]\right\}$$

$$= \bar{u}\frac{d(\rho h)}{dx} \tag{11.22}$$

Of course this model lacks a terminal regime; however, when G is very small ($<10^5$ Pa) the film thickness calculated using equations (11.21) and (11.22) can be accurate.

3. The Spriggs model incorporates a terminal regime.

$$\eta = \begin{cases} \mu, & \dfrac{\tau}{G} \le 1 \\ \mu \left|\dfrac{\tau}{G}\right|^{1-\frac{1}{n}}, & \dfrac{\tau}{G} > 1 \end{cases} \tag{11.23}$$

For pure rolling, $u_2 = u_1$, or $\Sigma = 0$, the one-dimensional Reynolds equation for the Spriggs model is [10]

$$\bar{u}\frac{d(\rho h)}{dx} = \begin{cases} \dfrac{d}{dx}\left(\dfrac{\rho h^3}{12\mu}\dfrac{dp}{dx}\right), & \left|\dfrac{h}{2G}\dfrac{dp}{dx}\right| < 1 \\ \dfrac{d}{dx}\left[\dfrac{\rho h G^2}{\mu \dfrac{dp}{dx}} \dfrac{\left|\dfrac{h}{2G}\dfrac{dp}{dx}\right|^{1+\frac{1}{n}}}{\left(2+\dfrac{1}{n}\right)} + \dfrac{2\rho G^3}{3\mu\left(\dfrac{dp}{dx}\right)^2} \dfrac{\left(\dfrac{1}{n}-1\right)}{\left(2+\dfrac{1}{n}\right)}\right], & \left|\dfrac{h}{2G}\dfrac{dp}{dx}\right| \ge 1 \end{cases} \tag{11.24}$$

4. The single-Newtonian Ellis model ($\mu_2 = 0$) is

$$\eta = \frac{\mu}{1 + \left|\dfrac{\tau}{G}\right|^{\left(\frac{1}{n}-1\right)}} \tag{11.25}$$

and the mid-plane stress, τ_m, may be found for $dp/dx \neq 0$ from [8]

$$\tau_m = \frac{\mu(u_2 - u_1)}{h} - \frac{G^{1-\frac{1}{n}}}{h\dfrac{dp}{dx}\left(1+\dfrac{1}{n}\right)}\left[\left|\tau_m + \frac{h}{2}\frac{dp}{dx}\right|^{1+\frac{1}{n}} - \left|\tau_m - \frac{h}{2}\frac{dp}{dx}\right|^{1+\frac{1}{n}}\right] \tag{11.26}$$

The one-dimensional Reynolds equation for the single-Newtonian Ellis model is [8]

$$\frac{d}{dx}\left\{\frac{\rho h^3}{12\mu}\frac{dp}{dx} + \frac{\rho G^{1-\frac{1}{n}}}{\mu\left(\dfrac{dp}{dx}\right)^2\left(1+\dfrac{1}{n}\right)}\left[\left(\frac{\tau_m - \dfrac{h}{2}\dfrac{dp}{dx}}{2+\dfrac{1}{n}} + \frac{h}{2}\frac{dp}{dx}\right)\left|\tau_m - \frac{h}{2}\frac{dp}{dx}\right|^{1+\frac{1}{n}}\right.\right.$$

$$\left.\left. - \left(\frac{\tau_m + \dfrac{h}{2}\dfrac{dp}{dx}}{2+\dfrac{1}{n}} - \frac{h}{2}\frac{dp}{dx}\right)\left|\tau_m + \frac{h}{2}\frac{dp}{dx}\right|^{1+\frac{1}{n}}\right]\right\} = \bar{u}\frac{d(\rho h)}{dx} \tag{11.27}$$

The Ellis model poses a very broad transition for low values of n.

5. The double-Newtonian Rabinowitsch model is

$$\eta = \mu_2 + \frac{\mu - \mu_2}{1 + \left(\frac{\tau}{G}\right)^2} \tag{11.28}$$

and the mid-plane stress, τ_m, may be found for $dp/dx \neq 0$ from [9]

$$\tau_m = \frac{\mu_2(u_2 - u_1)}{h} - \frac{G^2\left(1 - \frac{\mu}{\mu_2}\right)}{2h\frac{dp}{dx}} \ln\left[\frac{\frac{\mu}{\mu_2}G^2 + \left(\tau_m + \frac{h}{2}\frac{dp}{dx}\right)^2}{\frac{\mu}{\mu_2}G^2 + \left(\tau_m - \frac{h}{2}\frac{dp}{dx}\right)^2}\right] \tag{11.29}$$

The one-dimensional Reynolds equation for the double-Newtonian Rabinowitsch model is [9]

$$\begin{aligned}\frac{d}{dx}\Bigg\{\frac{\rho h^3}{12\mu_2}\frac{dp}{dx} + \frac{\rho G^2\left(\frac{\mu}{\mu_2} - 1\right)}{2\mu_2\left(\frac{dp}{dx}\right)^2}\Bigg\{\tau_m \ln\left[\frac{\frac{\mu}{\mu_2}G^2 + \left(\tau_m + \frac{h}{2}\frac{dp}{dx}\right)^2}{\frac{\mu}{\mu_2}G^2 + \left(\tau_m - \frac{h}{2}\frac{dp}{dx}\right)^2}\right] \\ -2h\frac{dp}{dx} + 2G\left(\frac{\mu}{\mu_2}\right)^{\frac{1}{2}}\left[\tan^{-1}\left(\frac{\tau_m + \frac{h}{2}\frac{dp}{dx}}{G\left(\frac{\mu}{\mu_2}\right)^{\frac{1}{2}}}\right) - \tan^{-1}\left(\frac{\tau_m - \frac{h}{2}\frac{dp}{dx}}{G\left(\frac{\mu}{\mu_2}\right)^{\frac{1}{2}}}\right)\right]\Bigg\}\Bigg\} \\ = \bar{u}\frac{d(\rho h)}{dx}\end{aligned} \tag{11.30}$$

The exponent, n, will not have the usual relation to the slope of a flow curve if μ/μ_2 is not large.

6. The double-Newtonian Ferry model has a broader transition from the first to the second Newtonian than does the double-Newtonian Rabinowitsch model.

$$\eta = \mu_2 + \frac{\mu - \mu_2}{1 + \left|\frac{\tau}{G}\right|} \tag{11.31}$$

The mid-plane stress, τ_m, may be found for $dp/dx \neq 0$ from [9]

$$\tau_m = \frac{\mu_2(u_2-u_1)}{h} - \frac{1}{h\dfrac{dp}{dx}}\left[G\left(1-\frac{\mu}{\mu_2}\right)\left(\left|\tau_m+\frac{h}{2}\frac{dp}{dx}\right|-\left|\tau_m-\frac{h}{2}\frac{dp}{dx}\right|\right)\right.$$
$$\left.+\frac{\mu}{\mu_2}G^2\left(\frac{\mu}{\mu_2}-1\right)\ln\left(\frac{\dfrac{\mu}{\mu_2}G+\left|\tau_m+\dfrac{h}{2}\dfrac{dp}{dx}\right|}{\dfrac{\mu}{\mu_2}G+\left|\tau_m-\dfrac{h}{2}\dfrac{dp}{dx}\right|}\right)\right] \tag{11.32}$$

The one-dimensional Reynolds equation for the double-Newtonian Ferry model is [9]

$$\frac{d}{dx}\left\{\frac{\rho h^3}{12\mu_2}\frac{dp}{dx}+\frac{\rho G\left(\dfrac{\mu}{\mu_2}-1\right)}{\mu_2^2\left(\dfrac{dp}{dx}\right)^2}\left[\frac{\mu}{\mu_2}G\,h\frac{dp}{dx}\right.\right.$$
$$+\left(\operatorname{sign}\left(\tau_m+\frac{h}{2}\frac{dp}{dx}\right)-\operatorname{sign}\left(\tau_m-\frac{h}{2}\frac{dp}{dx}\right)\right)$$
$$\times\left(\frac{\left(\tau_m+\dfrac{h}{2}\dfrac{dp}{dx}\right)\left(\tau_m-\dfrac{h}{2}\dfrac{dp}{dx}\right)}{2}+\left(G\frac{\mu}{\mu_2}\right)^2\ln\left(\frac{\dfrac{\mu}{\mu_2}G}{\dfrac{\mu}{\mu_2}G+\left|\tau_m-\dfrac{h}{2}\dfrac{dp}{dx}\right|}\right)\right)$$
$$\left.\left.-G\frac{\mu}{\mu_2}\ln\left(\frac{\dfrac{\mu}{\mu_2}G+\left|\tau_m+\dfrac{h}{2}\dfrac{dp}{dx}\right|}{\dfrac{\mu}{\mu_2}G+\left|\tau_m-\dfrac{h}{2}\dfrac{dp}{dx}\right|}\right)\left(\tau_m+G\frac{\mu}{\mu_2}\operatorname{sign}\left(\tau_m+\frac{h}{2}\frac{dp}{dx}\right)\right)\right]\right\}=\bar{u}\frac{d(\rho h)}{dx} \tag{11.33}$$

7. The single-Newtonian Carreau model is of the inappropriate form, $\eta(\dot{\gamma})$, for an exact Reynolds equation.

$$\eta = \frac{\mu}{\left[1+(\lambda\dot{\gamma})^2\right]^{\frac{1-n}{2}}} \tag{11.34}$$

This is an unfortunate situation since the Carreau model is the one most often found useful for the description of shear-thinning of single-component liquid lubricants. The present author has offered [10] an approximate Reynolds–Carreau equation by numerically solving for the flow rate, Q, over a large range of operating parameters. The reduced rolling velocity, $\bar{u}\mu/(hG)$, was investigated for values from 0.01 to 10. The velocity difference was investigated by variation of the slide-to-roll ratio $\Sigma = (u_1 - u_2)/\bar{u}$ from -0.5 to 1. The reduced pressure gradient was varied from $(dp/dx)h/G = -27$ to 27. Values of $n = 0.3, 0.5$ and 0.75 were investigated. An approximation for the mid-plane stress was found to be

$$\tau_m \approx \frac{\mu(u_2-u_1)}{h}\left[1+0.053\left(\frac{h}{G}\frac{dp}{dx}\right)^4\right]^{\frac{1-\frac{1}{n}}{4}}\left\{1+\left[\frac{\mu(u_2-u_1)}{Gh}\right]^2\right\}^{\frac{n-1}{2}} \tag{11.35}$$

The approximate one-dimensional Reynolds equation for the single-Newtonian Carreau model was found [10] to be

$$\frac{d}{dx}\left\{\frac{\rho h^3}{12\mu}\frac{dp}{dx} - 0.57\frac{(n-1)}{\left(\frac{1}{n}+1\right)}\left(\frac{\rho G^2 h}{\mu\frac{dp}{dx}}\right)\left|\frac{h}{2}\frac{dp}{dx}\right|^{\left(\frac{1}{n}+\frac{4}{3}\right)}\right\} = \bar{u}\frac{d(\rho h)}{dx} \tag{11.36}$$

The mid-plane stress, τ_m, does not appear in the Reynolds equation as before. This is due to the approximate nature of equation (11.36) and the relative ineffectiveness of sliding in changing the overall flow rate. A difference in surface velocities increases the stress near one surface while decreasing the stress near the opposite surface. Unless the departure from linearity of the constitutive equation is substantial, the effect of sliding on the flow is small and this effect was absorbed into the error in the above approximation.

11.2.2 Two-dimensional Reynolds equation for a generalized Newtonian fluid

The only exact two-dimensional Reynolds equation for a generalized Newtonian model was, at the time of this writing, the contribution by Greenwood [11] for the single Newtonian Rabinowitsch model,

$$\eta = \frac{\mu}{1+\left(\frac{\tau^*}{G}\right)^2} \tag{11.37}$$

where τ^* is given by equation (11.13). In the limit of high shear stress the response is power-law with $n = 1/3$, a relatively low value for a lubricant.

For simplicity, assume that the surface velocities are in the x-direction only, although Greenwood's result [11] was more general. Then the mid-plane stress, τ_m, is in the x-direction and is found from

$$\frac{\mu(u_2-u_1)}{h} = \tau_m\left[1+\frac{\tau_m^2}{G^2}+\frac{h^2}{4G^2}\left(\frac{\partial p}{\partial x}\right)^2+\frac{h^2}{12G^2}\left(\frac{\partial p}{\partial y}\right)^2\right] \tag{11.38}$$

The two-dimensional Reynolds equation for the single-Newtonian Rabinowitsch model is then

$$\begin{aligned}&\frac{\partial}{\partial x}\left\{\frac{\rho h^3}{\mu}\frac{\partial p}{\partial x}\left[1+3\frac{\tau_m^2}{G^2}+\frac{3h^2}{20G^2}\left[\left(\frac{\partial p}{\partial x}\right)^2+\left(\frac{\partial p}{\partial y}\right)^2\right]\right]\right\}\\&+\frac{\partial}{\partial y}\left\{\frac{\rho h^3}{\mu}\frac{\partial p}{\partial y}\left[1+\frac{\tau_m^2}{G^2}+\frac{3h^2}{20G^2}\left[\left(\frac{\partial p}{\partial x}\right)^2+\left(\frac{\partial p}{\partial y}\right)^2\right]\right]\right\}=12\bar{u}\frac{\partial(\rho h)}{\partial x}\end{aligned} \tag{11.39}$$

The pressure gradients in the x and y-directions are coupled. The Rabinowitsch model is the Ellis model with $n = 1/3$ and the equation above for $\partial p/\partial y = 0$ is equivalent to the single Newtonian Reynolds–Ellis equation (11.27) with $n = 1/3$.

The generalized Reynolds equations presented above assume the simplest description of non-Newtonian flow, the generalized Newtonian liquid. With the understanding that Reynolds equations, in general, cannot represent exactly the pressure generated in films of non-Newtonian liquids [12], viscoelastic liquids for example, and with the understanding that they may not be accurate for flows that combine a large shear stress with ordinary piezoviscous response [3,4], the generalized Reynolds equations can be useful for calculations of EHL film thickness which is essentially determined by the flow in the inlet zone where conditions are less severe than in the high-pressure, load-bearing region.

References

[1] Reynolds, O., "On the Theory of Lubrication and Its Application to Mr. Beauchamp Tower's Experiments Including an Experimental Determination of the Viscosity of Olive Oil", *Phil. Trans. Royal Soc. London*, Vol. 177, 1886, pp. 157–234.

[2] Dowson, D., "A Generalized Reynolds Equation for Fluid Film Lubrication", *Int. J. Mech. Sci.*, Vol. 4, 1962, pp. 159–170.

[3] Bair, S., Khonsari, M. and Winer, W.O., "High-Pressure Rheology of Lubricants and Limitations of the Reynolds Equation", *Tribol. Int.*, Vol. 31, No. 10, 1998, pp. 573–586.

[4] Rajagopal, K.R. and Szeri, A.Z., "On an Inconsistency in the Derivation of the Equations of Elastohydrodynamic Lubrication", *Proc. Roy. Soc. London A*, Vol. 459, 2003, pp. 2771–2786.

[5] Hron, J., Malek, J. and Rajagopal, K.R., "Simple Flows of Fluids with Pressure-Dependent Viscosities", *Proc. Roy. Soc. London A*, Vol. 457, 2001, pp. 1603–1622.

[6] Dowson, D. and Higginson, G.R., *Elastohydrodynamic Lubrication*, Pergamon Press, 1966, p. 25.

[7] Tanner, R.I., *Engineering Rheology*, second edition, Oxford University Press, Oxford, 2000, pp. 286–310.

[8] Bair, S., "A Reynolds-Ellis Equation for Line Contact with Shear-Thinning", *Tribol. Int.*, Vol. 39, 2006, pp. 310–316.

[9] Bair, S. and Khonsari, M., "Generalized Reynolds Equations for Line Contact with Double-Newtonian Shear-Thinning", *Tribol. Lett.*, Vol. 18, No. 4, 2005, pp. 513–520.

[10] Bair, S. and Khonsari, M., "Reynolds Equations for Common Generalized Newtonian Models and an Approximate Reynolds-Carreau Equation", *Proc. I. Mech. E.*, Vol. 220, Part J, 2006, pp. 365–374.

[11] Greenwood, J.A., "Two-Dimensional Flow of a Non-Newtonian Lubricant", *Proc. I. Mech. E.*, Vol. 214, Part J, 2000, pp. 29–41.

[12] Bair, S., Lockwood, F.E., Anderson, W.B., Zhang, Z. and Dotson, D., "Measurements of Elasticity in Multigrade Motor Oil at Elevated Pressure", submitted *STLE Tribol. Trans.*, 2006.

CHAPTER 12

Applications to Elastohydrodynamics

12.1 Introduction

Models have been introduced for the temperature and pressure dependence of density in Chapter 4, for the temperature and pressure dependence of the low-shear viscosity in Chapter 6, for the shear dependence of viscosity in Chapter 8 and a failure criterion was introduced to limit the shear stress at high pressures in Chapter 10. These models are directly applicable to elastohydrodynamic simulations and should provide a means to relate the tribological behavior of the lubricated contact to the properties of the liquid lubricant. It has already been mentioned that accurate calculations of film thickness may be obtained from the isothermal Newtonian assumption for neat low-molecular-weight liquids. Then, only a description of the pressure dependence of density and viscosity are necessary. Squalane, 2,6,10,15,19,23-hexamethyltetracosane, with a molecular weight of 423 kg/kmol is an example of a liquid for which a Newtonian film thickness calculation can be very accurate using measured properties.

Figure 12.1 presents some measurements of the film thickness in a circular point contact between a glass flat and a steel ball obtained at Brno University of Technology using thin film colorimetric interferometry [1] for squalane. The composite radius was $R = 12.7$ mm in the rolling and transverse directions. The reduced Young's modulus was $E' = 123.4$ GPa and the contact half-width was $b = 161$ μm. Other operating conditions are given in the figure. The volume and the low-shear viscosity were measured in the present author's laboratory at 25°C and fit to the isothermal Tait equation (4.6) and the Doolittle equation (6.44) respectively. This is an isothermal problem, $\mu_R = \mu_0$ and $V_\infty = V_{\infty R}$, and the Doolittle equation becomes

$$\mu = \mu_0 \exp\left[BR_0\left(\frac{1}{\dfrac{V}{V_R} - R_0} - \frac{1}{1 - R_0}\right)\right] \tag{12.1}$$

The Tait parameters were found to be $K_0' = 11.74$ and $K_0 = 1.312$ GPa. The Doolittle parameters are $\mu_0 = 0.0276$ Pa s, $B = 5.258$ and $R_0 = 0.663$.

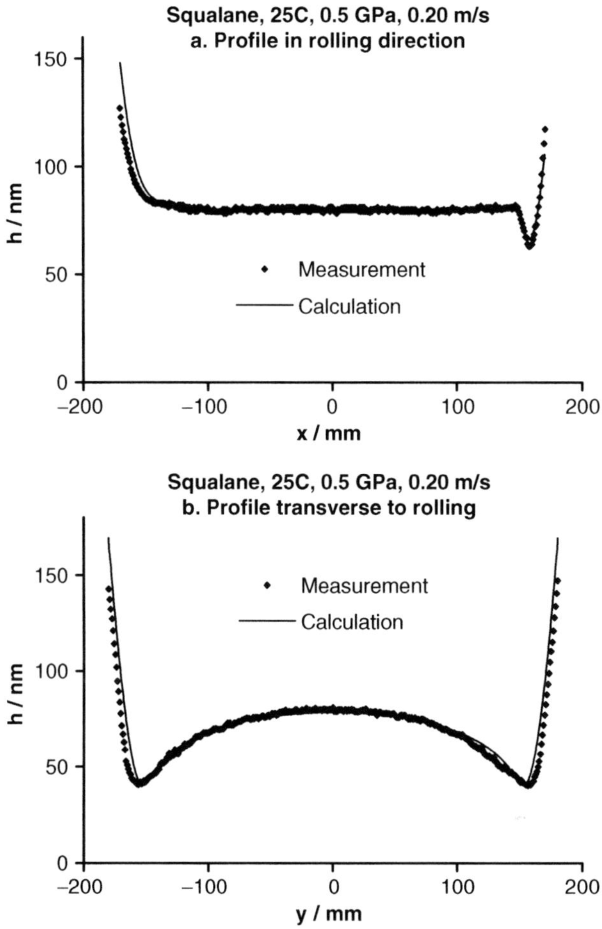

Fig. 12.1. The film thickness in circular point contact for squalane. Calculations by Liu and Wang at Northwestern University using properties of squalane measured at Georgia Institute of Technology compared with the measurements by Hartl and Krupka at Brno University of Technology.

Numerical simulations of EHL have been traditionally plagued by poor convergence at high-pressure when the viscosity depends strongly on pressure. A simulation code has been developed by Hu and Zhu [2] which apparently overcomes this difficulty, allowing numerical simulations with realistic piezoviscous response. Liu and Wang [3] at Northwestern University have used an improved version of this code to calculate the film thickness of squalane for the conditions of Fig. 12.1. They employed the Tait equation for the density and the Doolittle equation for the low-shear viscosity with the parameters given above to solve the Reynolds equation (1.17) coupled with the relation for the elastic deflection of the rollers. The resulting calculations are shown as the curves in Fig. 12.1(a) for the film profile in the rolling direction, x, and Fig. 12.1(b) for the direction transverse to the rolling velocity, y. The agreement is striking and did not involve the adjustment of properties to achieve it.

The success of the Newtonian assumption for the film thickness of squalane can be typical for neat, low-molecular-weight liquids that are used as lubricants such as naphthenic mineral oils, polybutenes and low-viscosity paraffinic mineral oils. We have seen in Fig. 8.7 that the Newtonian assumption is reasonable for squalane to a shear stress of at least 5 MPa. Apparently, this stress is not exceeded in the inlet zone. In Chapter 8, there are examples of lubricants for which shear-thinning begins at much lower stress. For these liquids, an accurate calculation must include a shear-dependent viscosity. As of this writing, there has been little success at solving the Reynolds equation for a generalized Newtonian liquid in combination with the elasticity equation. This is doubly unfortunate since an accurate calculation of traction requires an accurate film thickness to establish the shear rate in the Hertz region where the traction force is generated. In the next section a very simple Grubin-style inlet analysis is described which allows a calculation of the film thinning effect of shear-dependent viscosity.

12.2 Film Thickness for Shear-Thinning Liquids

It is possible to produce a somewhat accurate prediction of the central elastohydrodynamic film thickness without solving simultaneously the Reynolds equation and the elastic film shape equation (1.8). The technique was originated by Grubin [4] who made the insightful assumption that the film shape should be approximately the same as that of an unlubricated contact except that an additional separation, h_0, could be attributed to the presence of the liquid film. For a line contact then,

$$h = h_0 + \frac{128R}{\pi^2}\left(\frac{\bar{p}}{E'}\right)^2 \left\{ \frac{-x}{b}\left(\frac{x^2}{b^2}-1\right)^{\frac{1}{2}} - \ln\left[\frac{-x}{b} + \left(\frac{x^2}{b^2}-1\right)^{\frac{1}{2}}\right]\right\}, x < -b \quad (12.2)$$

for composite roller radius, R. The Hertz half-width is

$$b = \frac{16R}{\pi}\frac{\bar{p}}{E'} \quad (12.3)$$

where the average pressure is related to the load per unit length, W, by $\bar{p} = W/2b$ and the Hertz (maximum) pressure is $p_H = 4\bar{p}/\pi$. Of course, this shape would be exact if the pressure within the film were exactly Hertzian, equation (1.9), inside the contact and zero outside. The film shape is shown schematically in Fig. 12.2 where the inlet zone extends to the left of $x = -b$.

The second assumption made by Grubin placed the value of the pressure at the edge of the Hertz zone sufficiently high that the Poiseuille component of the flow was approximately zero there. The argument was made [4] that a Newtonian liquid with exponentially pressure-dependent viscosity, equation (6.20), would satisfy this boundary

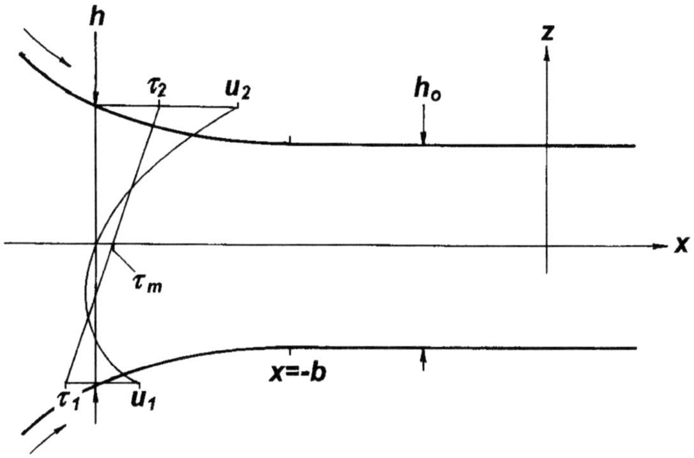

Fig. 12.2. The geometry of the Grubin-style inlet zone analysis for shear-thinning liquids. The inlet zone extends to the left of $x = -b$ where the velocity and shear stress distributions across the film are shown.

condition for

$$1 - \exp\left[-\alpha p\left(\frac{x}{b} = -1\right)\right] \approx 1 \tag{12.4}$$

and setting $p(x/b = -1) = 4/\alpha^*$ seems to satisfy this requirement.

The solution proceeds by numerically integrating dp/dx with respect to x starting from an arbitrary position in the inlet where the pressure is set to zero and continuing the integration to $x = -b$. The starting position must be sufficiently far upstream that the position has no effect upon the solution, that is to say, the inlet will not be "starved". Starting at $x = -8b$ seems to be sufficiently far upstream. It is advantageous to use a nonuniform grid spacing with narrower spacing near the edge of the Hertz zone. The value of h_0 is varied to obtain the pressure boundary condition, $p(x/b = -1) = 4/\alpha^*$. Whereas this procedure has been used for reasonably accurate Newtonian solutions to the central film thickness it can be applied to the shear-thinning case as well.

The one-dimensional Reynolds equations for generalized Newtonian liquids that were presented in Chapter 11, were of the form

$$\frac{d}{dx}[\rho\bar{u}h - \rho Q] = \bar{u}\frac{d}{dx}(\rho h) \tag{12.5}$$

where Q is the volume flow rate per unit width. Somewhere within the interval $-b < x < b$, $dp/dx = 0$ and at this location the flow must be $Q = \bar{u}h_0$. For the incompressible assumption, the one-dimensional Reynolds equations for generalized Newtonian liquids can be integrated to yield equations of the form,

$$Q\left(\frac{dp}{dx}, \mu, \mu_2, h, G, n, \tau_m\right) = \bar{u}h_0 \tag{12.6}$$

where the RHS is a constant of integration. Relationships for the midplane shear stress were also given in Chapter 11 of the form,

$$\tau_m = F\left(\frac{dp}{dx}, \Sigma, \bar{u}, \mu, \mu_2, h, G, n, \tau_m\right) \tag{12.7}$$

These equations (12.6) and (12.7) may be solved numerically to provide the value of *dp/dx* required for the numerical integration that yields the pressure.

There is another approach to finding *dp/dx* which requires more computing time but has some advantages. In this "direct integration" technique, the flow is found by numerical evaluation across the film of the double integral,

$$Q = \int_{-h/2}^{h/2} u dz = \int_{-h/2}^{h/2} \left[u_1 + \int_{-h/2}^{z} \dot{\gamma} dz\right] dz \tag{12.8}$$

using $\dot{\gamma}(\tau)$ that may be obtained algebraically from a viscosity function of the form $\eta(\tau)$ or obtained numerically from a viscosity function of the form $\eta(\dot{\gamma})$. The shear stress is distributed across the film as $\tau = \tau_m + z(dp/dx)$. At each x an iteration of τ_m proceeds until $u_1 + \int_{-h/2}^{h/2} \dot{\gamma} dz = u_2$. Also, at each x an iteration of *dp/dx* proceeds until $Q = \bar{u}h_0$, repeating the previous iteration if necessary. This technique does not require a viscosity function of the form $\eta(\tau)$ and the details of the inlet flow are revealed. Implicit in these methods is the assumption that the cross-film pressure gradient does not affect the viscosity distribution across the film due to the piezoviscous effect.

Some examples of the direct integration method follow. The inlet zone conditions for a line contact in rolling and sliding are shown in Fig. 12.3. Here it was assumed that $R = 0.02$ m, $p_H = 1.02$ GPa and $\bar{u} = 0.5$ m/s and that the viscosity varies exponentially with pressure according to equation (6.20) with $\mu_0 = 0.06$ Pa s and $\alpha = 15$ GPa^{-1}. Rollers are steel. Conditions are shown for a Newtonian liquid in Fig. 12.3(a) where the film thickness is $h_0 = 223$ nm. The Dowson and Toyoda formula, equation (1.15), yields $h_c = 275$ nm here and the velocity dependence for that formula is $d(\log h_c)/d(\log \bar{u}) = 0.69$ in contrast to the present Grubin-style analysis for which the velocity dependence is approximately 0.8. The absolute values of maximum shear stress, maximum shear rate and pressure are plotted against the ratio of local film thickness to h_0 in Fig. 12.3. For a Newtonian liquid, the peak in maximum shear rate occurs at $h = 2h_0$, but more importantly, the peak maximum shear stress, 5 MPa, occurs very much closer to the beginning of the Hertz zone, $h = h_0$.

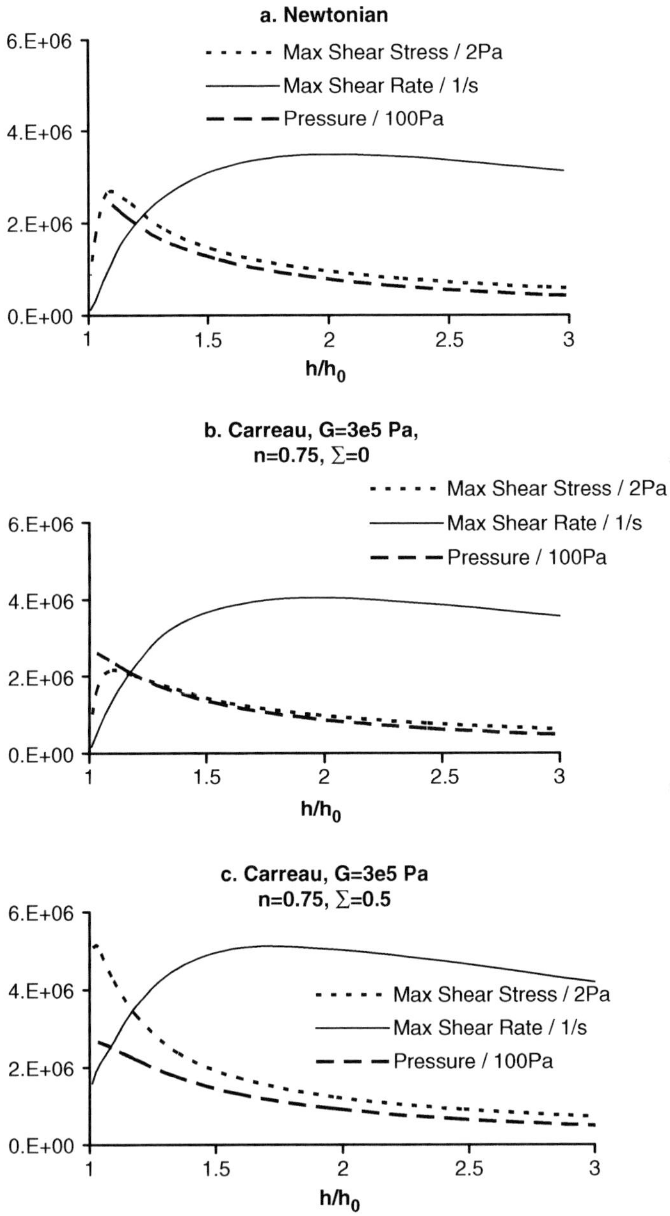

Fig. 12.3. (a) Inlet zone conditions for a Newtonian liquid. (b) Carreau liquid in pure rolling. (c) A Carreau liquid in a sliding contact.

The same calculations were performed for a single Newtonian Carreau liquid,

$$\eta = \frac{\mu}{\left[1 + (\lambda\dot{\gamma})^2\right]^{\frac{1-n}{2}}} \tag{11.34}$$

with $G = 0.3$ MPa and $n = 0.75$ for pure rolling in Fig. 12.3(b) where $h_0 = 199$ nm and for $\Sigma = (u_2 - u_1)/\bar{u} = 0.5$ in Fig. 12.3(c) where $h_0 = 195$ nm. Shear-thinning reduced the film thickness by 11% and sliding produced an additional 2% reduction. The introduction of shear-thinning to the inlet calculation increased the shear rate and moved the peak in maximum shear rate closer to the edge of the Hertz zone whereas the peak maximum shear stress decreased to 4 MPa. The introduction of sliding ($\Sigma = 0.5$) increased both the shear rate and the shear stress (to 10 MPa) and moved the peak in maximum shear rate closer still to the edge of the Hertz zone.

The general character of a film thickness calculation for a liquid following the single Newtonian Carreau model in a sliding contact [5] is illustrated in Fig. 12.4 where the power-law exponent, n, assumes values of 0.875, 0.625 and 0.375 and values for constant G are indicated in the figure. Here it was assumed that $R = 0.02$ m, $p_H = 1.0$ GPa and $\bar{u} = 1.0$ m/s and that the viscosity varies exponentially with pressure according to equation (6.20) with $\mu_0 = 0.05$ Pa s and $\alpha = 20$ GPa^{-1}. The trend is for film thickness at a given slide-to-roll ratio to decrease with decreasing G. The rate of decrease in h_0 with decreasing G, increases as Σ increases and as G is reduced. For large values of n, that is as n approaches unity, the influence of sliding on the film thickness is small. However, for the smaller values of n in combination with small values of G, the effect of sliding is pronounced.

It would be useful to have some means to know if a Newtonian calculation will be accurate. The value of G is plotted against n which resulted in a 5% reduction in film thickness in Fig. 12.5. Here it was assumed that $R = 0.02$ m, $p_H = 1.02$ GPa and $\bar{u} = 1.0$ m/s and that the viscosity varies exponentially with pressure according to equation (6.20) with $\alpha = 15$ GPa^{-1}. It was assumed that G is constant and Carreau behavior (11.34) was assumed. Film thicknesses of 430 nm and 90 nm were investigated by assuming $\mu_0 = 0.08$ Pa s and 0.01 Pa s, respectively. For pure rolling, if $G > 2$ MPa, then a Newtonian calculation should be accurate to better than 5%. Thicker films and sliding contacts are more restrictive on the value of G. For very low molecular weight liquids such as squalane, even pure sliding should not result in film-thinning from constitutive behavior.

The technique described above is not as accurate as a calculation of film thickness from a simultaneous solution of the Reynolds and elasticity equations. However, as of the writing of this book, there have been no such solutions for ordinary shear-thinning, although there were several laboratories actively engaged in their development. As a substitute for the accuracy of a full simulation this author has used the simple Grubin-style analysis described above to correct a classical Newtonian solution. The procedure has been to calculate the film thickness, h_0, for the Newtonian assumption, h_N, and the film thickness for the shear-thinning assumption, h_{NN}, for the same operating conditions. The ratio, $\kappa = h_N/h_{NN}$, can then be divided into any Newtonian film thickness to give a new corrected film thickness. This procedure can be used for point contacts as well, assuming that the shear-thinning affects point and line contacts in the same way.

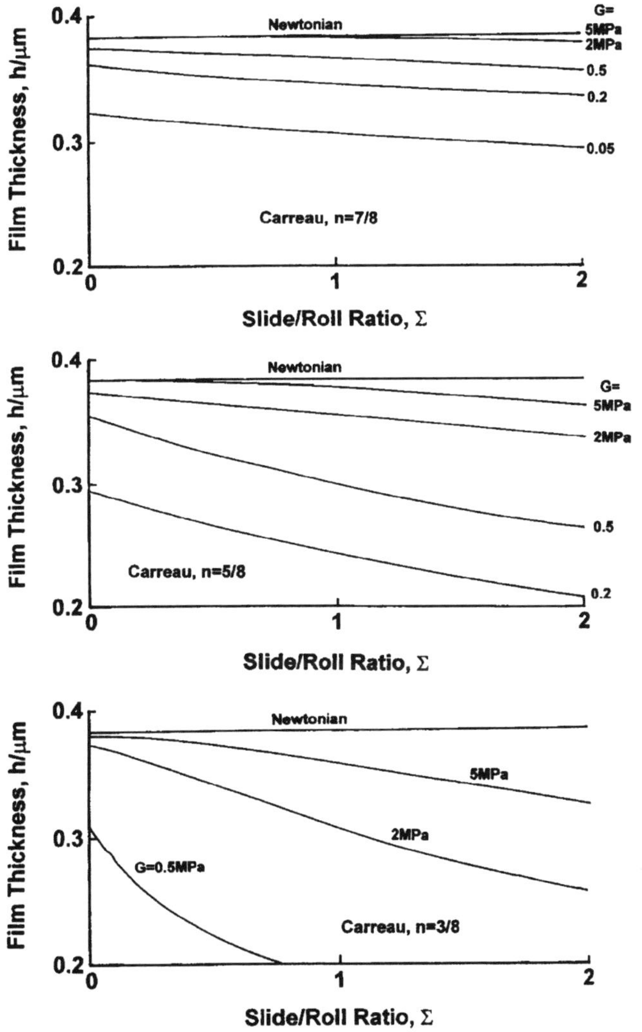

Fig. 12.4. Central film thicknesses for various values of n and G. Reprinted with permission of Elsevier from Ref. [5].

In Ref. [7], 99 combinations of R, $\bar{u}$, μ_0, Σ, n and G were used to calculate κ. The viscosity function used here was

$$\eta = \frac{\mu}{\left[1 + \left(\frac{\tau}{G}\right)^2\right]^{\frac{\frac{1}{n}-1}{2}}} \tag{12.9}$$

This model approximates the single Newtonian Carreau equation (11.34) reasonably well when n is not too small. It is useful to define an inlet Weissenberg number, $\Gamma = \mu_0\bar{u}/(h_N G)$. The 99 results for κ were found [7] to be reasonably well described

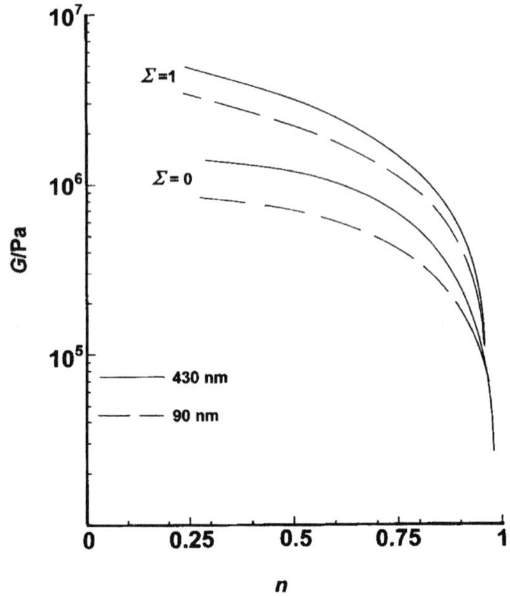

Fig. 12.5. The value of G that results in a 5% reduction in film thickness. Reproduced from Bair, S. and Khonsari, M., "Reynolds Equations for Common Generalized Newtonian Models and an Approximate Reynolds-Carreau Equation", *Proc. I. Mech. E.*, Vol. 220, Part J, 2006, pp. 365–374, with permission granted by the Council of the Institution of Mechanical Engineers.

by the relationship,

$$\kappa = \left\{1 + 0.79[(1+\Sigma)\Gamma]^{\frac{1}{1+0.2\Sigma}}\right\}^{3.6(1-n)^{1.7}} \tag{12.10}$$

Now, a prediction of the shear-thinned film thickness, h_N/κ, may be obtained from a Newtonian prediction, h_N.

To illustrate the usefulness of the film thickness correction given by equation (12.10), consider the film thickness measurements presented in Fig. 12.6 for squalane at 20°C and for a polyglycol gear oil at 65°C in an elliptical point contact between a sapphire flat and a steel roller. The composite radius was $R = 7.6$ mm in the rolling direction. The reduced Young's modulus was $E' = 290$ GPa and the contact width was $2b = 0.112$ mm in the rolling direction and 167 mm in the transverse direction, giving $p_H = 0.617$ GPa and ellipticity ratio, $e = 1.5$. The thickness of the film was measured by interferometry using monochromatic light.

The solid straight curves marked "H&D Newtonian" in Fig. 12.6 represent the Hamrock and Dowson [8] formula for central film thickness in elliptical contacts

$$h_c = 2.5\alpha^{*0.53}(\mu_0\bar{u})^{0.67}E'^{-0.073}R^{0.464}\left(eb^2p_H\right)^{-0.067}[1-0.61\exp(-0.73e)] \tag{12.11}$$

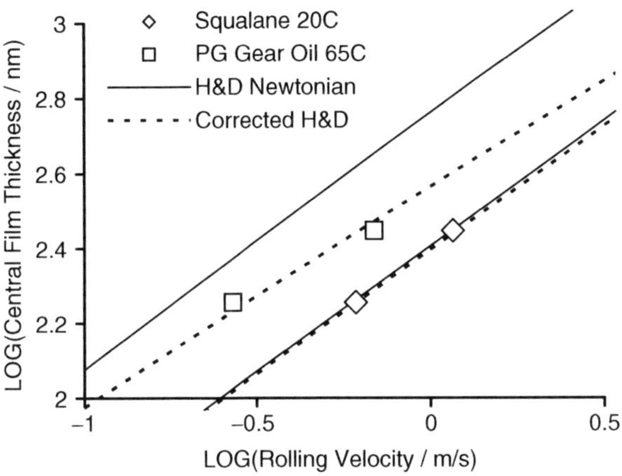

Fig. 12.6. The central film thickness of a polyglycol gear oil and squalane. Experimental measurement compared with the Newtonian calculation and the Newtonian calculation corrected for shear-thinning. The Newtonian slope is predicted to be 0.67, whereas the shear-thinning slope is 0.57 for the gear oil.

For squalane and the gear oil, $\mu_0 = 0.036$ and 0.18 Pa s and $\alpha^* = 19.7$ and 11.4 GPa^{-1}, respectively. As expected, the prediction for squalane is excellent, including the velocity dependence, $d \ln h_c / d \ln \bar{u}$, which apparently is 0.67 for Newtonian inlet behavior. For the gear oil, in contrast, the film thickness is overestimated by about 30% and the velocity dependence of the measured film is somewhat reduced from the Newtonian prediction.

The broken curves in Fig. 12.6 represent h_c/κ, using equation (12.10) with $G = 6.0$ and 0.22 MPa and $n = 0.46$ and 0.67 for squalane and the gear oil respectively. The correction has little effect on the prediction for squalane but for the gear oil, it brings the prediction into good agreement with the measurement. The velocity dependence of the corrected Newtonian prediction decreases with increasing velocity and is about $d \ln h_c / d \ln \bar{u} = 0.57$ in the neighborhood of the measurements.

Other examples of film thinning from generalized Newtonian behavior may be found in Fig. 1.6 and in Refs. [9–15]. The ability to calculate the film thickness from the properties of a liquid lubricant is important not only for the assessment of the durability of a lubricant film but also as an input for a calculation of traction. The film thickness establishes the Couette component of the shear rate in the Hertz zone of a sliding contact and, as was shown in Fig. 8.2, if the shear-thinning substantially reduces the film thickness, a calculation of traction based on the Newtonian film thickness will not be accurate.

12.3 The Calculation of Traction from Material Properties

The prediction of lubricated concentrated contact traction from the properties of the liquid lubricant has been a goal of elastohydrodynamics for at least 40 years. Progress here has proceeded more slowly than for film thickness and the accuracy of a traction calculation does not approach that of film thickness. A possible explanation for the slower progress may be the lack of a regime of conditions for which the Newtonian assumption yields reasonable predictions. The examples for squalane in Figs. 1.6, 12.1 and 12.6 indicate that, given a liquid with a large Newtonian limit, G, the film thickness can be predicted without considering shear-thinning. It was shown in Section 8.1, that for a high-molecular-weight PAO, the Newtonian portion of a traction curve must occur at slide-roll ratio smaller than the usual experimental limit of resolution. It may be possible to experimentally observe Newtonian traction at very low contact pressure and slide-roll ratio by selecting a liquid with the combination of relatively low pressure dependence of viscosity such as the polyol ester of Fig. 6.3 and relatively large Newtonian limit such as the dicylohexyl methylpentane of Table 8.1.

Traction is a topic for which full numerical solutions of the coupled Reynolds and elasticity equations have not proven useful. It is necessary to model the viscosity over a range of pressure for which the pressure dependence is usually greater-than-exponential and doing so has caused numerical stability problems. That is to say, it is necessary to include the pressure fragility property in a traction calculation.

For the purposes of discussion, it may be useful to classify traction behavior at small slide-to-roll ratio into low, moderate and high-pressure regimes. The classification of the operating regime of a contact should depend on both the contact pressure and the response of the shear properties of the liquid to pressure. At intermediate pressures the average shear stress, when plotted against the logarithm of sliding velocity, slide-roll ratio, or other measure of average shear rate, will show a somewhat straight portion that indicates logarithmic traction response which may be described by the equation (1.19),

$$\bar{\tau} = \tau_E \sinh^{-1}(C_E \Sigma) \tag{1.19}$$

or, more appropriately, the approximation for large Σ, $\bar{\tau} \approx \tau_E \ln(2C_E) + \tau_E \ln \Sigma$, where τ_E is the slope of the logarithmic portion. At low pressures the logarithmic portion often does not develop before thermal softening begins to affect the traction. At high pressures the logarithmic portion does not develop at all and a traction curve on double linear axes will show a linear increasing regime and a plateau separated by a short transition. Of course, the transition between pressure regimes is gradual.

The pressure dependence of the low-shear viscosity plays a key role in all aspects of traction. To illustrate the intermediate pressure behavior, the viscosities of two gear oil base stocks, a polyol ester and a poly alpha olefin are given in Table 12.1. These data were fitted to the Bair and Kottke pressure–viscosity relation in Ref. [16].

$$\mu = \mu_0 \exp\left(\frac{-P_-}{p_{-\infty}} + \frac{-P_+}{p_\infty}\right) \cdot \exp\left(\frac{P_+}{p_\infty - p} - \frac{P_-}{p - p_{-\infty}}\right) \tag{12.12}$$

Here, $\alpha_0 = P_-/p_{-\infty}^2 + P_+/p_\infty^2$. The five required parameters are listed in Table 12.1. The viscosities of the two liquids are plotted as functions of pressure in Fig. 12.7 showing equation (12.12) as the curves. The polyol ester could have been described accurately by the four-parameter version of the above relation, equation (6.40);

Table 12.1. Pressure dependent viscosities, in Pa s, for ISO 68 gear oil base stocks, a polyol ester, POE, and a poly alpha olefin, PAO, at 57°C.

p/MPa	POE	PAO
0.1	0.0169	0.0201
50	0.0382	
69		0.0722
100	0.0757	
146		0.217
200	0.243	
230	0.345	
300		1.11
349	1.256	1.79
467	4.36	4.99
586	15.6	12.7
610		16.62
705	54.9	34.2
765		52.3
823	227	77.9
942	938	197
1060	4430	506
1179		1402
1298		4380
α^*/GPa^{-1}	14.4	16.4
μ_0/Pa s	0.0172	0.0205
p_∞/GPa	3.38	3.38
$p_{-\infty}$/GPa	−0.381	−0.422
P_+/GPa	71.29	37.51
P_-/GPa	1.503	3.011

however, the poly alpha olefin with its broad exponential regime requires the additional parameter of equation (12.12). This pair of liquids is especially interesting since the liquid with the lower value of initial pressure–viscosity coefficient, α_0, possesses the greater viscosity and greater local pressure–viscosity coefficient, α, at pressures above 0.5 GPa. Clearly, it is impossible to predict the pressure–viscosity response at high-pressure from the low-pressure behavior.

A simple traction calculation can be made for point contact by assuming that the Hertz pressure distribution for point contact,

$$p = p_H(1 - r^2)^{\frac{1}{2}} \tag{12.13}$$

applies to the lubricant film as well. Here, r is the dimensionless contact radius, equal to one at the edge of the contact and zero at the center. The average contact shear stress

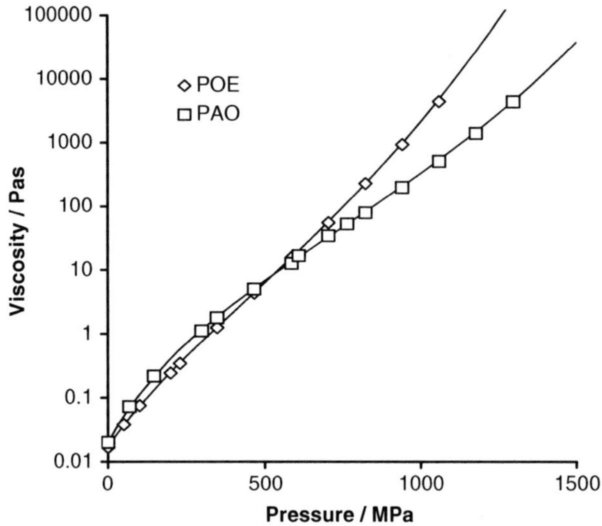

Fig. 12.7. The viscosities of two gear oil base stocks, a poly olester, POE and a poly alpha olefin, PAO, at 57°C. The curves are equation (12.12) with the parameters given in Table 12.1.

is computed from integration over the contact area, A.

$$\bar{\tau} = \frac{1}{A}\int_A \tau(p,\dot{\gamma})dA = 2\int_0^1 \tau(p,\dot{\gamma})rdr \tag{12.14}$$

In general, the stress is taken to be

$$\tau = \min\left[\eta(p,\dot{\gamma})\dot{\gamma},\ \Lambda p\right] \tag{12.15}$$

where Λ is the pressure-limiting stress coefficient (see Chapter 10). It has been customary to neglect the Poiseuille component of the shear rate in the Hertz region for an analysis of traction [17]. Then, in the absence of spin, the shear rate is

$$\dot{\gamma} = \frac{\Sigma_l \bar{u}}{h_c} \tag{12.16}$$

where $\Sigma_l = (u_2 - u_1)/\bar{u}$ is defined as the slide-roll ratio calculated from the surface velocities, u_i, within the contact patch. For traction analysis, a distinction must be made between this definition of slide-roll ratio and the Σ that is obtained using velocities measured distant from the contact or calculated from shaft rotation speeds as there may be a difference due to the elasticity of the roller material.

A great deal can be learned from a calculation that neglects shear-thinning if the pressure is in the intermediate range. Here, $\tau = \min\left[\mu(p)\dot{\gamma},\ \Lambda p\right]$ and assuming arbitrarily that $\Lambda = 0.07$, integration (equation (12.14)) using the pressure–viscosity relation of

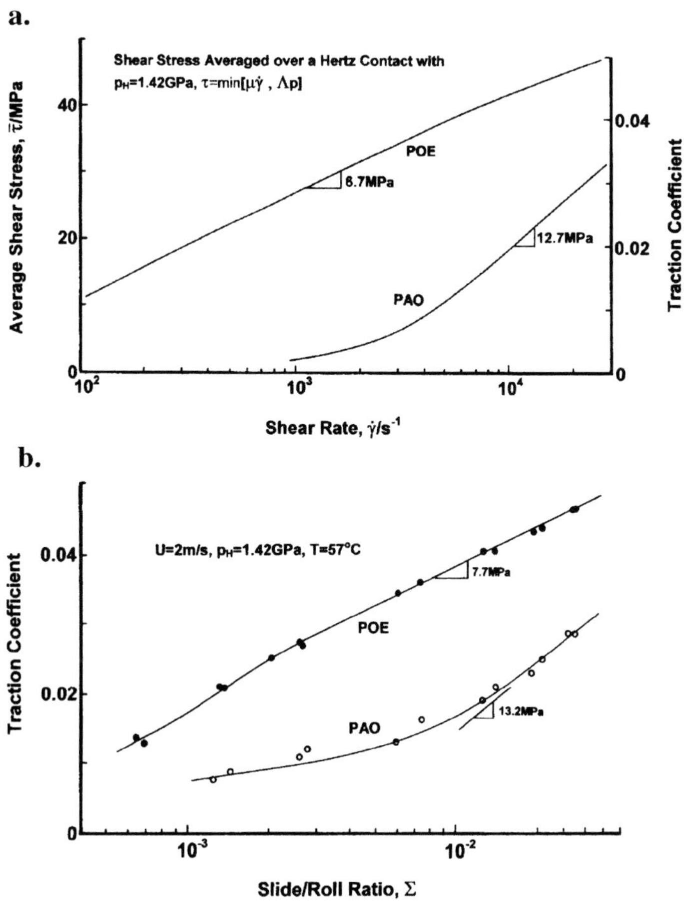

Fig. 12.8. The traction behavior of the two gear oils described in Table 12.1. (a) The calculated traction curve. (b) The measured traction. Reproduced from Bair, S. and Kottke, P., "Pressure-Viscosity Relationships for Elastohydrodynamics", *STLE Tribol. Trans.*, Vol. 46, 2003, pp. 289–295, with permission of the Society of Tribologists and Lubrication Engineers.

equation (12.12) and the pressure distribution of equation (12.13) yields the traction curves shown in the graph at the top of Fig. 12.8. The Hertz pressure is $p_H = 1.42$ GPa. Each curve may be seen to have a logarithmic portion that appears as a relatively straight line with slopes of 6.7 and 12.7 MPa for the POE and the PAO oils, respectively. The slope of the logarithmic traction gradient may be approximated [18] by

$$\tau_E \approx \frac{2\Lambda}{\alpha(p = p_H)} \tag{12.17}$$

The local pressure–viscosity coefficient at Hertz pressure can be calculated from the pressure–viscosity relation of equation (12.12) to give $\alpha(p = p_H) = 19.0$ and 10.6 GPa^{-1} for the POE and the PAO oils, respectively. Applying these values to equation (12.17) yields $\tau_E \approx 7.4$ and 13.1 MPa for the POE and the PAO oils, respectively. The pressure–viscosity relationship appropriate for Hertz pressures combined with the limiting stress will result in a prediction of logarithmic traction response. Consideration of shear-thinning is unnecessary for a point contact.

Measurements of the transverse traction coefficient, $\bar{\tau}/\bar{p}$, for the same oils at the same Hertz pressure using a rolling velocity of 2 m/s resulted in the traction curves shown in the graph at the bottom of Fig. 12.8. Newtonian predictions of central film thickness result in $h_C = 169$ and 204 nm for the POE and the PAO oils, respectively. The similar character of the traction calculation and the traction measurement is obvious in Fig. 12.8. The slopes of the logarithmic traction gradients are accurately estimated by equation (12.17), although a choice of a smaller value of Λ would have reduced the estimated slopes proportionately. The slide-roll ratio may be calculated from the shear rate in the graph at the top of Fig. 12.8, and that calculated slide-roll ratio is about an order of magnitude smaller than that for the measurement at equal values of traction. Two effects were missing from the exercise above, which tend to increase the apparent shear rate in a traction calculation; shear-thinning and roller compliance.

In Section 6.3 and Table 6.1, an inverse correlation was found for the pressure fragility, m_p(or α_g), and the representative stress, τ_E. The reason for this relationship may be understood from equation (12.17) if $\alpha(p = p_H)$ depends closely upon the local pressure-viscosity coefficient at the glass transition, α_g.

An example of experimental traction for a viscous poly alpha olefin in the low pressure regime ($p_H = 0.528$ GPa) was given in Fig. 8.2 and the calculation of traction using the Carreau law was discussed in Section 8.1. For these conditions, the local shear stress remained a small fraction of the local pressure (<0.015) and therefore the limit to the stress, Λp, had no effect for the usual values of Λ. The low-pressure traction response of squalane is shown in Fig. 12.9 from Ref. [19] where measurement and calculation are successfully compared. Here, the shear stress was obtained from equation (12.15) using the single-Newtonian Carreau model,

$$\eta = \mu \left[1 + \left(\frac{\mu \dot{\gamma}}{G} \right)^2 \right]^{\frac{(n-1)}{2}} \tag{8.1}$$

and the standard time–temperature–pressure shifting rule

$$G = G_R \frac{\rho T}{\rho_R T_R} \tag{8.16}$$

Again, the pressure dependence of viscosity was described by the Doolittle equation (12.1) with the relative volume provided by the Tait EOS (4.6). The value of Λ was estimated to be 0.75 from the traction coefficient along the plateau of a traction curve as shown in Fig. 10.1. For the largest slide-roll ratio, $\Sigma = 0.9$, in Fig. 12.9, the maximum value of the ratio τ/p is 0.052 and occurs at the contact center, although the traction coefficient here is $\bar{\tau}/\bar{p} = 0.035$. Therefore, the shear stress limit may be

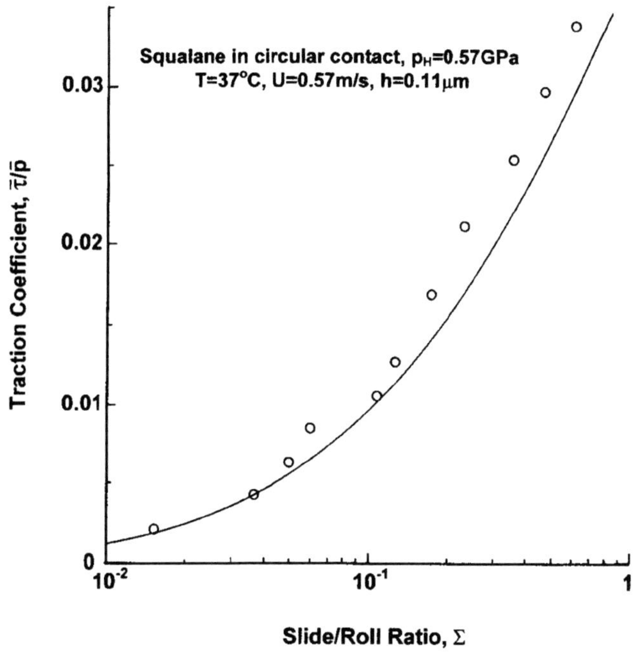

Fig. 12.9. Low-pressure traction curve for squalane. Reproduced from Bair, S., McCabe, C. and Cummings, P.T., "Calculation of Viscous EHL Traction for Squalane Using Molecular Simulation and Rheometry", *Tribol. Lett.*, Vol. 13, No. 4, 2002, pp. 251–254, with kind permission of Springer Science and Business Media.

reached locally before the traction coefficient reaches to Λ. This is what has occurred in the intermediate pressure traction curve for squalane in Fig. 12.10 from [19].

Multiple traction calculations are shown in Fig. 12.10; Newtonian, Newtonian with a stress limit, Carreau and Carreau with a stress limit. Comparison with measured traction indicates that in this regime, both shear-thinning and a stress limit are required for the best accuracy; however the Newtonian viscosity with a stress limit reproduces the logarithmic character well with $\tau_E = 11$ MPa.

A somewhat larger contact pressure has been investigated for squalane in Fig. 12.11 from Ref. [20]. These data show some aspects of the high-pressure regime. The traction coefficient at the largest slide-roll ratio has become relatively independent of Σ and, at low slide-roll ratio, the viscous calculation severely overestimates the stress. Values of dry traction have been included on the plot in Fig. 12.11 and they are surprisingly similar to the full film traction. A curve is shown indicating a linear gradient of $m_s = 40$, which is a reasonable description of the behavior of the dry contact and perhaps the lubricated contact as well to $\Sigma = 5 \times 10^{-4}$.

Bair and Kotzalas [20] investigated the traction of both dry and lubricated contacts at very high pressures, 1.43–2.65 GPa, varying the shear modulus of the rollers, G_s,

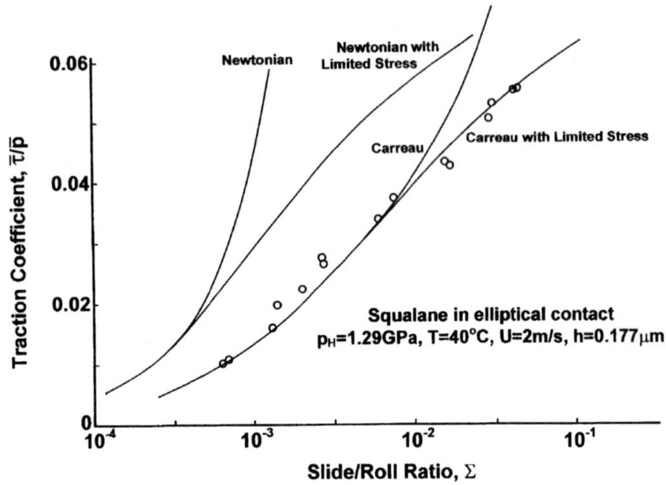

Fig. 12.10. Intermediate-pressure traction curve for squalane. Reproduced from Bair, S., McCabe, C. and Cummings, P.T., "Calculation of Viscous EHL Traction for Squalane Using Molecular Simulation and Rheometry", *Tribol. Lett.*, Vol. 13, No. 4, 2002, pp. 251–254, with kind permission of Springer Science and Business Media.

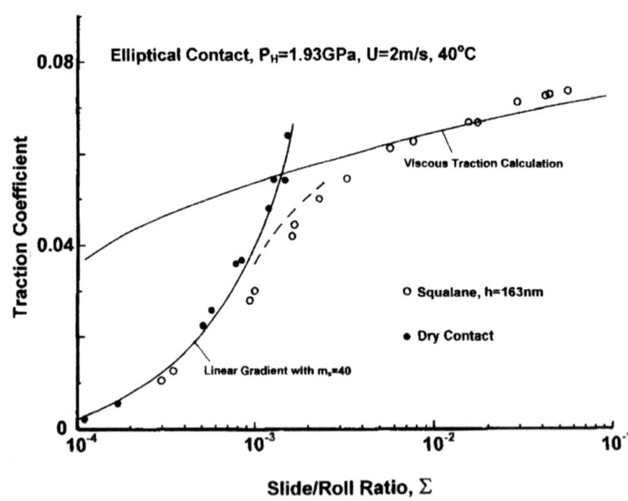

Fig. 12.11. Traction curve for squalane indicating the beginning of the high-pressure regime. The low slide-roll behavior is similar to dry contact. Reproduced from Bair, S. and Kotzalas, M., "The Contribution of Roller Compliance to Elastohydrodynamic Traction", *STLE Tribol. Trans.*, Vol. 49, No. 2, 2006, pp. 218–224, with permission of the Society of Tribologists and Lubrication Engineers.

from 45 to 179 GPa and using two different traction machines of different geometries. They found that the initial traction gradient, $m_s = (\bar{\tau}/\bar{p})/\Sigma$, in dry or lubricated contact, could be described by the very simple relation,

$$m_s = \frac{G_s}{p_H} \tag{12.18}$$

Kalker [21] developed a theory to relate the velocity difference of a pair of contacting elastic bodies to the tangential traction transferred between them. Even for perfect adhesion on the contact patch where the surface velocities must match, the velocities distant from the contact will differ due to elastic compliance. In a traction measurement, the slide-roll ratio is calculated from the roller shaft rotational velocities. Bair and Kotzalas [20] compared the experimental traction gradient with the theory of Kalker and found the experimental gradient was roughly two-thirds of theory over the experimental range. No explanation for the two-thirds factor has been found. It may then be assumed that the enhanced values of Σ at high pressure and low sliding are the result of elastic compliance of the rollers.

An obvious method to correct a traction calculation at high pressure and low slide-roll ratio is to simply add the contribution of roller compliance to the contribution of viscous shear of the film to arrive at the total slide-roll ratio

$$\Sigma = \Sigma_l(\bar{\tau}) + \frac{3}{2}\frac{\bar{\tau}}{G_s} \tag{12.19}$$

The broken curve in Fig. 12.11 is the equation (12.19) and is an obvious improvement over the purely viscous traction calculation.

The traction curves displayed in Fig. 10.1 for a polybutene are excellent examples of the high-pressure regime of traction where a pronounced plateau appears following a linear increase in traction coefficient with Σ. This linear increase may be described simply by $\bar{\tau}/\bar{p} = \Sigma G_s/p_H$. See Table 12.2 where the experimental and analytical gradients are compared.

The examples of isothermal, point-contact traction behavior, above, should not be used to set pressure ranges for the behavior of liquids in general. For a given liquid, an increase of temperature will increase the pressure at which a particular aspect of traction will be present. Fragile liquids (Chapter 6) will display each aspect of traction behavior at reduced pressure compared with strong liquids. Pressure fragility may be the most important property regarding traction. The model liquids introduced in Section 6.6 with

Table 12.2. The initial traction slope for the traction curves of Fig. 10.1 for the polybutene, L100, at 50°C and 2.5 m/s.

p_H/GPa	1.93	2.52	3.31
Measured m_s	39	32	26
G_s/p_H	40	31	24

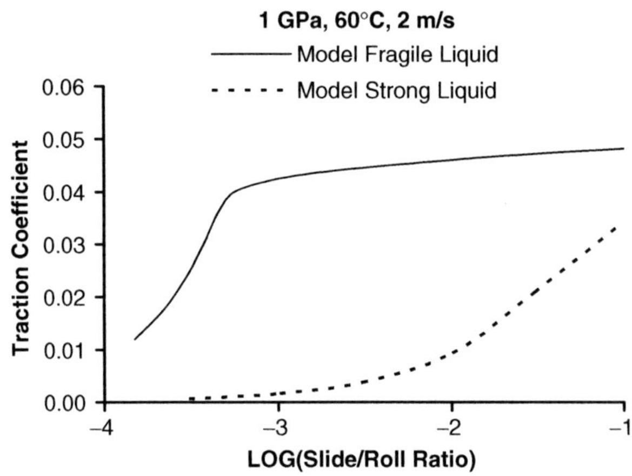

Fig. 12.12. Calculated traction curves for the model fragile and strong liquids of Chapter 6.

parameters given in Table 6.4 were used in a circular point contact traction calculation shown in Fig. 12.12. For both liquids the single Newtonian Carreau equation (11.34) was used to represent shear-thinning with the parameters of Fig. 12.3. For both liquids $\Lambda = 0.05$ and $\mu_R = 0.1$ Pa s. The rollers were steel with $R = 6$ mm. The film thickness was calculated from equation (1.18) corrected with equation (12.10). The traction response of the two model liquids, as seen in Fig. 12.12, is very different for the pressure of $p_H = 1.0$ GPa. The fragile liquid produces much greater traction at low slide-roll ratio and displays a linear gradient there. The strong liquid shows logarithmic behavior with $\tau_E = 7.4$ MPa. Reducing the Hertz pressure to $p_H = 0.5$ GPa results in logarithmic response over about two decades of Σ for the fragile liquid with $\tau_E = 2.3$ MPa.

The simple traction calculations described above have not considered the possibility of multiple transitions in the shear-thinning law (Section 8.6), the effect of the glass transition on the viscosity (see Section 9.4), the influence of the liquid film on the pressure distribution (requires a full simulation), and thermal softening. The thermal traction response at larger slide-roll ratio where the temperature varies along and across the film will probably require a full EHL simulation and the descriptions of temperature dependence given in the preceding chapters should be helpful to that end.

References

[1] Hartl, M., Krupka, I., Poliscuk, R., Liska, M., Molimard, J., Querry, M. and Vergne, P., "Thin Film Colorimetric Interferometry", *STLE Tribol. Trans.*, Vol. 44, No. 2, 2001, pp. 270–276.

[2] Hu, Y.Z. and Zhu, D., "A Full Numerical Solution to the Mixed Lubrication in Point Contacts", *ASME J. Tribol.*, Vol. 122, No. 1, 2000, pp. 1–9.

[3] Liu, Y.C., Wang, Q.J., Wang, W.Z., Hu, Y.Z. and Zhu, D., "Effect of Differential Scheme and Mesh Density on EHL Film Thickness in Point Contacts", *ASME J. Tribol.*, Vol. 128, No. 3, 2006, pp. 641–653.

[4] Dowson, D. and Higginson, G.R., *Elastohydrodynamic Lubrication*, Pergamon Press, Oxford 1966, pp. 70–71.

[5] Bair, S. and Qureshi, F., "Ordinary Shear Thinning Behavior and Its Effect Upon EHL Film Thickness", *Tribological Research and Design for Engineering Systems*, Dowson, D., Priest, M., Dalmaz, G. and Lubrecht, A.A., (Eds.), Elsevier, 2002, pp. 693–699.

[6] Bair, S. and Khonsari, M., "Reynolds Equations for Common Generalized Newtonian Models and an Approximate Reynolds-Carreau Equation", *Proc. I. Mech. E.*, Vol. 220, Part J, 2006, pp. 365–374.

[7] Bair, S., "Shear-Thinning Correction for Rolling/Sliding Elastohydrodynamic Film Thickness", *Proc. I. Mech. E.*, Vol. 219, Part J, 2005, pp. 69–74.

[8] Hamrock, B.J., "*Fluid Film Lubrication*", NASA Ref. Publ., 1255, Washington, D.C., 1991, p. 475.

[9] Bair, S. and Khonsari, M. "An EHD Inlet Zone Analysis Incorporating the Second Newtonian", *ASME J. Tribol.*, Vol. 118, 1996, pp. 341–343.

[10] Bair, S., "Elastohydrodynamic Film Forming with Shear Thinning Liquids", *ASME J. Tribol.*, Vol. 120, 1998, pp. 173–178.

[11] Greenwood, J.A. and Kauzlarich, J.J., "Elastohydrodynamic Film Thickness for Shear Thinning Lubricants", *Proc. Instn. Mech. Engrs.*, Vol. 212, Part J, 1998, pp. 179–191.

[12] Bair, S., Vergne, P. and Marchetti, M. "The Effect of Shear-Thinning on Film Thickness for Space Lubricants", *Tribol. Trans.*, Vol. 45, No. 3, 2002, pp. 330–333.

[13] Bair, S. and Qureshi, F., "The High-Pressure Rheology of Polymer-Oil Solutions", *Tribol. Int.*, Vol. 36, 2003, pp. 637–645.

[14] Bair, S., Qureshi, F. and Kotzalas, M., "The Low Shear Stress Rheology of a Traction Fluid and the Influence on Film Thickness", *Proc. Instn. Mech. Engrs.*, Vol. 218, Part J, 2004, pp. 95–98.

[15] Bair, S., Vergne, P. and Querry, M., "A Unified Shear-Thinning Treatment of both Film Thickness and Traction in EHD", *Tribol. Lett.*, Vol. 18, No. 2, 2005, pp. 145–152.

[16] Bair, S. and Kottke, P., "Pressure-Viscosity Relationships for Elastohydrodynamics", *STLE Tribol. Trans.*, Vol. 46, 2003, pp. 289–295.

[17] Johnson, K.L. and Tevaarwerk, J.L., "Shear Behaviour of Elastohydrodynamic Oil Films", *Proc. Roy. Soc. London A*, Vol. 356, 1977, pp. 215–236.

[18] Bair, S. and Winer, W.O., "The High Shear Stress Rheology of Liquid Lubricants at Pressures of 2 to 200 MPa", *Trans. ASME J. Tribol.*, Vol. 112, No. 2, 1990, pp. 246–253.

[19] Bair, S., McCabe, C. and Cummings, P.T., "Calculation of Viscous EHL Traction for Squalane Using Molecular Simulation and Rheometry", *Tribol. Lett.*, Vol. 13, No. 4, 2002, pp. 251–254.

[20] Bair, S. and Kotzalas, M., "The Contribution of Roller Compliance to Elastohydrodynamic Traction", *STLE Tribol. Trans.*, Vol. 49, No. 2, 2006, pp. 218–224.

[21] Kalker, J.J., *On the Rolling Contact of Two Elastic Bodies in the Presence of Dry Friction*, Dr.-Ing Thesis, TLT Delft, 1967, pp. 90–95.

Index